Single Element Semiconductors

# Single Element Semiconductors

Properties and Devices

*Professor Yi Shi*
*Professor Shancheng Yan*

WILEY VCH

**Authors**

***Professor Yi Shi***
Nanjing University
No. 163 Xianlin Da Dao
Qixia District
Nanjing
CH, 210023

***Professor Shancheng Yan***
Nanjing University of Posts and
Telecommunications
No. 9 Wenyuan Road
Qixia District
Nanjing
CH, 210023

Cover Design: Wiley
Cover Image: © asharkyu/Shutterstock



**Library of Congress Card No.:** has been applied for

**British Library Cataloguing-in-Publication Data**
A catalogue record for this book is available from the British Library.

**Bibliographic information published by the Deutsche Nationalbibliothek**
The Deutsche Nationalbibliothek lists this publication in the Deutsche Nationalbibliografie; detailed bibliographic data are available on the Internet at http://dnb.d-nb.de



The manufacturer's authorized representative according to the EU General Product Safety Regulation is Wiley-VCH GmbH, Boschstr. 12, 69469 Weinheim, Germany, e-mail: Product_Safety@wiley.com.



**Print ISBN: 9783527355037**
**ePDF ISBN: 9783527851775**
**ePub ISBN: 9783527851768**
**oBook ISBN: 9783527851782**

**Typesetting**: Lumina Datamatics
**Printing and Binding:** CPI Group (UK) Ltd, Croydon, CR0 4YY

C9783527355037_220525

## Contents

# Preface

In the rapidly developing field of nanotechnology, single-element semiconductor materials have become a research hotspot due to their unique physical and chemical properties and wide application prospects. The purpose of this book is to discuss in depth the preparation, properties, and applications of single-element semiconductor materials in modern electronic devices. Special attention is paid to the nanostructures formed by elements such as carbon, silicon, germanium, phosphorus, and tellurium, which show great potential in the fields of energy, electronics, and sensing due to their unique physicochemical properties. The book will provide the latest research progress on the representative structures of graphene, silicon nanowires, black phosphorus, and other elements, including their preparation methods, electronic and optical properties, as well as the application prospects in energy storage, optoelectronics, biomedicine, and other fields. The purpose of this book is to provide a comprehensive reference resource for researchers and engineers to advance the research and application development of single-element semiconductor materials. In this book, we not only review the historical development of single-element semiconductor materials but also focus on their synthesis strategies and application prospects. We believe that the unique properties of these materials will revolutionize future electronic devices. With the deepening of research, the potential of these materials will be further tapped and contribute to scientific and technological progress.

# I

## Carbon

*Yi Shi*

### 1.1 Introduction

Carbon (C) is a nonmetallic element of period 2, group 14 (group IVA) of the periodic table, and the fourth most abundant element in the universe [1]. There are four valence electrons ($2s^2 2p^2$) outside the carbon atom, meaning that all three types ($sp$, $sp^2$, $sp^3$) of hybridization between *s*- and *p*-orbitals could occur in carbon atoms during the formation of chemical bonds. Due to the bonding characteristics, millions of organic compounds in which carbon atoms usually bond with other nonmetallic atoms (e.g. hydrogen, oxygen, nitrogen, etc.) or the C atom itself (forming carbon chain, branch, or ring structures) have been discovered in nature or synthesized artificially.

For elementary substances of carbon (not regarded as organic molecules but inorganic ones), different hybridization modes between carbon atoms lead to the diversity of structure and properties of its allotropes. Consider two typical types of carbon bulk materials: diamond is a three-dimensional (3D) network composed of $sp^3$-hybridized carbon, known as the hardest natural material, while graphite is a 3D structure stacked by a $sp^2$-hybridized carbon monolayer with a low degree of hardness but good lubricity. Carbon nanomaterials include diverse low-dimensional allotropes of carbon, such as graphene [2], graphene nanoribbons (GNRs) [3], carbon nanotubes (CNTs) [4], graphyne (GY) [5], fullerenes [6], and carbon dots (CDs) [7]. In contrast to natural bulk materials, theoretical research, practical preparation, and potential applications of carbon nanomaterials have only been in development for about one century. Nevertheless, numerous studies have demonstrated these carbon nanomaterials' unlimited promise with retained properties of bulk materials and unique properties of low-dimensional materials. In particular, the primary fabrication of two-dimensional (2D) graphene monolayers triggered an unprecedented graphene "gold rush" while opening the door to the two-dimensional materials (2DM) system, which is a historic step for the carbon nanomaterial system [8].

In this chapter, we will take graphene, CNTs, and GY as representative carbon nanomaterials and illustrate their fabrication methods, essential properties, and applications through recent frontier research findings.

*Single Element Semiconductors: Properties and Devices*, First Edition. Yi Shi and Shancheng Yan.

## 1.2 Fabrication of Carbon Nanomaterials

### 1.2.1 Graphene

Graphene is a 2D building material for all other dimensions of $sp^2$-hybridized carbon materials, which can be transferred to zero-dimensional buckyballs (fullerenes), one-dimensional (1D) nanotubes, and 3D graphite (Figure 1.1) [8]. Although it was not until the early twenty-first century that graphene was isolated from bulk graphite, the concept of "graphene" or "monolayer graphite" had been proposed in the mid-twentieth century but initially used as a theoretical model that did not exist in a free state. This concept was derived from the conventional perspective proposed by Landau [9] and Peierls [10], who argued that 2D crystals could not exist independently because of their thermodynamic instability. For a long time afterward, atomic monolayer was considered to be obtained only by epitaxy on a 3D substrate. However, due to the interaction between the substrate and 2DM, epitaxial graphene may not entirely reflect the characteristics of "monolayer graphite." Hence, researchers are still looking for methods to prepare independent graphene [11–13].

In fact, since the groundbreaking isolation of graphene monolayer [2], various top-down methods, such as micromechanical cleavage, liquid-phase exfoliation, and graphene oxide (GO) reduction, have successfully proved the independent existence of 2D crystals. Furthermore, aiming for the scalable application for the post-Moore era, bottom-up methods, such as confinement-controlled sublimation (CCS) and chemical vapor deposition (CVD), are actively used in large-scale, high-quality synthesis of graphene.

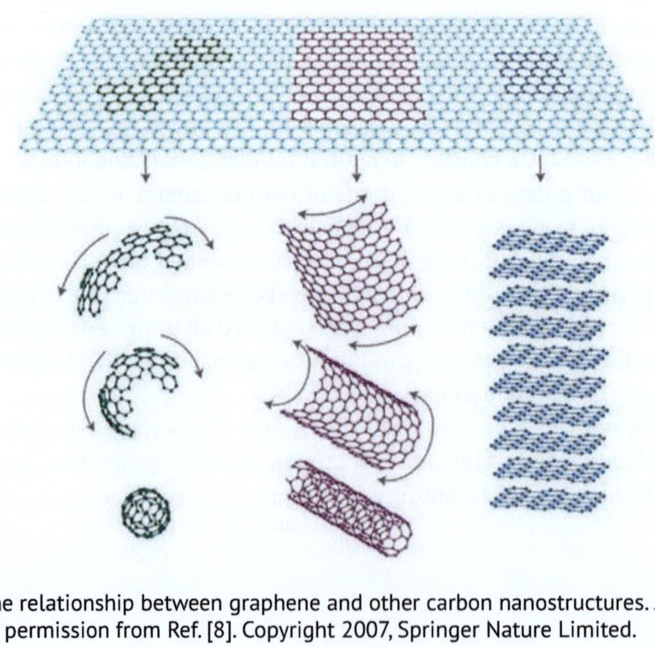

**Figure 1.1** The relationship between graphene and other carbon nanostructures. *Source:* Reprinted with permission from Ref. [8]. Copyright 2007, Springer Nature Limited.

#### 1.2.1.1 Top-down Methods

##### *1.2.1.1.1 Dry Exfoliation of Graphene*

In 2004, Novoselov et al. demonstrated the preparation of few-layer graphene (FLG) consisting of monolayer by micromechanical cleavage (repeated peeling by scotch tape) of highly oriented pyrolytic graphite [2]. This surprisingly simple method produced FLG films up to 10 μm in size with stability and reliability (Figure 1.2a), and the properties of FLG are almost identical to those of theoretical studies. With convenience but low yield, the original cleavage method is suitable for research or proof of concept but not practical for large-scale applications. A layer-engineered exfoliation (LEE) technique was introduced by Moon et al. for large-area, layer-controlled graphene exfoliation (Figure 1.2b) [14]. In the LEE process, bulk natural graphite was firstly cleaved on adhesive tape, and then a selective metal film was directly deposited on the graphite by e-beam evaporation. The lattice mismatch between metal and graphite produced tensile stress at the interface, which created a crack at the boundaries of graphite induced by external bending and eventually led to large-area exfoliation of graphene [15]. The thickness of exfoliated graphene was determined by the difference between the metal–graphene binding energy ($\gamma_{\text{metal-Gr}}$) and interlayer binding energy of graphite ($\gamma_{\text{Gr-Gr}}$), resulting in a thicker layer with a higher difference (Figure 1.2b,c) [15, 16]. Due to the slight difference between $\gamma_{\text{Au-Gr}}$ and $\gamma_{\text{Gr-Gr}}$, Au-assisted LEE graphene showed a defect-free monolayer with a large lateral size of 1 mm and could be repeatedly exfoliated from the same bulk graphite (Figure 1.2d,e). The large-area exfoliation strategy controlled by interface binding energy between metal and 2DM is also applicable in other 2DM systems, such as transition metal dichalcogenides [17].

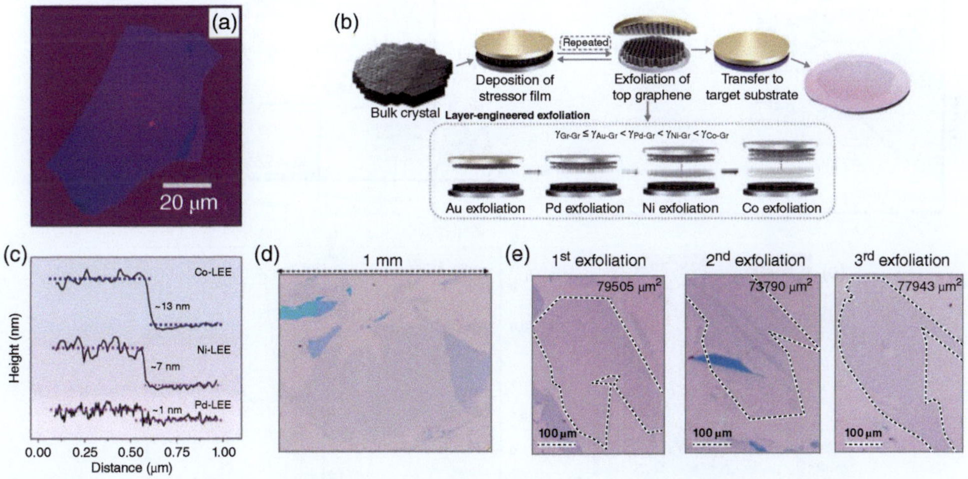


**Figure 1.2** (a) OM image of FLG produced using original micromechanical cleavage method. *Source:* Reprinted with permission from Ref. [2]. Copyright 2004, The American Association for the Advancement of Science. (b) Schematic of LEE process and the relation between $\gamma_{\text{metal-Gr}}$ and $\gamma_{\text{Gr-Gr}}$. (c) AFM profile of Co-, Ni-, and Pd-LEE graphene with the relation $\gamma_{\text{Pd-Gr}} < \gamma_{\text{Ni-Gr}} < \gamma_{\text{Co-Gr}}$. (d) Low-magnification OM image of LEE graphene with millimeter-size monolayer. (e) OM images of repeated LEE graphene. *Source:* Reprinted with permission from Ref. [14]. Copyright 2020, The American Association for the Advancement of Science.

#### 1.2.1.1.2 Liquid-phase Exfoliation of Graphene

Compared with air ambience, liquid immersion can significantly reduce the van der Waals (vdW) interaction between neighboring layers of graphite. With external forces induced by sonication [18], ball milling [19], or shear mixing [20], graphene nanosheets could be easily exfoliated in the liquid phase, generally provided by water or organic solvents. High tension at the solid/liquid interface hurts solid dispersion in a liquid medium. Therefore, the selection of a liquid medium can directly determine the quality of liquid-exfoliated graphene, and solvents with a lower surface tension ($\gamma$ ~40 mN·m$^{-2}$ [18]) could minimize the tension at the graphene/solvent interface [21]. Furthermore, adding surfactants or intercalation particles can also weaken the vdW interaction of graphite.

Supercritical fluid (SCF) is a special substance with a temperature and pressure above the critical point ($T_c$, $p_c$), where liquid and gas phases cannot be distinct (Figure 1.3a). SCFs have specific intermediate properties between liquid and gas, such as gas-like diffusivity, liquid-like solubility, and adjustable density and viscosity controlled by temperature and pressure [22, 23]. Supercritical carbon dioxide ($SCCO_2$), with an easily accessible critical point ($T_c = 31.1°C, p_c = 7.38 MPa$ [24]), is a suitable medium for graphene exfoliation owing to extremely low surface tension, shearing effect produced by complex hydrodynamics, and intercalation of $CO_2$ molecules. Zhu et al. characterized the $SCCO_2$-exfoliated graphene and explained the stress mechanism in $SCCO_2$ applying on graphite (Figure 1.3b) [25].

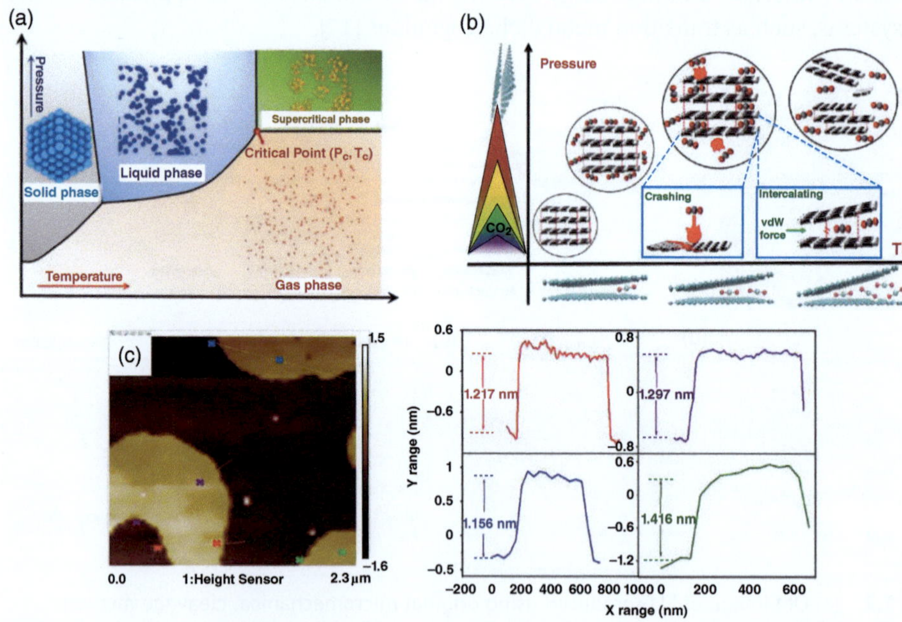


Figure 1.3 (a) Schematic of a pressure–temperature phase diagram showing the supercritical region. *Source:* Reprinted with permission from Ref. [22]. Copyright 2023, Wiley-VCH GmbH. (b) Demonstration of $SCCO_2$-assisted graphene exfoliation mechanism. (c) AFM image of graphene nanosheets with centrifugation posttreatment and corresponding height profiles along different labels. *Source:* Reprinted with permission from Ref. [25]. Copyright 2021, Elsevier Ltd.

Tangential stress could enhance the intercalation effect of $SCCO_2$ on graphite's vdW gap, while the direct impact of normal stress on the graphene surface could break C—C bonds and reduce the size of graphene. The increasing pressure makes $SCCO_2$ denser, enhancing the stress mechanism and leading to more efficient exfoliation. Longer processing time also contributes to complete exfoliation, but excessive processing will cause the agglomeration of graphene. Under the same posttreatment condition, graphene obtained at 45 MPa, 48 h showed the highest concentration with an average thickness of ~1.272 nm, which could be regarded as two or three layers (Figure 1.3c).

#### *1.2.1.1.3 Electrochemical Exfoliation*

Electrochemical exfoliation (ECE) is still a liquid-phase intercalation method. However, the strong electric field generated by the ECE device allows the intercalation of larger ions or molecules, resulting in fast expansion and exfoliation of 2DM. A typical dual-electrode ECE device (Figure 1.4a) consists of a working electrode (usually

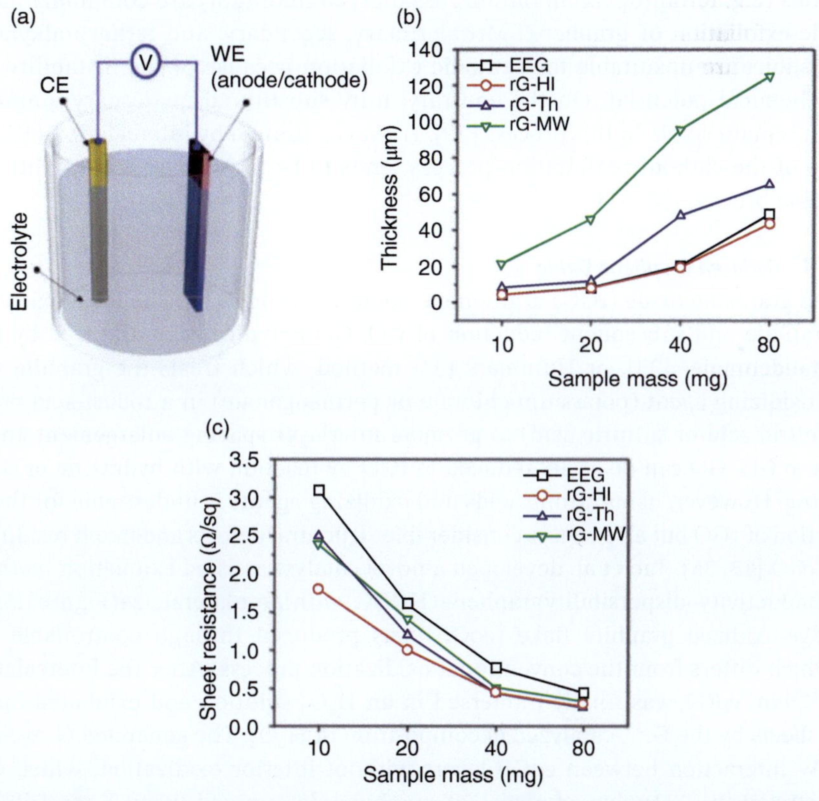


**Figure 1.4** (a) Schematic of a typical dual-electrode ECE setup. *Source:* Reprinted with permission from Ref. [26]. Copyright 2024, The Royal Society of Chemistry. (b) Thickness and (c) sheet resistance of electrochemically exfoliated graphene (EEG) and reduced graphene (rG) by HI, thermal (Th), and microwave (MW) treatment. *Source:* Reprinted with permission from Ref. [27]. Copyright 2021, American Chemical Society.

bulk materials for exfoliation such as graphite), a counter electrode (usually a metal plate), an electrolyte, and a power source [26]. According to charge characteristics of ions involved in intercalation, ECE can be classified into two types: anodic exfoliation, in which graphite acts as an anode to attract anions, and cathodic exfoliation, in which graphite acts as a cathode to attract cations. Anodic exfoliation is usually generated in an aqueous electrolyte containing anions such as $SO_4^{2-}$, $ClO_4^-$, $PO_4^{3-}$, $OH^-$, and halide ($Cl^-$, $Br^-$, $I^-$). Due to the positive voltage applied to graphite, anodic exfoliation provides higher productivity but inevitably leads to oxidation and defects in the exfoliated graphene [26, 27]. Mirkhani et al. proposed a high-temperature vapor reduction process to improve the morphology and electrical properties of graphene produced by anodic exfoliation [28]. Compared with thermal and microwave treatment, graphene with HI vapor treatment indicated a more compacted structure with more metallic luster and lower sheet resistance (Figure 1.4b,c). Cathodic exfoliation is a relatively mild process since the applied negative voltage could prevent graphene's oxidation, preserving its intrinsic properties. Nonaqueous media (e.g. *N*-methyl-2-pyrrolidone, dimethylformamide) containing cations such as alkali ions ($Li^+$, $K^+$, $Na^+$) and quaternary ammonium ions (e.g. tetrapropylammonium, tetrabutylammonium) are commonly used for cathodic exfoliation of graphene. Most primary, secondary, and tertiary alkylammonium cations are unsuitable for cathodic exfoliation because of the instability under electrochemical potential. On the contrary, fully substituted quaternary ammonium ions can remain stable in this process [29]. However, limited by intercalation efficiency, the rate of the cathodic exfoliation process tends to be lower than that of the anodic exfoliation process [30, 31].

#### *1.2.1.1.4 Reduced Graphene Oxide*

Reduced graphene oxide (rGO) is graphene obtained through the chemical oxidation of bulk graphite and subsequent reduction of GO. GO is typically synthesized by Brodie [32], Staudenmaier [33], or Hummers [34] method, which treats the graphite with a potent oxidizing agent (potassium chlorate or permanganate) in a robust acid environment (nitric acid or sulfuric acid) to promote interlayer spacing enlargement and conversion to GO. GO can be easily reduced to rGO by reaction with hydrazine or thermal annealing. However, using strong acids and oxidizing agents is undesirable for the mass production of rGO but also causes considerable structural defects and much residual oxygen in rGO [13, 35]. Tao et al. developed a novel catalyst-assisted exfoliation method for high-conductivity–dispersibility graphene (HCDG) with large lateral size (Figure 1.5a) [36]. The edge-oxidized graphite flake (eoGF) was produced through controllable oxidation, which differs from the conventional oxidization process. After the intercalation of the $Fe^{3+}$ ion, eoGF was finally immersed in an $H_2O_2$ solution and exfoliated into graphene sheets by the $Fe^{3+}$-catalyzed decomposition of $H_2O_2$. The generated $O_2$ weakened the vdW interaction between eoGF layers without interior oxidization, which can be demonstrated in (002) characteristics of graphene determined using X-ray Diffraction (XRD), high-resolution transmission electron microscopy (Figure 1.5b,c), and localized, low-level oxidization determined using Raman spectra and X-ray Photoelectron Spectroscopy (XPS) (Figure 1.5d,e).

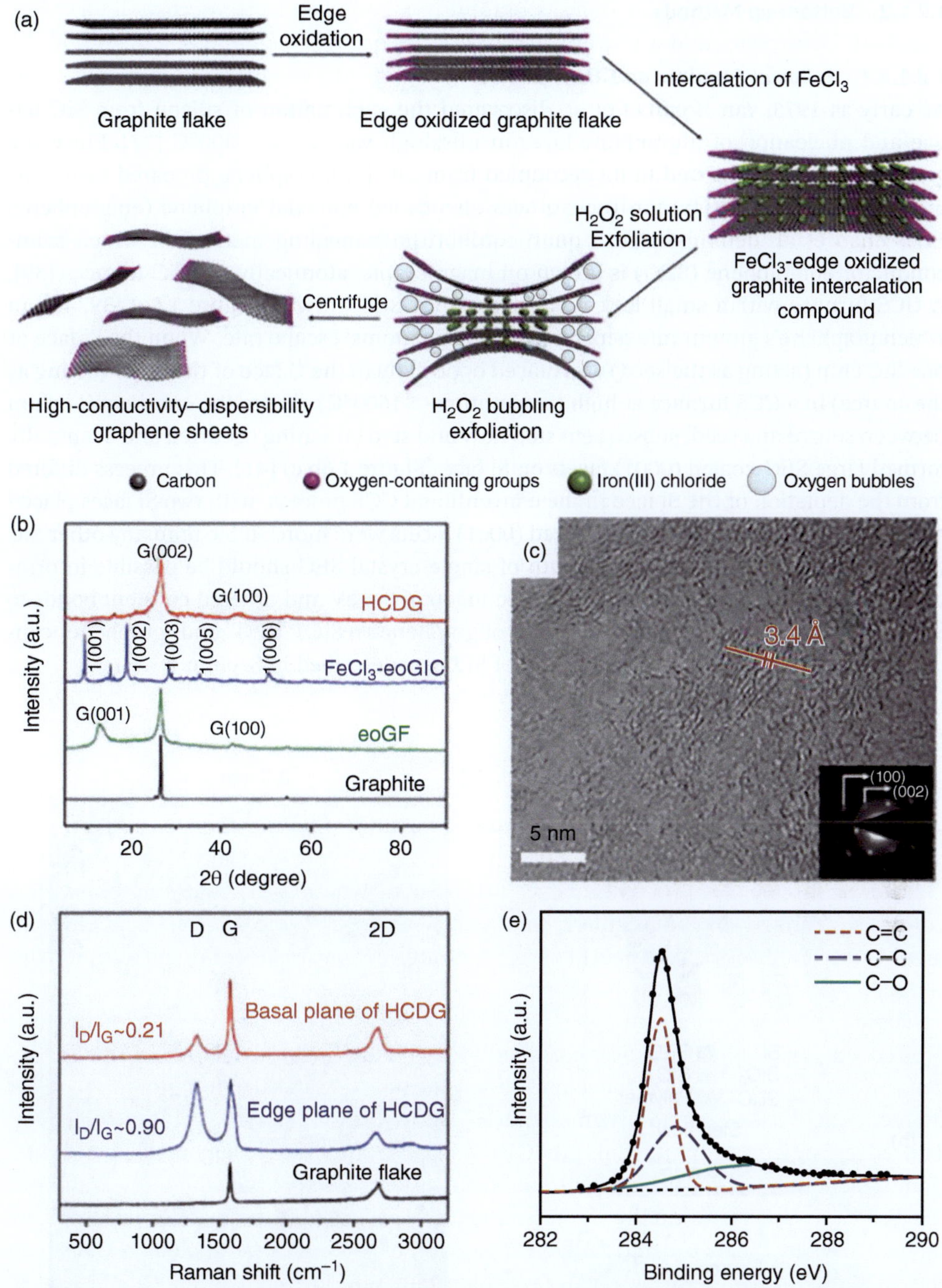


**Figure 1.5** (a) Schematics of catalyst-assisted exfoliation of HCDG. (b) XRD patterns of HCDG, $FeCl_3$-eoGIC, eoGF, and pristine graphite. (c) Transmission electron microscopy (TEM) image and selected area electron diffraction (SAED) pattern of the HCDG basal plane. (d) Raman spectra of HCDG and pristine graphite. (e) C1s spectra of XPS and deconvoluted peaks of HCDG. *Source:* Reprinted with permission from Ref. [36]. Copyright 2020, Wiley-VCH GmbH.

### 1.2.1.2 Bottom-up Methods

#### *1.2.1.2.1 Epitaxial Graphene on Silicon Carbide*

As early as 1975, van Bommel et al. discovered the sublimation of silicon from SiC lattice and nucleation of a graphene layer in ultrahigh vacuum at ~800°C [37]. Since the graphene layer was proved to be decoupled from SiC [38], graphene prepared using SiC high-temperature sublimation process was also called epitaxial graphene (epigraphene, EG). Zhao et al. demonstrated a quasi-equilibrium annealing method in which semiconductor epigraphene (SEG) is grown on macroscopic, atomically flat SiC terraces [39]. A CCS furnace with a small leak was designed for SEG growth (Figure 1.6a) [39, 40], in which graphene's growth rate depended on silicon atoms' escape rate. When the Si face of one SiC chip (acting as the seed) was placed oppositely to the C face of the other (acting as the source) in a CCS furnace at high temperature (>1600°C), due to the quasi-equilibrium between source and seed, subsequent step flow and step bunching occurred and eventually formed large SEG-coated (0001) facets on Si face (Figure 1.6b,c) [41]. This process differed from the depletion of the Si face in the conventional CCS process, with two Si faces placed oppositely. In addition, the SEG-covered (0001) facets were more stable than any other SiC facets, implying that wafer-level growth of single-crystal SEG should be possible in principle. Characterization of SEG showed a bandgap of 0.6 eV and ordered covalent bonds to SiC substrate, while there was no evidence of graphene on SiC ("SEG" and "graphene" cannot be equated). The electrical properties of SEG are described in Section 1.3.1.1.

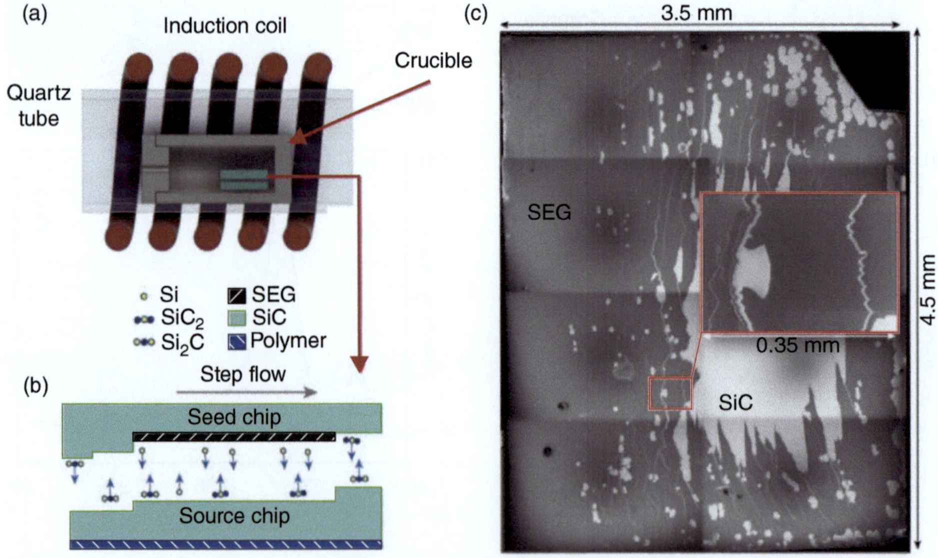


**Figure 1.6** (a) Schematic of a CCS furnace with two SiC chips inside a graphite crucible. (b) Schematic of two SiC chips stacked with the C face of the source chip facing the Si face of the seed chip. (c) Composite electron microscope image of a 3.5 mm × 4.5 mm SiC wafer with 80% SEG coverage. The localized area shows the contrast between SiC and SEG. *Source:* Reprinted with permission from Ref. [39]. Copyright 2024, Springer Nature Limited.

#### *1.2.1.2.2 Chemical Vapor Deposition*

CVD is a widely used process for the deposition of various kinds of material in microelectronics, such as metals, elemental or compound semiconductors, gate oxides, dielectrics, etc. In the graphene CVD process, gaseous precursors like hydrocarbons are primarily used for deposition, while solid [42, 43] and liquid [44] precursors are also feasible. Transition metal is a well-known substrate for graphene CVD, acting as a catalyst for dehydrogenating hydrocarbon precursors. Nucleation and growth of graphene ensue after the dehydrogenation process, and growth mechanisms on metal substrates can be divided into two categories according to the solubility of carbon in a specific metal. Ni and Cu are the two most commonly used metal substrates for graphene CVD, corresponding to the segregation growth mechanism on metals with higher carbon solubility and the surface growth mechanism on metals with lower carbon solubility. The segregation mechanism on Ni tends to grow into multilayer graphene, in which the thickness, crystalline state, and defect of Ni substrate will influence the morphology of graphene [45, 46]. The surface mechanism on Cu refers to a self-limiting and robust reaction process, which is more favorable for monolayer graphene growth [47]. However, due to the misorientation among different initial nuclei, the merging of crystal domains originating from these nuclei can only finally form polycrystalline graphene. Two strategies have been derived from this issue for the growth of large-scale, monocrystalline graphene on Cu: one is to reduce the nucleation density and maximize the expansion of a single nucleus, and the other is to achieve the perfect coalescence of domains by controlling the consistent orientation of multiple nuclei.

Through elaborate condition control and optimization, bilayer graphene (BLG) with AB (Bernal) stacking can be further produced on Cu. Zhang et al. achieved rapid CVD growth of large-area continuous BLG on Cu foil by introducing trace $CO_2$ (Figure 1.7a) [48]. With the assistance of $CO_2$, ~50% BLG coverage could be achieved within 10 min based on the whole-covered monolayer graphene (MLG), and continuous BLG was obtained within 20 min. The growth mechanism of BLG was investigated by isotopic labeling with alternating introduction of $^{13}CH_4$ and $^{12}CH_4$, which demonstrated that the second-layer graphene grew below the first layer and showed concentric rings composed of alternating $^{12}C$ and $^{13}C$ under Raman intensity mapping (Figure 1.7b–d). The introduced $CO_2$ etched the first-layer graphene to form point defects, providing diffusion sites for the carbon source to form the second layer (Figure 1.7a), which differs from the conventional self-limiting growth mechanism on Cu. The AB-stacking structure was dominant in BLG, accounting for as high as 61% on Cu (100)-dominated polycrystalline Cu foils and 100% on super flat epitaxial single-crystal Cu (111) on annealed c-surface sapphire (Figure 1.7e,f). The $CO_2$-assisted strategy also showed high compatibility in submeter-scale (0.3 m × 0.1 m) substrates and roll-to-roll mass production, achieving more than 90% coverage of BLG.

### 1.2.2 Carbon Nanotubes

CNT is a type of seamless tubular graphite structure first prepared by Iijima in 1991 [4]. Although such 1D cylindrical tubular structure was rare in inorganic crystals at that time, Iijima predicted that structure engineering of carbon nanomaterials would be

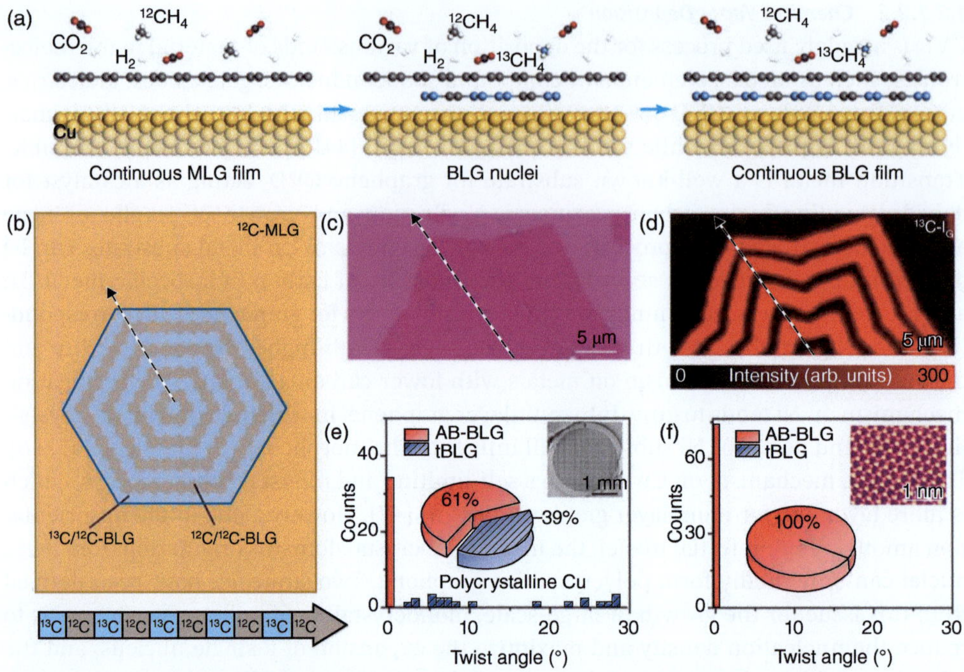


**Figure 1.7** (a) Schematic of $CO_2$-assisted BLG growth through the alternating supplement of $^{12}CH_4$ and $^{13}CH_4$. (b) Schematic of alternating $^{12}C$ and $^{13}C$ pattern of graphene. (c) OM image and (d) Raman intensity mapping of $^{13}C$ G-band of half BLG domain. (e, f) Distribution of twist angles based on SAED patterns of the BLG grown on (e) polycrystalline and (f) single-crystal Cu. *Source:* Reprinted with permission from Ref. [48]. Copyright 2023, Springer Nature Limited.

possible at much larger scales than fullerenes, which was confirmed by the first isolation of graphene in 2004 [2, 4]. According to the thickness of walls, CNTs can be classified into single-walled carbon nanotubes (SWCNTs) and multi-walled carbon nanotubes (MWCNTs). A SWCNT with a specific diameter can be regarded as the structure formed by the rotation and enclosure of a GNR monolayer around a particular vector, which can be described by a pair of chiral indices $(n, m)$ (Figure 1.8) [49]. The relationship between $n$ and $m$ values reflects the configuration of SWCNTs [50, 51]. To be specific, SWCNTs with indices $(n, 0)$ and $(n, n)$ are called zigzag and armchair, which are two types with high symmetry. For $n > m > 0$, the indices $(n, m)$ refer to a type of SWCNTs with chiral isomers. Chirality has an essential influence on the electrical properties of SWCNTs [52], which will be discussed in Section 1.3.2.

The main synthetic methods of CNTs are arc discharge (AD), laser ablation (LA), CVD, etc., which allow controlling and adjusting the morphology and parameters of CNTs by optimizing the growth conditions. In addition, since transition metal catalysts are necessary for the methods mentioned previously, the rational design of catalysts is also an effective strategy for controlled synthesis of CNTs [53].

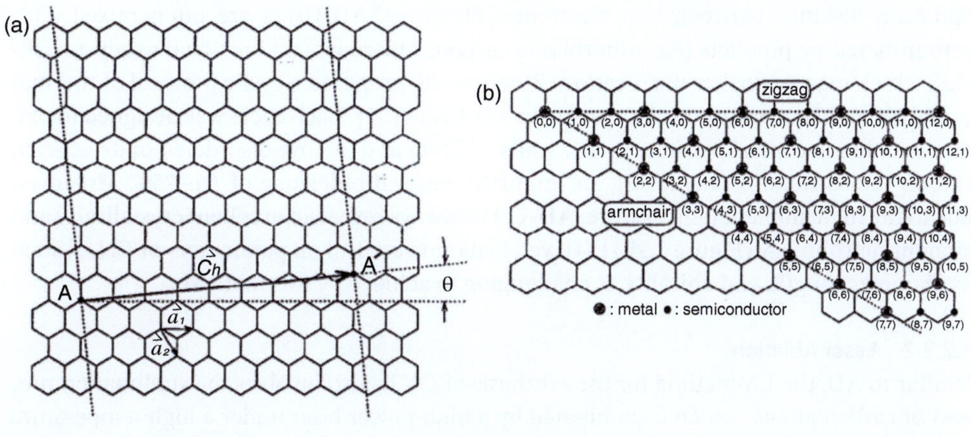


**Figure 1.8** (a) Chiral vector $\vec{C}_h$ of a SWCNT in the unfolded 2D graphene layer. The unit vectors $\vec{a}_1$ and $\vec{a}_2$, and chiral angle $\theta$ are also shown. (b) Possible chiral indices with $\theta$ ranging from 0° to 30°. The circled dots and dots denote each SWCNT's metallic and semiconducting behavior. *Source:* Reprinted with permission from Ref. [49]. Copyright 1992, American Institute of Physics.

#### 1.2.2.1 Arc Discharge

The phenomenon of ADis usually generated by the approach of two electrodes under an inert ambiance. This principle was initially applied for the preparation of fullerenes, while CNTs were occasionally synthesized in this way by Iijima [4]. For AD synthesis of CNTs, graphite is used as both electrodes and high-temperature (above 1500°C) plasma is generated by the arc under a large current, which causes the sublimation of carbon atoms from the anode and coagulation into CNTs on the cathode (Figure 1.9a) [54]. Therefore, the anode graphite is continuously consumed and designed to be mobile to maintain the

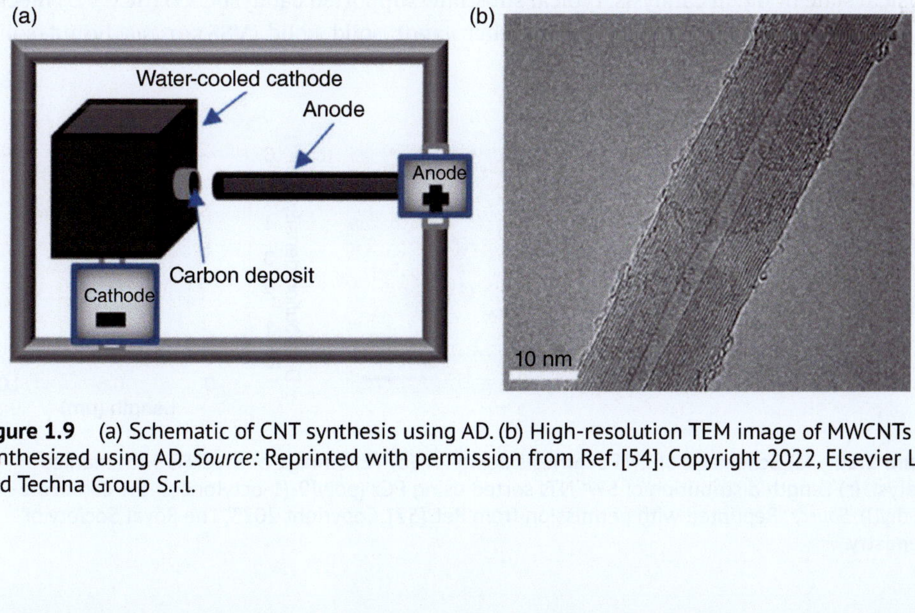


**Figure 1.9** (a) Schematic of CNT synthesis using AD. (b) High-resolution TEM image of MWCNTs synthesized using AD. *Source:* Reprinted with permission from Ref. [54]. Copyright 2022, Elsevier Ltd and Techna Group S.r.l.

optimum distance between two electrodes. However, AD-CNTs are often mixed with carbon-based by-products (e.g. amorphous carbon, fullerenes) and residual catalyst particles, which require further purification. Ribeiro et al. proposed a cyclic protocol of thermal oxidation and HCl solution reflux to purify AD-CNTs [55]. Each cycle was designed to oxidize and dissolve metal particles on the walls of CNTs and remove disordered or defect-rich carbon nanostructures, achieving an impurity removal efficiency of 66–75%. Moreover, due to the high ambient temperature, AD-CNTs always reveal an excellent crystallinity and straight morphology (Figure 1.9b) [54]; yet, elaborate control on parameters of CNTs, such as diameter, length, and chirality, is challenging to achieve by this method.

#### 1.2.2.2 Laser Ablation

Similar to AD, the LA method for the synthesis of CNTs also involves the sublimation process of carbon atoms, which is dominated by a high-power laser under a high temperature (~1200°C) (Figure 1.10a) [56]. Unlike the intermittent discharge process of AD, LA enables a continuous and efficient synthesis of CNTs with no requirement for the conductivity of carbon targets [56, 57]. Conventional LA process of CNTs used graphite target as a carbon source, while Chen et al. recently reported high-yield synthesis of SWCNT bundles using low-graphitized coal as a carbon source (Figure 1.10b) [57]. Despite complex carbon components and various impurity elements in coal, it is more easily ablated than graphite under the same LA conditions, resulting in a higher carbon evaporation rate. LA-SWCNTs synthesized from coal present an ideal distribution of diameter (~1.1–1.3 nm at 1198 K) for electronic applications, which can be fabricated as the active layer of carbon nanotubes field effect transistors (CNT-FETs) (Figure 1.10c). However, it should be pointed out that the high-power laser in the LA setup prohibits its large-scale application to CNT synthesis [53].

#### 1.2.2.3 Chemical Vapor Deposition

Due to the unacceptable condition of high temperatures in the AD and LA process, CVD is a suitable and mature alternative for synthesizing CNTs. On the one hand, according to the physical state of metal catalysts, typical substrate-supported catalyst CVD (SCCVD) mechanisms of CNTs can be classified into either vapor–solid–solid (VSS) or gas–liquid–solid

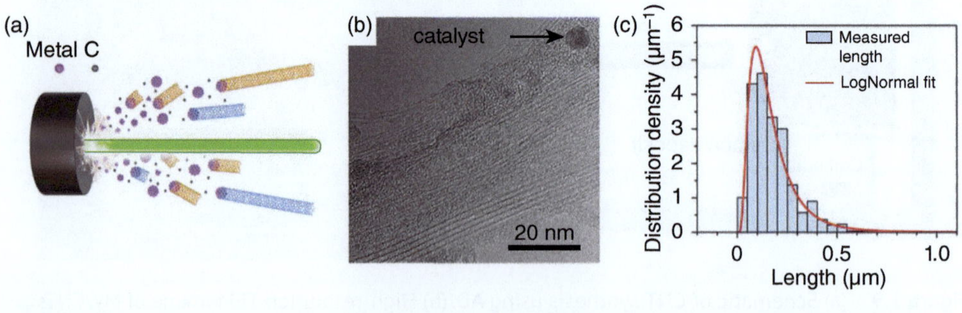


**Figure 1.10** (a) Schematic of CNT synthesis using LA. (b) TEM image of SWCNTs with marked catalyst. (c) Length distribution of SWCNTs sorted using PCz (poly[9-(1-octylonoyl)-9*H*-carbazole-2,7-diyl]). *Source:* Reprinted with permission from Ref. [57]. Copyright 2023, The Royal Society of Chemistry.

(VLS). In the VSS growth mechanism, the solid catalyst usually maintains a relatively stable morphology, and the precipitation of carbon atoms only happens on the catalyst's surface. The structure and components of the catalyst can be manipulated to narrow the diameter and chirality distribution of CNTs [58, 59]. In the VLS mechanism, the liquid state of the catalyst will lead to the dissolution of carbon and the movement of the catalyst droplet. Hence, the chirality of the synthesized CNT is randomly distributed.

On the other hand, as CVD growth progresses, the catalyst may eventually be located at the base or the tip of CNTs. VSS growth of CNTs may follow a base- or tip-growth mechanism, which is related to the strength of the interaction between catalyst metal and substrate, and the growth direction (vertical or parallel to substrates) of CNTs can be adjusted by plasma or airflow [60, 61]. VLS growth of CNTs is generally regarded as a tip-growth mechanism [62], as carbon atoms are only precipitated at the migrating catalyst droplet.

Different from metal catalyst pre-deposition in SCCVD methods, floating catalyst CVD (FCCVD) introduces catalyst precursor and gaseous carbon source together into the reaction chamber, followed by catalyst formation, carbon nucleation, and CNT growth [63, 64]. FCCVD-CNTs are usually floating in the chamber, which can be blown out with airflow and collected easily [65]. However, due to the short reaction time in the flowing atmosphere, CNTs are blown out without sufficient growth, exhibiting short lengths and random alignment. Jiang et al. proposed a substrate interception and orientation strategy (SIDS) to fabricate high-density ultralong CNT arrays by combining the so-called "kite-mechanism" (tip-growth mechanism) of ultralong CNTs with an improvement of FCCVD (Figure 1.11a,b) [66]. A substrate was placed horizontally into the center of the FCCVD chamber and used to intercept and reorient the floating short CNTs, which is absent in conventional FCCVD (Figure 1.11c,d). Local turbulence at the edge of the substrate caused one end of the CNTs to be anchored to the vertical step of the substrate, which sequentially followed the kite-like growth mechanism with an adequate flowing supply of carbon source (Figure 1.11e–g). Conventional growth of ultralong CNTs was based on VSS methods. In this case, the tip-growing CNTs initially need to overcome external forces such as vdW and drag forces to grow vertically without lying on the substrate [61, 67]. Actually, these dead CNTs account for a considerable portion of the total, resulting in a low catalyst utilization rate and CNT density [67].

### 1.2.3 Graphyne

GY is a type of 2D carbon nanostructure predicted by Baughman et al. in 1987, in which *sp*- and $sp^2$-hybridized carbon atoms coexist [68]. In GY structures, benzene rings are connected by acetylene bonds and form a hexagonal symmetric network, resulting in four types of chemical bonds with different lengths between carbon atoms: (i) $C_{sp2}—C_{sp2}$ bonds in benzene rings; (ii) $C_{sp} \equiv C_{sp}$ acetylene bonds; (iii) $C_{sp2}—C_{sp}$ bonds between benzene ring and acetylene bond; and (iv) $C_{sp}—C_{sp}$ bonds between two acetylene bonds [69]. According to the number ($n$) of acetylene bonds between two benzene rings, GY family can be typically classified into GY ($n = 1$), graphdiyne (GDY) ($n = 2$), graphtriyne ($n = 3$), and graphtetrayne ($n = 4$), or uniformly indicated by GY-$n$ (Figure 1.12) [70]. Among them, GDY is one of the most widely studied GY structures, which was first synthesized by Li et al. in 2010 [5].

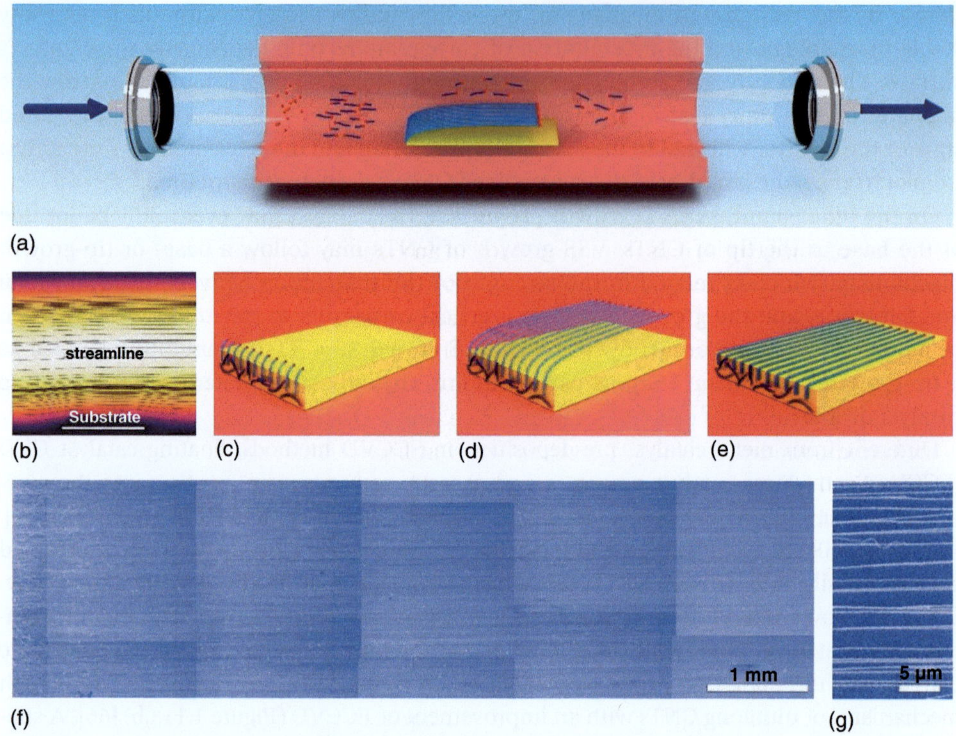


**Figure 1.11** (a) Schematic of ultralong CNT synthesis using SIDS. (b) CFD simulation of the velocity distribution and streamlines adjacent to the substrate. (c–e) Schematics of (c) interception, (d) reorientation and growth, and (e) the resultant array of ultralong CNTs during the FCCVD process with SIDS. (f, g) Scanning electron microscopy (SEM) image of the high-density ultralong CNT arrays with different scale bars. *Source:* Reprinted with permission from Ref. [66]. Copyright 2023, American Chemical Society.

GDY synthesis largely depends on the formation of diacetylene bonds, which can be achieved through an alkyne coupling reaction between two aryl monomers. In the early stages of exploring the synthesis of GDY, studies mainly focused on synthesizing several substructures of GDY or nanoGDY due to the difficulty in controlling the conditions for large ordered couplings [71]. The large-area synthesis of GDY was not extensively studied

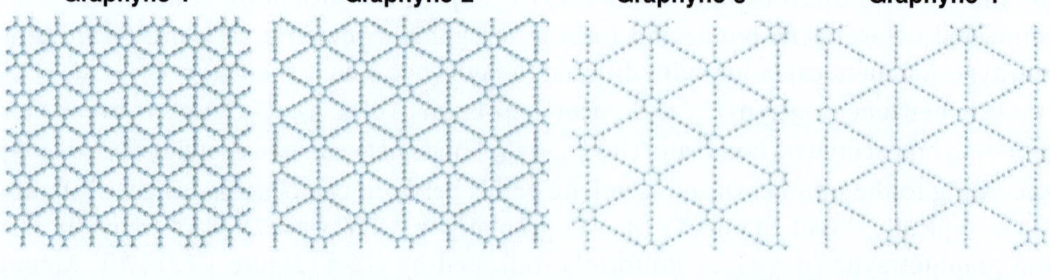


**Figure 1.12** GY containing different numbers ($n = 1, 2, 3, 4$) of inserted acetylene bonds. *Source:* Reprinted with permission from Ref. [70]. Copyright 2023, American Chemical Society.

until Li et al. introduced the cross-coupling reaction of hexaethylbenzene (HEB) and demonstrated the obtained nanoscale GDY films [5]. According to the physical state of the substrate, GDY synthesis can be mainly classified into liquid-phase and solid-phase synthesis.

#### 1.2.3.1 Liquid-phase Synthesis

The initial synthesis of GDY developed by Li et al. was carried out in a liquid-phase environment provided by tetrahydrofuran [5]. The cross-coupling reaction of HEB was driven by the release of trace Cu(II) ions from the Cu foil in an alkaline environment provided by pyridine, which ultimately formed a GDY thin film on Cu. Hence, the Cu foil served a dual role as catalyst and substrate [72]. Li et al. synthesized crystalline GDY (6L, ~2.19 nm) through a modified Glaser–Hay coupling reaction of HEB and directly observed the rhombohedral (ABC) stacking in GDY by low-voltage TEM (Figure 1.13a,b) [73]. Direct characterization of few-layered GDY crystals is difficult due to structural fragility. Apart from growth at the liquid/solid interface, GDY can also grow at liquid/liquid and liquid/gas interfaces [74, 75]. Liquid/liquid interface is usually provided by water and water-immiscible organic solvents. During the reaction process, the catalyst in the aqueous phase and monomers in the organic phase undergo a coupling reaction at the interface and gradually grow into a suspended GDY layer, which can be easily transferred to other substrates using Langmuir–Schäfer method (Figure 1.13c,d). The stability of the liquid/liquid interface can effectively avoid the random interaction between monomer and catalyst [74]. The thickness of GDY can be adjusted by changing the concentrations of monomer and catalyst. GDY modification can be achieved by adding metal ions to the aqueous phase to form metal-GDY composite structures [76].

Besides interface-mediated synthesis, GDY can be synthesized directly in the liquid phase through coupling reaction. He et al. proposed a mild, convenient, and tunable one-pot method for the synthesis of GDY through Pd-catalyzed decarboxylation coupling reaction (Figure 1.14a) [77]. The synthesized GDY was characterized by typical 2D structure, high degree of crystallinity, and exhibited localized folding behaviors (Figure 1.14b,c). Li et al. proposed a space-confined synthesis method using MXene as a

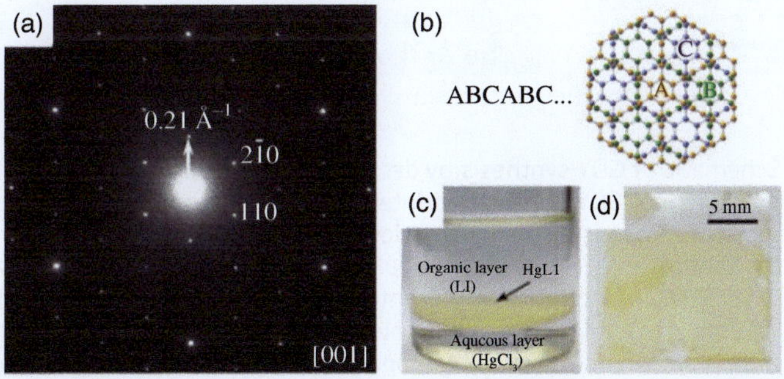


**Figure 1.13** (a) Experimental SAED pattern and (b) schematic stacking mode of ABC-stacked GDY. *Source:* Reprinted with permission from Ref. [73]. Copyright 2018, Tsinghua University Press and Springer-Verlag GmbH Germany. (c, d) Multilayer **HgL1** (L1=tris(4-ethynylphenyl)amine) nanosheets (c) at the liquid/liquid interface and (d) on a slide glass. *Source:* Reprinted with permission from Ref. [75]. Copyright 2021, Wiley-VCH GmbH.)

template for precise control of GDY's thickness and long-range ordering (Figure 1.14d) [78]. The sub-nanometer-sized gap between MXene layers allows monomers to be intercalated and coupled in situ and effectively constrains the out-of-plane growth or vertical stacking of GDY, ultimately yielding ML-GDY within MXene layers (Figure 1.14e). By liquid-phase exfoliation, 2DM mixtures containing free-standing ML-GDY with a lateral size distribution of 0.3–2.4 μm could be obtained, in which the monolayer thickness and ordered in-plane structure of GDY were preserved (Figure 1.14f). Yan et al. directly exfoliated GDY bulk materials into monolayer or multilayer sheets by the assistance of $Li_2SiF_6$ inorganic salt (Figure 1.14g) [79]. The most challenging aspect of GDY exfoliation is the potential introduction of structural defects or internal oxidation, which may affect the properties and application value of the eGDY. The noncovalent interactions generated by $SiF_6^{2-}$ and the intercalation of small-radius $Li^+$ together contributed to the efficient exfoliation and the preservation of GDY's original crystal structure without generating additional oxides (Figure 1.14h,i).

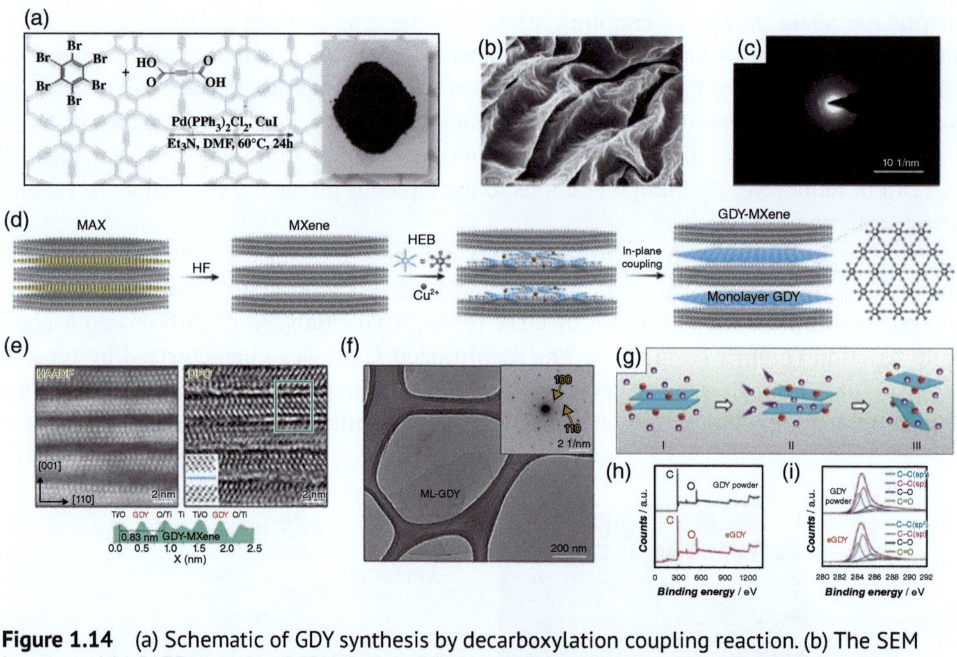


**Figure 1.14** (a) Schematic of GDY synthesis by decarboxylation coupling reaction. (b) The SEM image and (c) SAED pattern of GDY. *Source:* Reprinted with permission from Ref. [77]. Copyright 2023, The Royal Society of Chemistry. (d) Schematics of MXene template strategy for multilayer graphdiyne (MLGDY) synthesis. (e) High Angle Angular Dark Field-Scanning Transmission Electron Microscopy (HAADF-STEM) and Integrated Differential Phase Contrast Transmission Electron Microscopy (iDPC-STEM) of GDY-MXene structure along the [001] axis. The image intensity profile from the marked area is shown below. (f) TEM image and SAED pattern (inset) of free-standing monolayer GDY (ML-GDY). *Source:* Reprinted with permission from Ref. [78]. Copyright 2023, Wiley-VCH GmbH. (g) Salt-assisted exfoliation mechanism of GDY by three steps. Step I: adsorption of anions and cations on GDY; Step II: intercalation of cations into the charged interlayer; Step III: dispersed GDY in salt solution. (h) Wide-range XPS and (i) C1s spectra of bulk GDY powder and exfoliated GDY (eGDY). *Source:* Reprinted with permission from Ref. [79]. Copyright 2019, Wiley-VCH Verlag GmbH & Co. KGaA, Weinheim.

#### 1.2.3.2 Solid-phase Synthesis

Metal-catalyzed, self-limited CVD is also suitable for GDY synthesis (Figure 1.15a,b). Although Cu maintains a high catalytic activity in the liquid-phase synthesis of GDY, in the case of CVD, Ag is more advisable than Cu as a catalyst and substrate for coupling. Unlike graphene CVD using gaseous carbon precursors such as $CH_4$, the CVD process of GDY must be kept at a low temperature to prevent the destruction of the aryl precursor [80]. Liu et al. obtained large ML-GDY films on Ag foils with hexaethynylbenzene (HEB) or 1,3,5-triethynylbenzene (TEB) as precursor at 150°C, which could be transferred by a standard PMMA process for further characterization (Figure 1.15c,d) [80]. Mechanical forces such as ball milling have also been proven to be an effective option for driving solvent-free, alkyne-based coupling reactions without catalyst assistance. Li et al. prepared pure GDY by ball-milling-driven reaction of calcium carbide ($CaC_2$) and hexabromobenzene and confirmed the 1 : 1 $sp/sp^2$-hybridized carbon ratio of GDY by characterization (Figure 1.15e,f) [81]. Previously, the group had successfully prepared hydrogen-substituted GDY by mechanochemical coupling of $CaC_2$ and 1,3,5-tribromobenzene [82].

## 1.3 Properties and Applications of Carbon Nanomaterials

### 1.3.1 Graphene

According to the *1995 IUPAC Recommended Terminology*, the term "graphene" was initially used to describe a single carbon layer with a graphitic structure, which should be used only when discussing the reactions, structural relations, or other properties of individual layers [83, 84]. In fact, with the increase in the number of stacked graphene monolayers, band

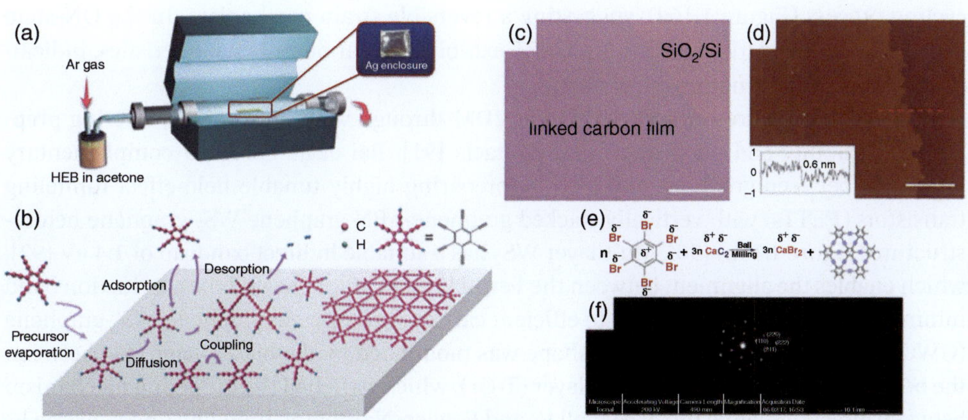


**Figure 1.15** (a, b) Schematics of (a) CVD system for GDY growth on Ag and (b) the surface growth process using HEB as a precursor. (c) Opticalmicroscopy (OM) and (d) Atomic force microscopy (AFM) images of GDY film transferred on $SiO_2$/Si substrate. Inset in (d): height profile along the dashed line. Scale bars: (i) 50 μm and (ii) 1 μm. *Source:* Reprinted with permission from Ref. [80]. Copyright 2017, Wiley-VCH Verlag GmbH & Co. KGaA, Weinheim. (e) Ball-milling-driven reaction for GDY preparation. (f) SAED pattern of as-prepared GDY. *Source:* Reprinted with permission from Ref. [81]. Copyright 2018, Elsevier Ltd.

structure evolves rapidly and approaches the 3D limit of graphite at 10 layers [8, 85]. Thus, the meaning of "graphene" extends from monolayer to 2D stacked systems of up to 10 layers, and it is crucial to make a distinction when focusing on such materials. When discussing specific properties, especially electrical properties, it is necessary to classify graphene into MLG, BLG, trilayer (TLG), and multilayer (<10 layers), as the difference in properties due to the variation in the number of layers will lead to different applications for these graphene structures.

#### 1.3.1.1 Electrical Properties and Applications

Under the tight-binding approximation, intrinsic MLG shows a linear dispersion relation in its Brillouin zone at the K and K′ points (Dirac points) [86, 87]. MLG's conduction and valence bands intersect at Dirac points; hence, it behaves as a zero-bandgap semimetal. Due to the ballistic transport properties of carriers, graphene can exhibit high mobility (theoretical maximum of ~200 000 $cm^2/(V{\cdot}s)$ at room temperature (RT) [88]), while the lack of bandgap becomes a major stumbling block for carrier modulation, which creates difficulties in the application of graphene field effect transistors (GFETs). Therefore, effective carrier modulation methods can significantly improve the performance of GFETs and highlight the advantage of high mobility. Zheng et al. introduced a graphene strain-effect transistor (GSET), which restricts and recovers current conduction through reversible nanocrack in the source/drain metal contacts (Figure 1.16a,b), resulting in a vast $I_{ON}/I_{OFF}$ ratio of over $10^7$ [89]. When a back-gate voltage exceeding the switching threshold ($|V_{BG}| > |V_{ST}|$) was applied to GSET, the piezoelectric lead zirconate titanate back-gate dielectric strained on the Ni/Au contacts and eventually switched off the GSET. The contrary ambipolar characteristics (OFF–ON–OFF) of GSET were quite different from the conventional ambipolar characteristics (ON–OFF–ON) (Figure 1.16c). Furthermore, the switching threshold voltages ($V_{ST,\,P}$ and $V_{ST,\,N}$) of GSET tended to stabilize after a 17-strain cycling process (Figure 1.16d), suggesting a reversible strain mechanism. In the ON-state ($|V_{BG}| < |V_{ST,\,N}|$ and $||V_{BG}| < |V_{ST,\,P}|$), GSET exhibited linear output characteristics, indicating the semimetallic nature of graphene.

vdW heterostructure combines different 2DM through vdW interactions, offering preparation flexibility and good interfacial contacts [91]. Bai et al. achieved complementary properties between graphene and $WS_2$ by preparing highly tunable field-effect tunneling transistors (FETTs) with vertically stacked graphene-hBN-graphene-$WS_2$-graphene heterostructure (Figure 1.16h) [90]. Multilayer $WS_2$ has a suitable indirect bandgap of 1.4 eV [92], which enables the alignment between the Fermi level of graphene and the conduction band minimum of $WS_2$ and contributes to efficient carrier tunneling. In graphene-$WS_2$-graphene (GWG) heterostructure, the barrier shape was modulated by the bias voltage ($V_b$) between the bottom (B-Gr) and top graphene layer (T-Gr), which switched the tunneling mechanism between direct tunneling under a small $V_b$ and Fowler–Nordheim tunneling under a large $V_b$ (Figure 1.16i–k). GWG-FETTs also changed the operating states through $V_b$ between n-type and bipolar complementary metal oxide semiconductor-like tunneling devices (Figure 1.16). Moreover, the thermionic emission mechanism was proved to contribute significantly to the on-current of FETTs (Figure 1.16l), as evidenced by high $I_{ON}/I_{OFF}$ ratios of $1.5 \times 10^6$ and $5 \times 10^8$ at RT (300 K) and low temperature (5 K), respectively, under small $V_b$.

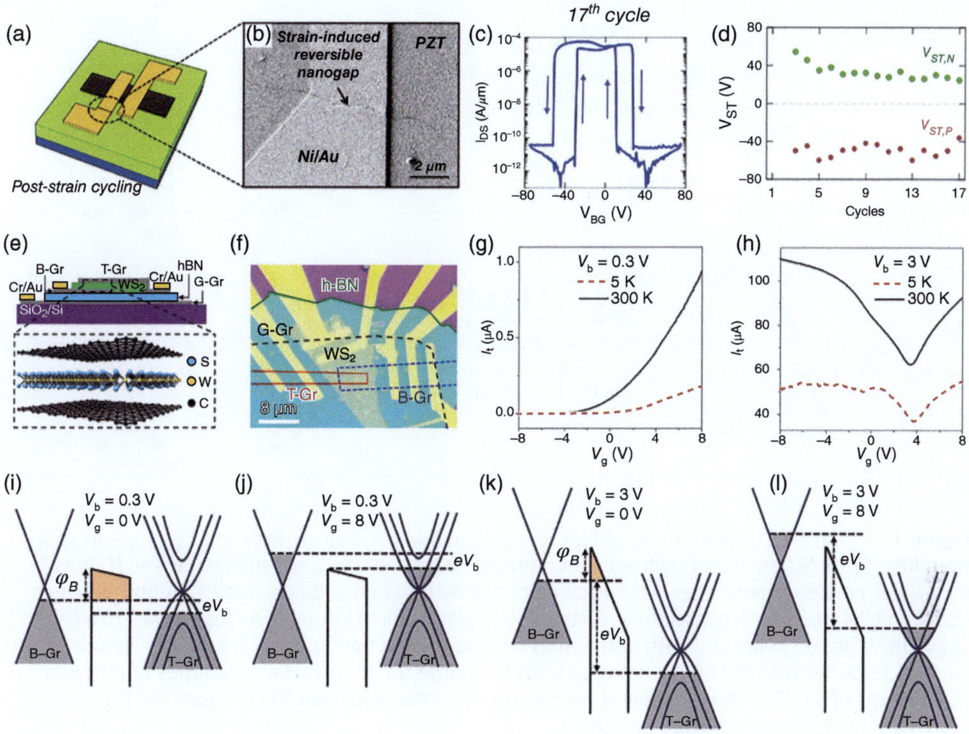

**Figure 1.16** (a) 3D schematic of the post-strain cycling process of GSET. (b) SEM image of the strain-induced nanogap across Ni/Au contact. (c) Transfer characteristics of GSET at the 17th strain cycle. (d) Switching threshold voltages ($V_{ST,P}$ and $V_{ST,N}$) of GSET in 17 strain cycles. *Source:* Reprinted with permission from Ref. [89]. Copyright 2023, American Chemical Society. (e) Schematics of GWG-FETT with the crystalline structure of each layer shown in the enlarged view. (f) OM image of GWG-FETT. (g, h) Corresponding transfer curves of GWG-FETT at (g) $V_b = 0.3V$ and (h) $V_b = 3V$. (i–l) Energy band alignment diagrams at (i, j) $V_b = 0.3V$ and (k, l) $V_b = 3V$ under zero and positive $V_g$. *Source:* Reprinted with permission from Ref. [90]. Copyright 2022, American Chemical Society.

The switching performance of GFETs can also be improved by modifying the band structure of graphene. In the pioneering report by Novoselov et al., the degree of energy band overlap ($\delta\varepsilon$) varied in FLGs with different thicknesses, and the thinnest FLG sample was considered to possess a zero bandgap [2]. The $\delta\varepsilon$ overlap could be used to account for the inversion of Hall coefficients ($R_H$) in FLG with the variation of gate voltage ($V_G$) (Figure 1.17a), suggesting that FLG could be transformed from mixed-carrier material to either fully electronic or entirely hole conductor. Li et al. introduced a bandgap by changing the hybridization form of carbon atoms in graphene (Figure 1.17b) [93]. A reversible hydrogenation process realized this transition in a highly concentrated hydrogen-ion electrolyte, in which the electric field generated by $V_G$ induced the hydrogen adsorption and desorption on GFETs. When the positive $V_G$ increased above the hydrogenation potential, the graphene lattice became highly activated and transformed into $sp^3$-hybridized insulating hydrogenated graphene. As $V_G$ was swept back to negative, the gradual increase in $I_{DS}$ indicated the recovery of graphene

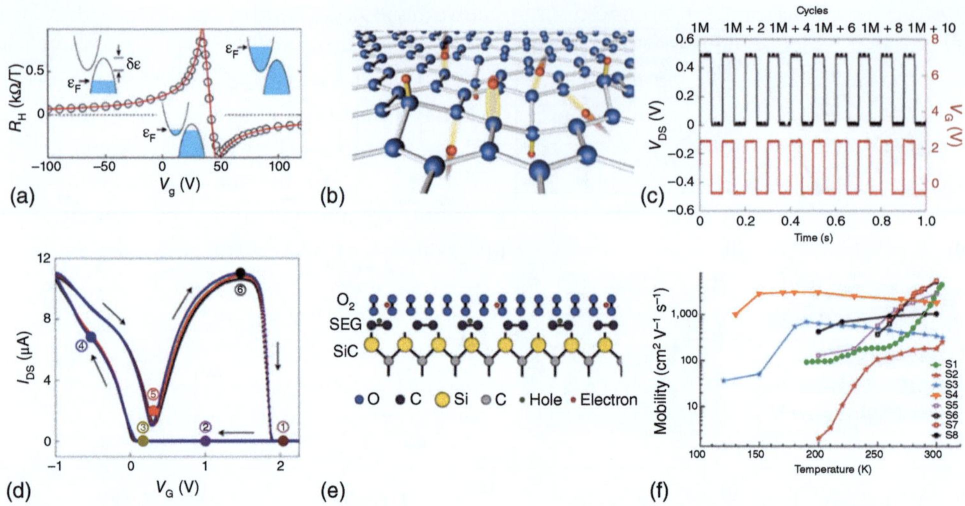

**Figure 1.17** (a) Hall coefficient $R_H$ of FLG versus $V_g$ at $T = 5$ K. *Source:* Reprinted with permission from Ref. [2]. (b) Schematic of hydrogenation process between the graphene lattice and $H^+$ ions. (c) $V_{DS}$ (left) measurement after 1M switching cycles with periodically changed $V_G$ (right) between −0.5V and 2.4V. (d) Three consecutive cycles of $I_{DS}$ measurements in MLG versus $V_G$ with fixed $V_{DS} = 5$mV. *Source:* Reprinted with permission from Ref. [93]. Copyright 2021, Springer Nature Limited. (e) Schematic of SEG p-doping process by charge transfer. (f) Hall mobilities of different SEG devices of S1–S8 versus temperature. *Source:* Reprinted with permission from Ref. [39]. Copyright 2024, Springer Nature Limited.

properties due to dehydrogenation. This reversible, hydrogen-induced conductor–insulator transition was completely observed in MLG (Figure 1.17d), BLG, and TLG with stable $I_{DS}$-$V_G$ curves. The hydrogenation strategy of graphene showed reliable gate-controlled switching capability, remaining operative without any degradation of $V_{DS}$ after 1 million (1M) switching cycles of MLG-FETs (Figure 1.17c). Zhao et al. p-doped the SEG devices by charge transfer of adsorbed oxygen to exhibit the intrinsic transport properties of SEG on SiC (Figure 1.17e; see Section 1.2.1.2 for SEG growth method) [39]. With the gradual temperature increase in the 100–300 K range, the conductivity and Hall mobility of the samples also showed an increasing trend with maximum mobility of 5500 $cm^2/(V \cdot s)$ (Figure 1.17f). The mobilities of S2, S3, and S4 increased rapidly and then tended to saturate with the temperature increase, which reflected the transition of the transport mechanism from localized defect states in the bandgap to high-mobility band transport, confirming the semiconductor nature of SEG. The $I_{ON}/I_{OFF}$ ratio of ~$10^6$ and *SS* of ~60 mV/dec obtained by the density of states calculations demonstrated that the properties of SEG were sufficient for digital electronic applications.

#### 1.3.1.2 Optoelectronic Properties and Applications

The unique properties of graphene have likewise attracted extensive interest in optoelectronics, prompting the construction of a range of graphene-based optoelectronic devices such as photodetectors (PDs), modulators, and hybrids. However, a number of factors, such as low light absorption (~2.3%), large dark currents arising from zero bandgap, and difficulty of separation and collection of ultrafast photocarriers, have limited further development of graphene in optoelectronics [94]. Koepfli et al. adopted a metamaterial integration strategy to maximize light absorption and photogenerated carrier extraction and realized high bandwidth (>500 GHz) and wide operating window (from <1400 nm to >4200 nm) for metamaterial graphene PD [95]. The metamaterial absorber consisted of a relatively simple stack of metal-insulator-MLG-metal-insulator layers to achieve an almost perfect light absorption (Figure 1.18a–e) [96–98]. The dipole resonators were connected with interdigitated metal contacts, and a thin silver layer was added underneath the thicker gold contact layer on every other line, which is the key to more efficient carrier extraction (Figure 1.18f). Under 1550-nm illumination, despite the low responsivity of PD ($R_{ext} = 0.75$ mA/W and $R_{int} = 1.57$ mA/W), the linear behavior of response allowed the PD to maximize the input power up to 100 mW (Figure 1.18), giving rise to high photocurrent and the highest graphene data rate of 132 Gbit/s so far. The

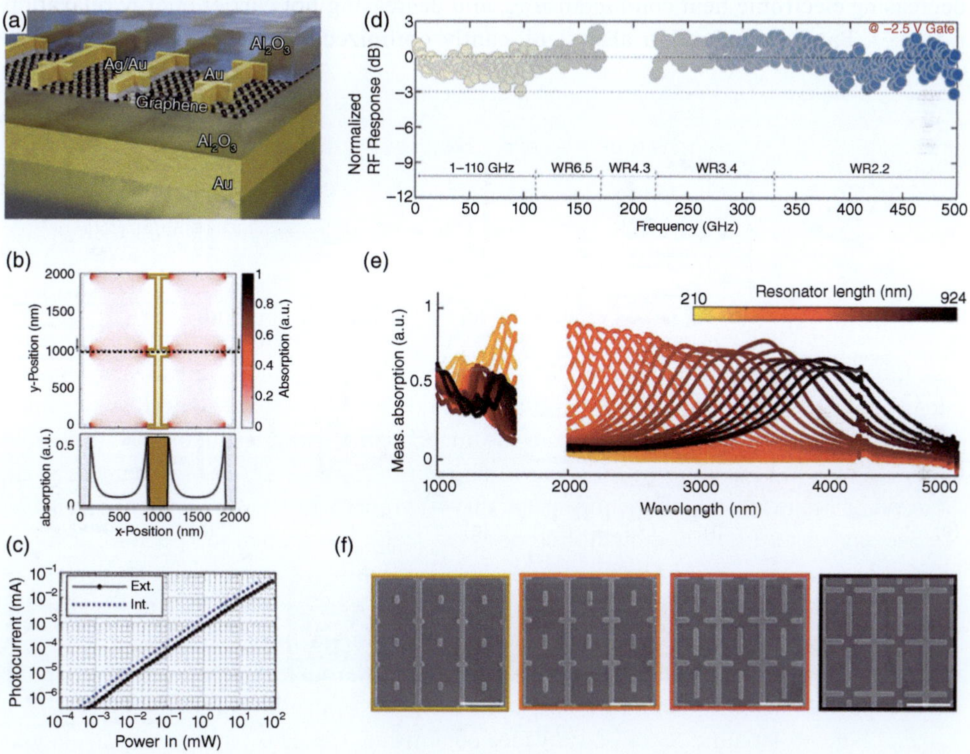


**Figure 1.18** (a) Schematic of Au-$Al_2O_3$-MLG-Ag/Au-$Al_2O_3$ stacked metamaterial layers. (b) Absorption distribution with strong confinement close to the dipole resonators. (c) Extracted photocurrent for an optical input power sweep over five orders of magnitude (from $10^{-3}$ to $10^2$ mW). (d) Normalized frequency response of the graphene PD over a 2- to 500-GHz range. (e) Measured spectral absorption of different metamaterial dipole lengths, which shows the wavelength tunability. (f) Four example SEM images of polarization-independent design, where the colors correspond to the resonator-length scale bar shown in (e). Scale bar: 1 μm. *Source:* Reprinted with permission from Ref. [95]. Copyright 2023, The American Association for the Advancement of Science.

high bandwidth of >500 GHz was mainly attributed to shorter carrier transit time ($\tau_{tr}$) (Figure 1.18), which was related to the structure design of the metamaterial. Through the direct alteration of dipole resonator length, the absorption spectra of PD revealed a tunable range of >3000 nm (Figure 1.18), having potential in both sensing and telecommunication applications.

Stacking of monolayers enriches the optoelectronic applications of graphene. Construction of a homogeneous p-n junction is a common strategy to enhance the performance of graphene PDs since the bandgap of BLG could yield some optoelectronic properties beyond MLG [99]. Titova et al. demonstrated the enhancement of the sub-terahertz ($f = 130$ GHz) response of split-gate BLG p-n junction PDs by an electrically induced bandgap (Figure 1.19a,b) [100, 101]. Bandgap induction at cryogenic temperature (~25 K) could improve the responsivity by enhancing the photo thermoelectric effect and emergence of tunneling rectification [102]. A larger bandgap was conducive to a higher thermoelectric optovoltage by increasing Seebeck coefficient $S$ (Figure 1.19c), decreasing electronic heat conductance $\chi_e$, and decreasing hot carrier energy relaxation rate $\tau\varepsilon^{-1}$. Bandgap induction also significantly optimized the noise equivalent power

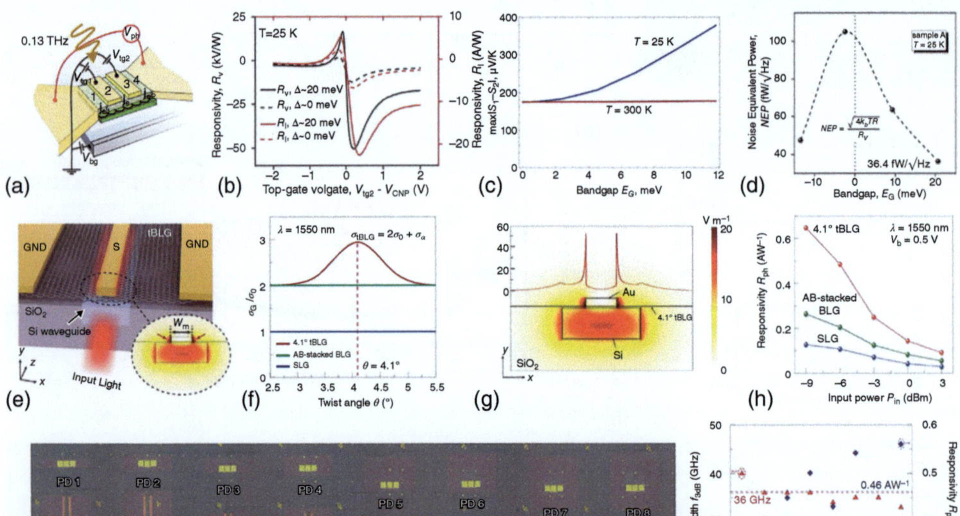


**Figure 1.19** (a) Schematic of split-gate BLG THz PD. (b) Comparison of current and voltage responsivity in the gapless (dashed lines) and gapped (solid lines) states. (c) Max $|S_2 - S_1|$ as a function of $E_G$ at charge neutral point. (d) Minimum attainable NEP for sample A versus $E_G$ at $T = 25$ K. *Source:* Reprinted with permission from [101]. Copyright 2023, American Chemical Society. (e) Schematic of Si waveguide-integrated tBLG PD. (f) Optical photoconductivity $\sigma_G$ of tBLG versus twist angle $\theta$. $\Delta E_{vHS}$ causes the transition $\sigma_\alpha$. (g) Strong field localization around the metal/tBLG interface. (h) Power dependence of responsivity for tBLG, BLG, and SLG PDs at $V_b = 0.5$V. (i) Optical image of tBLG PDs array. Scale bar: 1mm. (j) Statistical plot of responsivity and bandwidth from eight devices in (i), with average bandwidth and responsivity of 36GHz and 0.46A/W, respectively. *Source:* Reprinted with permission from Ref. [106]. Copyright 2024, Springer Nature Limited.

(NEP) of BLG p-n junctions from 100 fW/Hz$^{1/2}$ at zero bandgap down to 36 fW/Hz$^{1/2}$ at the maximum bandgap (~25 meV) (Figure 1.19d). The NEP achieved a comparable level to commercially available detectors at a much lower degree of bandgap [103]. Furthermore, when two graphene layers are stacked with a twist angle $\theta$, the overlap and reconstruction of the Dirac cones produce van Hove singularities (vHS), which is quite different from the case of AB-stacked BLG [104]. The introduction of the vHS significantly enhances the light-matter interactions in twisted bilayer graphene (tBLG), and the energy gap between vHS ($\Delta E_{vHS}$) can be tuned with the change of $\theta$ [105]. Wu et al. demonstrated the potential of tBLG for applications of optical communications utilizing heterogeneous integration with silicon photonics (Figure 1.19e) [106]. Large-area controllable tBLG growth yielded a twist angle of ~4.1° demonstrated by $\Delta E_{vHS}$-induced photoconductivity transition (Figure 1.19f). The incident electric field distribution simulation showed a strong localization at the electrode/tBLG interface, which is favorable for tBLG absorption (Figure 1.19g). Under 1550 nm incident wavelength, tBLG-integrated PD exhibited a high responsivity of 0.65 A/W, which was superior to that of AB-stacked BLG and MLG (Figure 1.19h), and showed uniform performance across eight individual devices and an 8-device array on silicon waveguide (Figure 1.19i,j). The bandwidth analysis showed a bandwidth of over 65 GHz and a high data rate of 50 Gbit/s, indicating the high sensitivity of tBLG PD.

#### 1.3.1.3 Spintronic Properties

Unlike the parabolic *E-k* relationship in conventional semiconductors such as Si, the linear *E-k* relationship in graphene enables its carriers to exhibit Dirac fermion-like behavior with near-zero mass and high Fermi velocity ($v_F$ ~10$^6$ m/s) [86, 107]. Spin-orbit coupling (SOC) in graphene triggers many novel physical effects, such as the quantum Hall effect, spin Hall effect, spin quantum Hall effect, and quantum anomalous Hall effect (QAHE) [108]. Recently, Sha et al. observed the ferromagnetic state in rhombohedral (ABCA)-stacked tetralayer graphene (4LG) through graphene/$WSe_2$ heterostructure (Figure 1.20a) [109]. The proximity-induced SOC by $WSe_2$ and spontaneous time-reversal symmetry breaking in ABCA-stacked 4LG led to the observation of QAHE in graphene. At a vertical displacement field of $D = -0.1$ V/nm, transport measurements on ABCA-4LG/$WSe_2$ showed ferromagnetism and Hall resistance quantization (Figure 1.20b,c), which was further demonstrated as a Chern insulator with Chern number of four. Besides, ABCA-4LG with $WSe_2$ exhibited a sequential transition of layer-antiferromagnetic (LAF) insulator, Chern insulator (CI), and layer-polarized insulator (LPI) states when $|D|$ was increased from zero. In contrast, the intermediate CI state was not observed in 4LG without $WSe_2$, suggesting that the $WSe_2$-induced SOC had a crucial role in the ferromagnetism generation of 4LG (Figure 1.20d,e). More interestingly, the ferromagnetism in ABCA-4LG/$WSe_2$ heterostructure could be modulated not only by magnetic field but also by displacement field and carrier concentration.

#### 1.3.1.4 Superconductive Properties

The unconventional superconductivity of graphene was experimentally observed by Cao et al. in a tBG system near a slight "magic angle" ($\theta$ ~1.1°) (Figure 1.21a) [110, 111]. Previously, Bistritzer et al. had predicted the electronic structure of tBG varying

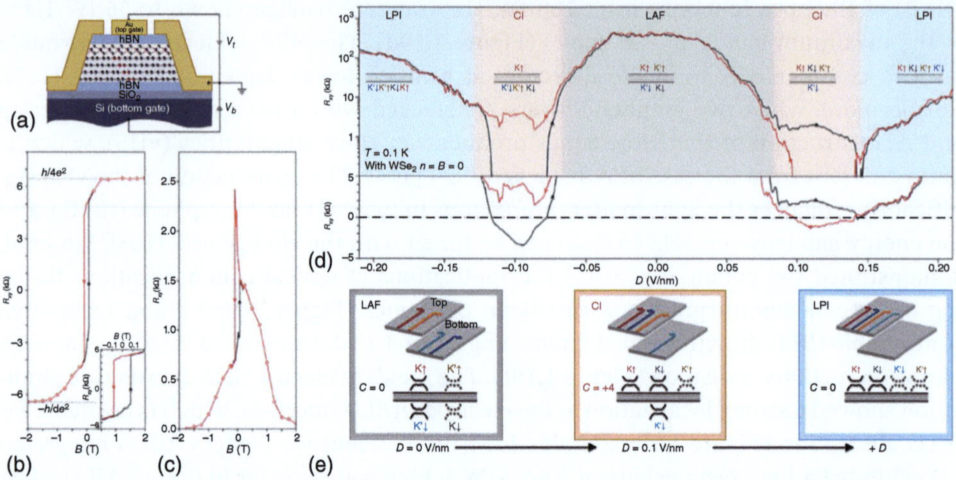


**Figure 1.20** (a) Schematic of a dual-gate $WSe_2$/ABCA-4LG device. (b, c) Magnitude field-dependent (b) Hall resistance $R_{xy}$ and (c) longitudinal resistance $R_{xx}$ at $D = -0.1$ V/nm, $T = 0.1$ K. (d) $R_{xy}$ as a function of $D$ at $B = 0$ in ABCA-4LG with SOC ($WSe_2$). Insets in (d) indicate the layer polarization of spin-valley flavors for different broken-symmetry states. (e) Schematics of the Chern numbers and Hall conductivity contributions of four spin-valley flavor pairs for different broken-symmetry states. *Source:* Reprinted with permission from Ref. [109]. Copyright 2024, The American Association for the Advancement of Science.

with periodic Moiré pattern, in which strong interlayer coupling leads to a flat energy band structure as $\theta$ approaches the "magic angle" [112]. The $v_F$ of electrons in these flat bands is significantly reduced and macroscopically manifests as a Mott-like insulating state of tBG, as confirmed in the experiments of Cao et al. (Figure 1.21c) [111]. Moreover, measurements of the conductance of magic-angle tBG (MATBG) at 70 mK showed a typical V-shaped conductance caused by the renormalized Dirac cones near the charge neutrality point ($n = 0$). However, they indicated the suppression of a superconducting state by a magnetic field in the vicinity of −2 electrons per Moiré unit cell (Figure 1.21b). In Figure 1.21c, a superconducting region was also observed near the Mott state with the highest critical temperature $T_c$ ~1.7 K. More peculiarly, the carrier concentration leading to this $T_c$ was merely $1.5 \times 10^{11}$ cm$^{-2}$, which could be only explained by strong electron–electron interactions. Besides tBLG, the superconductivity can be realized in other stacked systems of graphene as well, such as ABC-TLG/h-BN heterostructure [113].

Researchers have been actively exploring applications of superconductivity in MATBG. An example given by Klein et al. is the reproducible bistability switching of MATBG between superconductor, metal, and correlated insulator states (Figure 1.21d,e) [114]. This bistability was accomplished through the gate hysteresis in the top-BN/MATBG/bottom-BN dual-gate structure [115], which could be readily controlled by gate voltage regulation (Figure 1.21 f,g). Characterization of the superconducting state with filling factor

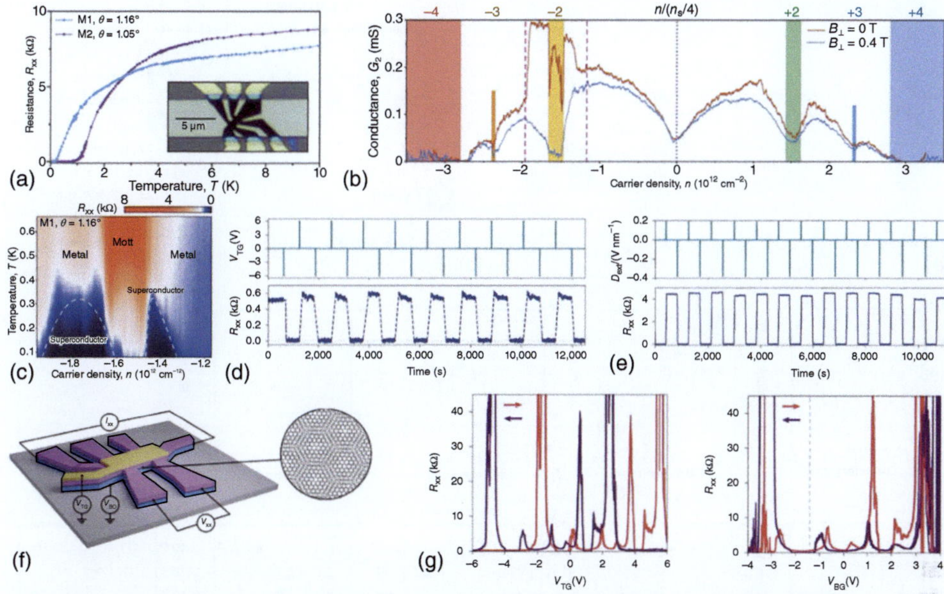


**Figure 1.21** (a) Four-probe resistance $R_{xx}$ of two devices, M1 and M2, with $\theta = 1.16°$ and $\theta = 1.05°$, respectively. Inset: optical image of M1. (b) Two-probe conductance $G_2$ of M1 measured in $B_\perp = 0$ (red) and $B_\perp = 0.4$ T (blue). (c) Measured $R_{xx}$ of M1 versus carrier density ($n$) and temperature ($T$). Metal, Mott, and superconductor regions are labeled. *Source:* Reprinted with permission from Ref. [111]. Copyright 2018, Springer Nature Limited. (d) Switching between metal and superconducting states with top-gate voltage $V_{TG} = \pm 6$V and fixed bottom-gate voltage $V_{BG} = -1.4V$ under $T = 65$mK. (e) Switching between correlated insulator and superconducting states with alternating displacement field $D_{ext}$ and fixed carrier density $n_{ext}$ under $T = 65$mK. (f) Schematic of dual-gate MATBG device in a Hall bar geometry. Inset: moiré pattern of the MATBG heterostructure. (g) $R_{xx}$ versus $V_{TG}$ and $V_{BG}$ with sweep directions up and down. *Source:* Reprinted with permission from Ref. [114]. Copyright 2023, Springer Nature Limited.

$\nu = \pm(2 + \delta)$ demonstrated its robustness on both electron and hole sides, leading to highly tunable MATBG superconducting devices.

### 1.3.2 Carbon Nanotubes

Since CNTs can be considered as rolled-up graphene layers, the carrier transport properties of CNTs are similar to those of graphene. However, unlike the zero-bandgap characteristics of graphene, the metallic or semiconductive properties of CNTs are determined by whether or not the chiral vector $(n, m)$ crosses the Dirac K-points [49], which is equivalent to judge whether mod $(2n + m, 3)$ is equal to zero. Metallic CNTs are satisfied with the relation mod $(2n + m, 3) = 0$, while semiconductive CNTs are not (Figure 1.8b). Therefore, CNTs exhibiting metallic and semiconductive properties should account for 1/3 and 2/3 of the total [49, 116].

Semiconductive SWCNTs can also be grouped into Type I or II with mod $(2n + m, 3) = 1$ or 2, and those with the same $2n + m$ value are classified as the same family [117].

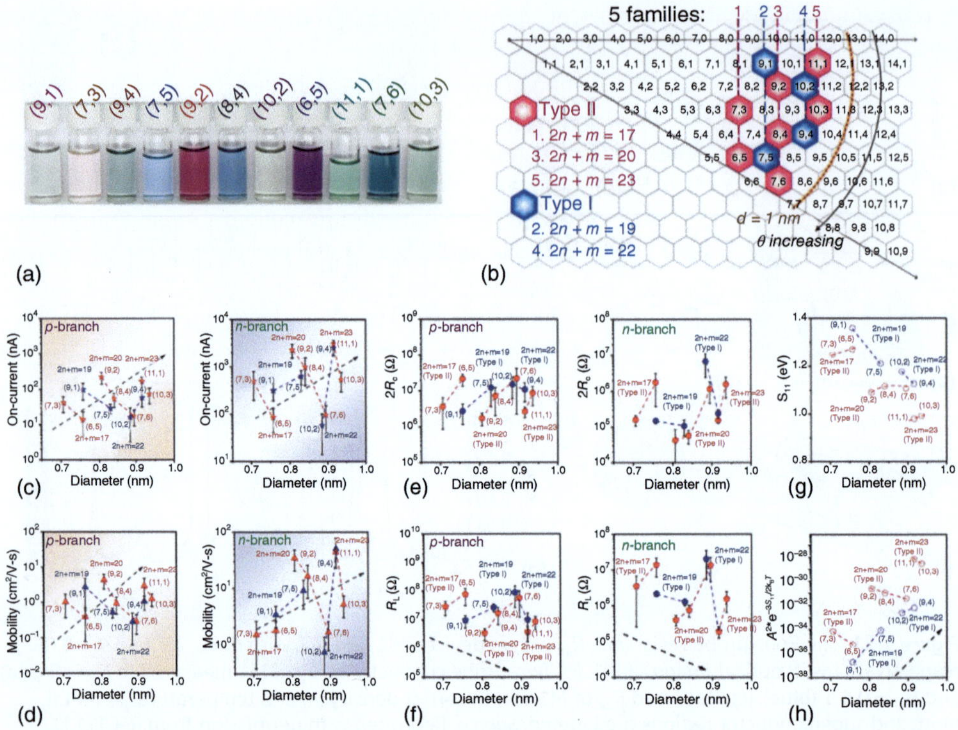


**Figure 1.22** (a) Photograph of single-chirality SWCNT solutions. (b) SWCNTs with different types and families are on the chiral map. (c) On-current, (d) mobility, (e) $2R_C$, and (f) $R_L$ of both p- and n-branch of SWCNTs as a function of diameter from statistical data. (g) $S_{11}$ and (h) $A^2e^{-3S_{11}/2k_BT}$ of SWCNTs as a function of diameter from further analysis. *Source:* Reprinted with permission from Ref. [118]. Copyright 2023, Springer Nature Limited.

Su et al. prepared 11 different types of single-chirality SWCNTs for electrical measurements by effective separation and purification (Figure 1.22a,b) [118, 119]. Electrical measurements, along with resistance analysis of SWCNTs, were carried out based on density-controlled SWCNT thin film transistors. Different types of semiconductive SWCNTs showed opposite trends of transport properties as the chiral angle increased. Specifically, for Type II SWCNTs, the on-current and mobility of both p- and n-branch within the same family decreased with increasing chiral angle, while it was contrary for Type I (Figure 1.22c,d). Measurements of various resistance components in SWCNT films, such as contact resistance ($R_C$) and channel resistance ($R_L$), also showed a similar dependence on the chiral angle (Figure 1.22e,f). By further analysis, $R_C$ and $R_L$ can be respectively attributed to the first bandgap ($S_{11}$) and relative junction conductance $G_{ij}^{NT} = A^2\exp[(-3S_{11})/2k_BT]$ of SWCNTs, showing a chiral dependence consistent with $R_C$ and $R_L$ (Figure 1.22g,h). However, experimental results in the p-branch of $2n + m = 19$ family were not by theoretical predictions, possibly due to interference from impurities or interfacial states. The systematic study on the relationship between chirality and electrical properties of semiconductive SWCNTs is instructive for the design of SWCNT devices.

The 1D morphology of CNTs facilitates the assembly of high-density, highly aligned arrays, which is a significant advantage for high-performance CNT-FETs, especially in the post-Moore era with shrinking feature sizes. Lin et al. reported aligned CNT-FETs with promising applications when scaled down to a sub-10 nm node [120]. With a contact gate pitch (CGP) of 175 nm (Figure 1.23a), CNT-FETs showed excellent p-type performance with an on current ($I_{ON}$) of 2.24 mA/μm and a peak transconductance ($g_m$) of 1.64 mS/μm, which is superior to the 45 nm silicon technology node (CGP = 160 nm). A CNT-FET-based six-transistor (6T) static random access memory (SRAM) was also fabricated at 175 nm CGP with an area of 0.976 μm$^2$ (Figure 1.23b), which improved the integration density while avoiding complex silicon-based back-end interconnections. To further scale down to a 10 nm node (CGP ~54 nm), a full-contact structure was employed for CNT-FETs, which significantly reduced the contact resistance while combining the benefits of side-contact and end-contact structures (Figure 1.23c,d). By scaling the gate length ($L_g$) and contact length ($L_{con}$) down to 35 nm and 20 nm (CGP ~61 nm), an $I_{ON}$ of 2.43 mA/μm and a $g_m$ of 2.45 mS/μm were achieved, which are comparable to the performance of Si- P-Channel Metal Oxide Semiconductor Field Effect Transistors (PMOSFETs) at 10 nm node (Figure 1.23e) [121].

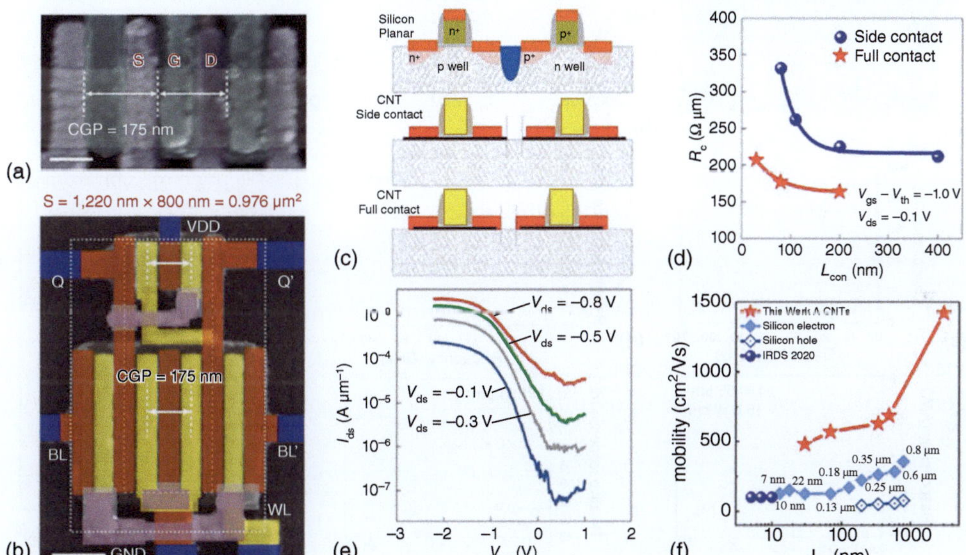


**Figure 1.23** (a) False-color SEM image of three top-gated CNT-FETs in series with shared source/drain contacts with 175 nm CGP. Scale bar: 100 nm. (b) False-color SEM image of CNT 6T SRAM cell with 175 nm CGP. Scale bar: 200 nm. (c) Schematic of the silicon planar structure, CNT side-contact structure, and CNT full-contact structure. (d) $R_c$ versus $L_{con}$ for side-contact and full-contact structure. (e) Transfer characteristics of the ultrascaled CNT-FET with 61 nm CGP. (f) Benchmarking the mobility of CNT-FETs at different $L_g$ with that of Si FETs from Intel and IRDS 2020. *Source:* Reprinted with permission from Ref. [120]. Copyright 2023, Springer Nature Limited.

More attractively, the mobility of CNT-FETs is at least four times higher than the electron or hole mobility of Si transistors at any gate length (Figure 1.23f), highlighting the great value for applications in digital integrated circuits.

Unlike 2D graphene, 1D CNT has a direct bandgap that is inversely proportional to its diameter ($E_g$ ~0.7 eV/$d$ (nm) [122, 123]), which enables arrays or thin films of CNTs with different diameters to cover a wide spectrum for photodetection [124]. Also, photo-excited electron-hole pairs interact strongly in the confined space of 1D CNTs, leading to large exciton binding energies. Hence, exciton behavior significantly impacts the performance of CNT optoelectronic devices [125]. Wu et al. demonstrated a high-density array CNT PD operational in 2-μm band photodetection, which provides a viable solution for high-speed optical communications (Figure 1.24a,b) [122]. The CNT arrays showed high alignment with the degree of polarization $P$ ~89%, and the estimated $E_g$ ~0.48 eV of CNTs demonstrated the ability for 2-μm band communications. To reduce the bandwidth limitation by resistance–conductance time constant, quartz was selected as the substrate of CNT PD, which effectively reduced parasitic capacitance and achieved a low total capacitance of ~12 fF(Figure 1.24c). At an incident wavelength of

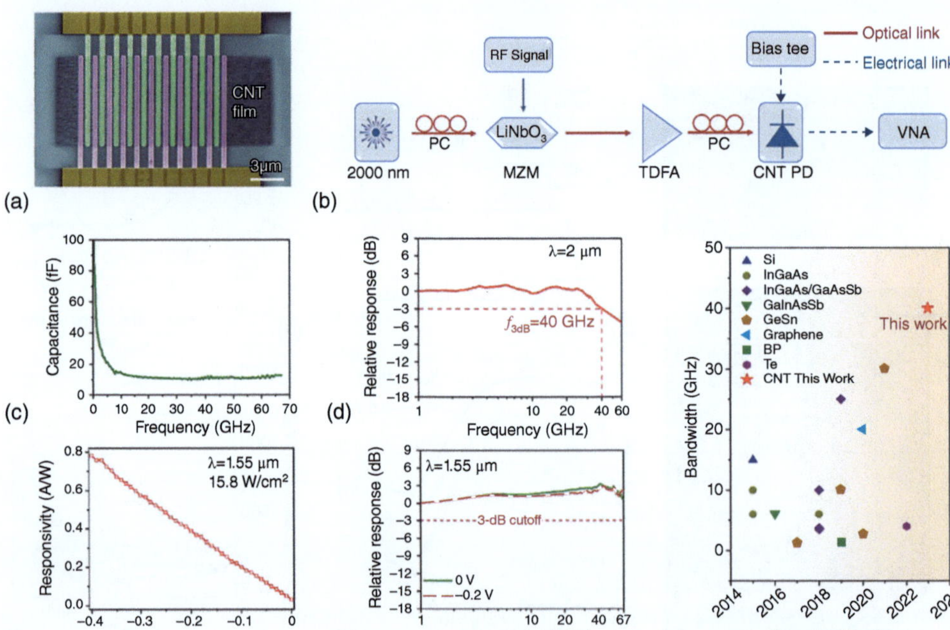


**Figure 1.24** (a) False-colored SEM image of aligned CNT PD with a channel length of 200 nm and a total channel width of 200 μm. (b) Frequency response measurement setup of CNT PD at 2 μm. (c) The capacitance of CNT PD as a function of frequency. (d) At $V = -0.2$ V, the relative response of CNT PD versus modulation frequency with $\lambda = 2$ μm. (e) Benchmarking of bandwidth of state-of-the-art high-speed PDs at the 2 μm band based on different materials. (f) The current responsivity of CNT PD versus $V$ with $\lambda = 1.55$ μm. (g) At different $V = 0$ and $V = -0.2$ V, the relative response of CNT PD versus modulation frequency with $\lambda = 1.55$ μm. *Source:* Reprinted with permission from Ref. [122]. Copyright 2023, American Chemical Society.

2 μm, CNT PD achieved a current responsivity of up to 0.62 A/W (Figure 1.24d), an NEP of <10 $pW/Hz^{1/2}$, and a bandwidth of >40 GHz, which is higher than any other state-of-the-art high-speed PDs in recent years (Figure 1.24e,f). In the commonly used band of 1.55 μm, CNT PD also achieved an outstanding performance of >67 GHz bandwidth and 0.78 A/W responsivity (Figure 1.24g), revealing the great potential of CNT-based optoelectronic devices for high-speed, low-power optical applications.

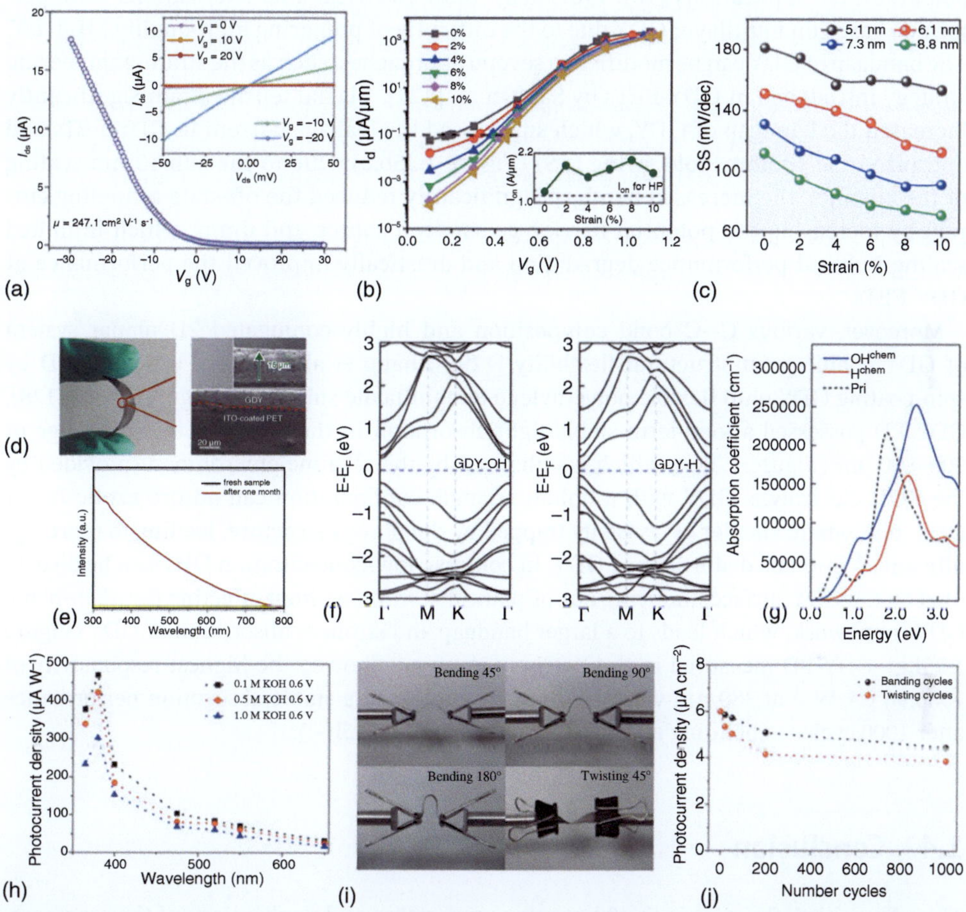


**Figure 1.25** (a) Transfer characteristics of ML-GDY FET at $V_{ds} = 0.05$ V. Inset: Output characteristics under various $V_g$ from −20 V to 20 V. *Source:* Reprinted with permission from Ref. [78]. Copyright 2023, Wiley-VCH GmbH. (b) Strain-dependent transfer characteristics of GDY-FET with a channel length of 7.3nm. Inset: $I_{ON}$ under different strains. (c) *SS* of GDY-FET with different channel lengths versus strain. *Source:* Reprinted with permission from Ref. [127]. Copyright 2020, Elsevier B.V. (d) Digital macrograph and SEM image of the PD with GDY thickness of 15μm. (e) UV–vis absorption spectra of GDY PD. (f) Band structures of $OH^-$ and $H^+$ chemically adsorbed GDY. (g) Comparison of the absorption spectra between pristine GDY and GDY with $OH^-$ and $H^+$ adsorption. (h) Responsivity of GDY PD versus incident wavelength at different KOH concentrations. (i) Photographs of GDY PD bending and twisting tests. (j) Photocurrent density of GDY PD versus number cycles of bending (black) and twisting (red) tests. *Source:* Reprinted with permission from Ref. [128]. Copyright 2020, Wiley-VCH Verlag GmbH & Co. KGaA, Weinheim.

### 1.3.3 Graphyne

Unlike graphene, GDY has a bandgap of ~1.04 eV, which is close to the bandgap value of silicon. However, the direct nature of bandgap may allow GDY to yield superior electronic and optoelectronic performance to silicon [78, 126]. Li et al. applied the prepared ML-GDY (see Section 1.2.3.1 for synthesis method) to FETs and measured an average mobility of 231.4 $cm^2/(V{\cdot}s)$ and a switching ratio of $1.2 \times 10^3$ (Figure 1.25a), indicating potential FET applications [78]. Previously, most electrical characterizations were performed based on multilayer GDY due to the difficulty of preparing high-quality ML-GDY. The bandgap of GDY can be modified in several approaches, such as the strain-engineering strategy introduced in GDY-FETs by Sang et al. [127]. Biaxial tensile strain significantly increased the bandgap of GDY, which suppressed the leakage current in GDY-FETs and optimized the subthreshold swing (*SS*) (Figure 1.25b,c). Due to the sub-10 nm scaling of the channel, the increased bandgap significantly reduced the off-state tunneling current by increasing the potential barrier $\varphi_B$ between source and drain, which inhibited scaling-induced performance degradation and drastically improved the performance of GDY-FETs.

Moreover, various C–C bond composition and highly conjugated 2D planar system of GDY enable great structural flexibility [129]. Zhang et al. prepared a flexible PD by spin-coating GDY onto flexible polyethylene terephthalate substrates (Figure 1.25d) [128]. GDY PD possessed a long-term stable light absorption in the short-wavelength range of 300–800 μm (Figure 1.25e), which is enhanced by the alkaline environment provided by the KOH electrolyte. GDY with a hollow triangle configuration can adsorb exotic $H^+$ or $OH^-$ ions, where smaller $H^+$ is easily trapped by the porous structure, leading to more significant *ab* in-plane distortions of GDY. In contrast, low-concentration $OH^-$ can be distributed on the GDY surface, inducing out-of-plane distortion without affecting the *ab* in-plane GDY framework, which leads to a larger bandgap and stronger absorption of GDY (Figure 1.25f,g). GDY PD measured in the KOH environment showed the highest responsivity of 1086.96 $\mu A{\cdot}W^{-1}$ at 380 nm wavelength and maintained a stable absorption performance after 1000 cycles of bending and twisting tests (Figure 1.25h–j).

## 1.4 Conclusion

This chapter briefly reviews the fabrication, properties, and applications of the representative carbon nanomaterials graphene, CNTs, and GY. Relevant research advances demonstrate the impressive properties and prospects of carbon nanomaterials. Represented by graphene, carbon nanomaterials are considered to be one of the candidates for replacing silicon and perpetuating Moore's law in transistor-based microelectronics. However, these materials do not yet have the mass production capability comparable to silicon-based technology. As graphene celebrates its $20^{th}$ anniversary, while researchers continue to explore the novel phenomena induced by carbon-based nanomaterials, it is still necessary to keep looking for ways to build a bridge from proof-of-concept to large-scale applications.

## References


1 Ehrenfreund, P., Irvine, W., Becker, L., Blank, J., Brucato, J.R., Colangeli, L., Derenne, S., Despois, D., Dutrey, A., Fraaije, H., Lazcano, A., Owen, T., Robert, F., ISSI-Team, A.I.S.S.I. Astrophysical and astrochemical insights into the origin of life. *Rep. Prog. Phys.* 2002, 65 (10): 1427–1487.

2 Novoselov, K.S., Geim, A.K., Morozov, S.V., Jiang, D., Zhang, Y., Dubonos, S.V., Grigorieva, I.V., Firsov, A.A. Electric field effect in atomically thin carbon films. *Science* 2004, 306 (5696): 666–669.

3 Li, X., Wang, X., Zhang, L., Lee, S., Dai, H. Chemically derived, ultrasmooth graphene nanoribbon semiconductors. *Science* 2008, 319 (5867): 1229–1232.

4 Iijima, S. Helical microtubules of graphitic carbon. *Nature* 1991, 354 (6348): 56–58.

5 Li, G., Li, Y., Liu, H., Guo, Y., Li, Y., Zhu, D. Architecture of graphdiyne nanoscale films. *Chem. Commun.* 2010, 46 (19): 3256.

6 Kroto, H.W., Heath, J.R., O'Brien, S.C., Curl, R.F., Smalley, R.E. C60: Buckminsterfullerene. *Nature* 1985, 318 (6042): 162–163.

7 Xu, X., Ray, R., Gu, Y., Ploehn, H.J., Gearheart, L., Raker, K., Scrivens, W.A. Electrophoretic analysis and purification of fluorescent single-walled carbon nanotube fragments. *J. Am. Chem. Soc.* 2004, 126 (40): 12736–12737.

8 Geim, A.K., Novoselov, K.S. The rise of graphene. *Nat. Mater.* 2007, 6 (3): 183–191.

9 Landau, L.D. Zur Theorie Der Phasenumwandlungen II. *Phys. Z Sowjetunion* 1937, 11 (545): 26–35.

10 Peierls, R. Quelques Propriétés Typiques Des corps solides. In: *Annales de l'institut Henri Poincaré* 1935, 5: 177–222.

11 Land, T., Michely, T., Behm, R., Hemminger, J., Comsa, G. STM investigation of single layer graphite structures produced on Pt (111) by hydrocarbon decomposition. *Surf. Sci.* 1992, 264 (3): 261–270.

12 Dresselhaus, M.S., Dresselhaus, G. Intercalation compounds of graphite. *Adv. Phys.* 2002, 51 (1): 1–186.

13 Wang, X.-Y., Narita, A., Müllen, K. Precision synthesis versus bulk-scale fabrication of graphenes. *Nat. Rev. Chem.* 2017, 2 (1): 0100.

14 Moon, J.-Y., Kim, M., Kim, S.-I., Xu, S., Choi, J.-H., Whang, D., Watanabe, K., Taniguchi, T., Park, D.S., Seo, J., Cho, S.H., Son, S.-K., Lee, J.-H. Layer-engineered large-area exfoliation of graphene. *Sci. Adv.* 2020, 6 (44): eabc6601.

15 Shim, J., Bae, S.-H., Kong, W., Lee, D., Qiao, K., Nezich, D., Park, Y.J., Zhao, R., Sundaram, S., Li, X., Yeon, H., Choi, C., Kum, H., Yue, R., Zhou, G., Ou, Y., Lee, K., Moodera, J., Zhao, X., Ahn, J.-H., Hinkle, C., Ougazzaden, A., Kim, J. Controlled crack propagation for atomic precision handling of wafer-scale two-dimensional materials. *Science* 2018, 362 (6415): 665–670.

16 Ming-Yuan, H., Hutchinson, J.W. Crack deflection at an interface between dissimilar elastic materials. *Int. J. Solids Struct.* 1989, 25 (9): 1053–1067.

17 Huang, Y., Pan, Y.-H., Yang, R., Bao, L.-H., Meng, L., Luo, H.-L., Cai, Y.-Q., Liu, G.-D., Zhao, W.-J., Zhou, Z., Wu, L.-M., Zhu, Z.-L., Huang, M., Liu, L.-W., Liu, L., Cheng, P., Wu, K.-H., Tian, S.-B., Gu, C.-Z., Shi, Y.-G., Guo, Y.-F., Cheng, Z.G., Hu, J.-P., Zhao, L.,

Yang, G.-H., Sutter, E., Sutter, P., Wang, Y.-L., Ji, W., Zhou, X.-J., Gao, H.-J. Universal mechanical exfoliation of large-area 2D crystals. *Nat. Commun.* 2020, 11 (1): 2453.

**18** Hernandez, Y., Nicolosi, V., Lotya, M., Blighe, F.M., Sun, Z., De, S., McGovern, I.T., Holland, B., Byrne, M., Gun'Ko, Y.K., Boland, J.J., Niraj, P., Duesberg, G., Krishnamurthy, S., Goodhue, R., Hutchison, J., Scardaci, V., Ferrari, A.C., Coleman, J.N. High-yield production of graphene by liquid-phase exfoliation of graphite. *Nat. Nanotechnol.* 2008, 3 (9): 563–568.

**19** León, V., Rodriguez, A.M., Prieto, P., Prato, M., Vázquez, E. Exfoliation of graphite with triazine derivatives under ball-milling conditions: preparation of few-layer graphene via selective noncovalent interactions. *ACS Nano* 2014, 8 (1): 563–571.

**20** Paton, K.R., Varrla, E., Backes, C., Smith, R.J., Khan, U., O'Neill, A., Boland, C., Lotya, M., Istrate, O.M., King, P., Higgins, T., Barwich, S., May, P., Puczkarski, P., Ahmed, I., Moebius, M., Pettersson, H., Long, E., Coelho, J., O'Brien, S.E., McGuire, E.K., Sanchez, B.M., Duesberg, G.S., McEvoy, N., Pennycook, T.J., Downing, C., Crossley, A., Nicolosi, V., Coleman, J.N. Scalable production of large quantities of defect-free few-layer graphene by shear exfoliation in liquids. *Nat. Mater.* 2014, 13 (6): 624–630.

**21** Ciesielski, A., Samorì, P. Graphene via sonication assisted liquid-phase exfoliation. *Chem. Soc. Rev.* 2014, 43 (1): 381–398.

**22** Liu, W., Zheng, X., Xu, Q. Supercritical $CO_2$ directional-assisted synthesis of low-dimensional materials for functional applications. *Small* 2023, 19 (33): 2301097.

**23** Romang, A.H., Watkins, J.J. Supercritical fluids for the fabrication of semiconductor devices: emerging or missed opportunities? *Chem. Rev.* 2010, 110 (1): 459–478.

**24** Jones, C.A., Zweber, A., DeYoung, J.P., McClain, J.B., Carbonell, R., DeSimone, J.M. Applications of "dry" processing in the microelectronics industry using carbon dioxide. *Crit. Rev. Solid State Mater. Sci.* 2004, 29 (3–4): 97–109.

**25** Zhu, H.-Y., Wang, Q.-B., Yin, J.-Z. High-pressure induced exfoliation for regulating the morphology of graphene in supercritical $CO_2$ system. *Carbon* 2021, 178: 211–222.

**26** Zhao, M., Casiraghi, C., Parvez, K. Electrochemical exfoliation of 2D materials beyond graphene. *Chem. Soc. Rev.* 2024, 53 (6): 3036–3064.

**27** Lee, H., Jin, S., Chung, J., Kim, M., Lee, S.W. Electrochemical production of two-dimensional atomic layer materials and their application for energy storage devices. *Chem. Phys. Rev.* 2023, 4 (1): 011306.

**28** Mirkhani, S.A., Iqbal, A., Kwon, T., Chae, A., Kim, D., Kim, H., Kim, S.J., Kim, M.-K., Koo, C.M. Reduction of electrochemically exfoliated graphene films for high-performance electromagnetic interference shielding. *ACS Appl. Mater. Interfaces* 2021, 13 (13): 15827–15836.

**29** Wang, S., Xue, J., Xu, D., He, J., Dai, Y., Xia, T., Huang, Y., He, Q., Duan, X., Lin, Z. Electrochemical molecular intercalation and exfoliation of solution-processable two-dimensional crystals. *Nat. Protoc.* 2023, 18 (9): 2814–2837.

**30** Yang, S., Zhang, P., Nia, A.S., Feng, X. Emerging 2D materials produced via electrochemistry. *Adv. Mater.* 2020, 32 (10): 1907857.

**31** Lu, X., Cai, M., Wu, X., Zhang, Y., Li, S., Liao, S., Lu, X. Controllable synthesis of 2D materials by electrochemical exfoliation for energy storage and conversion application. *Small* 2023, 19 (9): 2206702.

**32** Brodie, B.C. Sur Le Poids Atomique Du Graphite. *Ann. Chim. Phys.* 1860, 59: 466–472.

**33** Staudenmaier, L. Verfahren Zur Darstellung Der Graphitsäure. *Berichte Dtsch. Chem. Ges.* 1898, 31 (2): 1481–1487.

**34** Hummers, W.S., Offeman, R.E. Preparation of graphitic oxide. *J. Am. Chem. Soc.* 1958, 80 (6): 1339–1339.

**35** Zelenák, F., Kováčová, M., Moravec, Z., Černák, M., Krumpolec, R. Fast, scalable, and environmentally friendly method for production of stand-alone ultrathin reduced graphene oxide paper. *Carbon* 2023, 215: 118436.

**36** Tao, R., Li, F., Lu, X., Liu, F., Xu, J., Kong, D., Zhang, C., Tan, X., Ma, S., Shi, W., Mo, R., Lu, Y. High-conductivity–dispersibility graphene made by catalytic exfoliation of graphite for lithium-ion battery. *Adv. Funct. Mater.* 2021, 31 (6): 2007630.

**37** Van Bommel, A.J., Crombeen, J.E., Van Tooren, A. LEED and auger electron observations of the SiC (0001) surface. *Surf. Sci.* 1975, 48 (2): 463–472.

**38** Forbeaux, I., Themlin, J.-M., Debever, J.-M. Heteroepitaxial graphite on 6 H – SiC (0001): interface formation through conduction-band electronic structure. *Phys. Rev. B* 1998, 58 (24): 16396–16406.

**39** Zhao, J., Ji, P., Li, Y., Li, R., Zhang, K., Tian, H., Yu, K., Bian, B., Hao, L., Xiao, X., Griffin, W., Dudeck, N., Moro, R., Ma, L., De Heer, W.A. Ultrahigh-mobility semiconducting epitaxial graphene on silicon carbide. *Nature* 2024, 625 (7993): 60–65.

**40** De Heer, W.A., Berger, C., Ruan, M., Sprinkle, M., Li, X., Hu, Y., Zhang, B., Hankinson, J., Conrad, E. Large area and structured epitaxial graphene produced by confinement controlled sublimation of silicon carbide. *Proc. Natl. Acad. Sci.* 2011, 108 (41): 16900–16905.

**41** Bao, J., Yasui, O., Norimatsu, W., Matsuda, K., Kusunoki, M. Sequential control of step-bunching during graphene growth on SiC (0001). *Appl. Phys. Lett.* 2016, 109 (8): 081602.

**42** Sun, Z., Yan, Z., Yao, J., Beitler, E., Zhu, Y., Tour, J.M. Growth of graphene from solid carbon sources. *Nature* 2010, 468 (7323): 549–552.

**43** Qi, T., Shao, Q., Qi, H., Xiao, J., Zhang, Y., Wang, Y., Jin, X., Chen, Y., Li, D., Zang, X., Cao, N. CVD growth of graphene on copper-plated scrap steel without external carbon source. *Nano Res.* 2023, 16 (10): 12263–12269.

**44** Yi, X., Song, Q., Chen, Q., Zhao, C., Watanabe, K., Taniguchi, T., Chen, Q., Yan, C., Wang, S. Quantum transport in CVD graphene synthesized with liquid carbon precursor. *Nanotechnology* 2022, 33 (35): 355601.

**45** Yu, Q., Lian, J., Siriponglert, S., Li, H., Chen, Y.P., Pei, S.-S. Graphene segregated on Ni surfaces and transferred to insulators. *Appl. Phys. Lett.* 2008, 93 (11): 113103.

**46** Kim, K.S., Zhao, Y., Jang, H., Lee, S.Y., Kim, J.M., Kim, K.S., Ahn, J.-H., Kim, P., Choi, J.-Y., Hong, B.H. Large-scale pattern growth of graphene films for stretchable transparent electrodes. *Nature* 2009, 457 (7230): 706–710.

**47** Li, X., Cai, W., An, J., Kim, S., Nah, J., Yang, D., Piner, R., Velamakanni, A., Jung, I., Tutuc, E., Banerjee, S.K., Colombo, L., Ruoff, R.S. Large-area synthesis of high-quality and uniform graphene films on copper foils. *Science* 2009, 324 (5932): 1312–1314.

**48** Zhang, J., Liu, X., Zhang, M., Zhang, R., Ta, H.Q., Sun, J., Wang, W., Zhu, W., Fang, T., Jia, K., Sun, X., Zhang, X., Zhu, Y., Shao, J., Liu, Y., Gao, X., Yang, Q., Sun, L., Li, Q., Liang, F., Chen, H., Zheng, L., Wang, F., Yin, W., Wei, X., Yin, J., Gemming, T., Rummeli, M.H., Liu, H., Peng, H., Lin, L., Liu, Z. Fast synthesis of large-area bilayer graphene film on Cu. *Nat. Commun.* 2023, 14 (1): 3199.

**49** Saito, R., Fujita, M., Dresselhaus, G., Dresselhaus, M.S. Electronic structure of chiral graphene tubules. *Appl. Phys. Lett.* 1992, 60 (18): 2204–2206.

**50** Muhulet, A., Miculescu, F., Voicu, S.I., Schütt, F., Thakur, V.K., Mishra, Y.K. Fundamentals and scopes of doped carbon nanotubes towards energy and biosensing applications. *Mater. Today Energy* 2018, 9: 154–186.

**51** Venkataraman, A., Amadi, E.V., Chen, Y., Papadopoulos, C. Carbon nanotube assembly and integration for applications. *Nanoscale Res. Lett.* 2019, 14 (1): 220.

**52** Dresselhaus, M.S., Dresselhaus, G., Saito, R. Physics of carbon nanotubes. *Carbon* 1995, 33 (7): 883–891.

**53** Liu, Y., Ge, Z., Li, Z., Chen, Y. High-power instant-synthesis technology of carbon nanomaterials and nanocomposites. *Nano Energy* 2021, 80: 105500.

**54** Zain Mehdi, S.M., Shin, T.H., Abbas, S.Z., Kwon, H., Seo, Y., Kim, D., Hong, S.J., Goak, J.C., Lee, N. Superior field emission characteristics of highly crystalline and thermally stable carbon nanotubes grown in N2 and O2 by arc discharge. *Ceram. Int.* 2023, 49 (3): 4668–4676.

**55** Ribeiro, H., Schnitzler, M.C., Da Silva, W.M., Santos, A.P. Purification of carbon nanotubes produced by the electric arc-discharge method. *Surf. Interfaces* 2021, 26: 101389.

**56** Guo, T., Nikolaev, P., Thess, A., Colbert, D.T., Smalley, R.E. Catalytic growth of single-walled nanotubes by laser vaporization. *Chem. Phys. Lett.* 1995, 243 (1–2): 49–54.

**57** Chen, S., Chen, Y., Xu, H., Lyu, M., Zhang, X., Han, Z., Liu, H., Yao, Y., Xu, C., Sheng, J., Xu, Y., Gao, L., Gao, N., Zhang, Z., Peng, L., Li, Y. Single-walled carbon nanotubes synthesized by laser ablation from coal for field-effect transistors. *Mater. Horiz.* 2023, 10 (11): 5185–5191.

**58** Qin, X., Li, Y., Chernov, A.I., He, M., Zhao, W., Gao, Y., Li, Y., Qiu, S., Li, Q. Towards the preparation of single-walled carbon nanotubes with average diameter of 1.2 nm. *Carbon* 2024, 224: 119059.

**59** Li, Y., Kim, W., Zhang, Y., Rolandi, M., Wang, D., Dai, H. Growth of single-walled carbon nanotubes from discrete catalytic nanoparticles of various sizes. *J. Phys. Chem. B* 2001, 105 (46): 11424–11431.

**60** Zheng, J., Yang, R., Xie, L., Qu, J., Liu, Y., Li, X. Plasma-assisted approaches in inorganic nanostructure fabrication. *Adv. Mater.* 2010, 22 (13): 1451–1473.

**61** Huang, S., Woodson, M., Smalley, R., Liu, J. Growth mechanism of oriented long single walled carbon nanotubes using "fast-heating" chemical vapor deposition process. *Nano Lett.* 2004, 4 (6): 1025–1028.

**62** Zhu, Z., Cui, C., Bai, Y., Gao, J., Jiang, Y., Li, B., Wang, Y., Zhang, Q., Qian, W., Wei, F. Advances in precise structure control and assembly toward the carbon nanotube industry. *Adv. Funct. Mater.* 2022, 32 (11): 2109401.

**63** Liu, F., Yuan, M., He, Y., Zhai, G., Xiang, H., Wu, Q., Zhu, M. Insights into the nucleation and structure of lignin-based carbon nanotubes synthesized using iron via floating catalyst chemical vapor deposition. *ACS Appl. Nano Mater.* 2024, 7 (7): 7184–7194.

**64** Xu, Y., Miao, X., Chen, H., Gu, X., Liu, Y., Ma, C., Zhang, X., Zou, J. Fe complex-based catalyst system for the chirality-selected continuous synthesis of single-walled carbon nanotubes. *ACS Appl. Nano Mater.* 2024, 7 (5): 5525–5533.

**65** Hou, P., Zhang, F., Zhang, L., Liu, C., Cheng, H. Synthesis of carbon nanotubes by floating catalyst chemical vapor deposition and their applications. *Adv. Funct. Mater.* 2022, 32 (11): 2108541.

**66** Jiang, Q., Wang, F., Li, R., Li, B., Wei, N., Gao, N., Xu, H., Zhao, S., Huang, Y., Wang, B., Zhang, W., Wu, X., Zhang, S., Zhao, Y., Shi, E., Zhang, R. Synthesis of ultralong carbon nanotubes with ultrahigh yields. *Nano Lett.* 2023, 23 (2): 523–532.

**67** Zhang, R., Zhang, Y., Zhang, Q., Xie, H., Qian, W., Wei, F. Growth of half-meter long carbon nanotubes based on Schulz–Flory distribution. *ACS Nano* 2013, 7 (7): 6156–6161.

**68** Baughman, R.H., Eckhardt, H., Kertesz, M. Structure-property predictions for new planar forms of carbon: layered phases containing $sp^2$ and *sp* atoms. *J. Chem. Phys.* 1987, 87 (11): 6687–6699.

**69** Zheng, X., Chen, S., Li, J., Wu, H., Zhang, C., Zhang, D., Chen, X., Gao, Y., He, F., Hui, L., Liu, H., Jiu, T., Wang, N., Li, G., Xu, J., Xue, Y., Huang, C., Chen, C., Guo, Y., Lu, T.-B., Wang, D., Mao, L., Zhang, J., Zhang, Y., Chi, L., Guo, W., Bu, X.-H., Zhang, H., Dai, L., Zhao, Y., Li, Y. Two-dimensional carbon graphdiyne: advances in fundamental and application research. *ACS Nano* 2023, 17 (15): 14309–14346.

**70** Li, H., Lim, J.H., Lv, Y., Li, N., Kang, B., Lee, J.Y. Graphynes and graphdiynes for energy storage and catalytic utilization: theoretical insights into recent advances. *Chem. Rev.* 2023, 123 (8): 4795–4854.

**71** Hu, G., He, J., Chen, J., Li, Y. Synthesis of a wheel-shaped nanographdiyne. *J. Am. Chem. Soc.* 2023, 145 (9): 5400–5409.

**72** Zhou, J., Gao, X., Liu, R., Xie, Z., Yang, J., Zhang, S., Zhang, G., Liu, H., Li, Y., Zhang, J., Liu, Z. Synthesis of graphdiyne nanowalls using acetylenic coupling reaction. *J. Am. Chem. Soc.* 2015, 137 (24): 7596–7599.

**73** Li, C., Lu, X., Han, Y., Tang, S., Ding, Y., Liu, R., Bao, H., Li, Y., Luo, J., Lu, T. Direct imaging and determination of the crystal structure of six-layered graphdiyne. *Nano Res.* 2018, 11 (3): 1714–1721.

**74** Matsuoka, R., Sakamoto, R., Hoshiko, K., Sasaki, S., Masunaga, H., Nagashio, K., Nishihara, H. Crystalline graphdiyne nanosheets produced at a gas/liquid or liquid/liquid interface. *J. Am. Chem. Soc.* 2017, 139 (8): 3145–3152.

**75** Xu, L., Sun, J., Tang, T., Zhang, H., Sun, M., Zhang, J., Li, J., Huang, B., Wang, Z., Xie, Z., Wong, W. Metallated graphynes as a new class of photofunctional 2D organometallic nanosheets. *Angew. Chem. Int. Ed.* 2021, 60 (20): 11326–11334.

**76** Huang, R., Kou, Z., Rodriguez, R.D., Wang, X., Hou, Y., Wang, T., Zhang, T. Two-dimensional silver-metalated graphdiyne nanosheets for photocatalytic degradation of antibiotics. *ACS Appl. Nano Mater.* 2023, 6 (9): 7395–7403.

**77** He, S., Wu, B., Xia, Z., Guo, P., Li, Y., Song, S. One-pot synthesis of gamma-graphyne supported pd nanoparticles with high catalytic activity. *Nanoscale Adv.* 2023, 5 (9): 2487–2492.

**78** Li, J., Cao, H., Wang, Q., Zhang, H., Liu, Q., Chen, C., Shi, Z., Li, G., Kong, Y., Cai, Y., Shen, J., Wu, Y., Lai, Z., Han, Y., Zhang, J. Space-confined synthesis of monolayer graphdiyne in MXene interlayer. *Adv. Mater.* 2024, 36 (7): 2308429.

**79** Yan, H., Yu, P., Han, G., Zhang, Q., Gu, L., Yi, Y., Liu, H., Li, Y., Mao, L. High-yield and damage-free exfoliation of layered graphdiyne in aqueous phase. *Angew. Chem. Int. Ed.* 2019, 58 (3): 746–750.

**80** Liu, R., Gao, X., Zhou, J., Xu, H., Li, Z., Zhang, S., Xie, Z., Zhang, J., Liu, Z. Chemical vapor deposition growth of linked carbon monolayers with acetylenic scaffoldings on silver foil. *Adv. Mater.* 2017, 29 (18): 1604665.

**81** Li, Q., Li, Y., Chen, Y., Wu, L., Yang, C., Cui, X. Synthesis of γ-graphyne by mechanochemistry and its electronic structure. *Carbon* 2018, 136: 248–254.

**82** Lee, J., Li, Y., Tang, J., Cui, X. Synthesis of hydrogen substituted graphyne through mechanochemistry and its electrocatalytic properties. *Acta Phys.-Chim. Sin.* 2018, 34 (9): 1080–1087.

**83** Fitzer, E., Kochling, K.-H., Boehm, H.P., Marsh, H. Recommended terminology for the description of carbon as a solid (IUPAC Recommendations 1995). *Pure Appl. Chem.* 1995, 67 (3): 473–506.

**84** Boehm, H. Graphene—How a laboratory curiosity suddenly became extremely interesting. *Angew. Chem. Int. Ed.* 2010, 49 (49): 9332–9335.

**85** Partoens, B., Peeters, F.M. From graphene to graphite: electronic structure around the K point. *Phys. Rev. B* 2006, 74 (7): 075404.

**86** Wallace, P.R. The band theory of graphite. *Phys. Rev.* 1947, 71 (9): 622–634.

**87** Castro Neto, A.H., Guinea, F., Peres, N.M.R., Novoselov, K.S., Geim, A.K. The electronic properties of graphene. *Rev. Mod. Phys.* 2009, 81 (1): 109–162.

**88** Morozov, S.V., Novoselov, K.S., Katsnelson, M.I., Schedin, F., Elias, D.C., Jaszczak, J.A., Geim, A.K. Giant intrinsic carrier mobilities in graphene and its bilayer. *Phys. Rev. Lett.* 2008, 100 (1): 016602.

**89** Zheng, Y., Sen, D., Das, S., Das, S. Graphene strain-effect transistor with colossal ON/OFF current ratio enabled by reversible nanocrack formation in metal electrodes on piezoelectric substrates. *Nano Lett.* 2023, 23 (7): 2536–2543.

**90** Bai, Z., Xiao, Y., Luo, Q., Li, M., Peng, G., Zhu, Z., Luo, F., Zhu, M., Qin, S., Novoselov, K. Highly tunable carrier tunneling in vertical graphene–$WS_2$ –graphene van Der Waals heterostructures. *ACS Nano* 2022, 16 (5): 7880–7889.

**91** Novoselov, K.S., Mishchenko, A., Carvalho, A., Castro Neto, A.H. 2D Materials and van Der Waals heterostructures. *Science* 2016, 353 (6298): aac9439.

**92** Braga, D., Gutiérrez Lezama, I., Berger, H., Morpurgo, A.F. Quantitative determination of the band gap of $WS_2$ with ambipolar ionic liquid-gated transistors. *Nano Lett.* 2012, 12 (10): 5218–5223.

**93** Li, S., Li, J., Wang, Y., Yu, C., Li, Y., Duan, W., Wang, Y., Zhang, J. Large transport gap modulation in graphene via electric-field-controlled reversible hydrogenation. *Nat. Electron.* 2021, 4 (4): 254–260.

**94** Chen, X., Shehzad, K., Gao, L., Long, M., Guo, H., Qin, S., Wang, X., Wang, F., Shi, Y., Hu, W., Xu, Y., Wang, X. Graphene hybrid structures for integrated and flexible optoelectronics. *Adv. Mater.* 2020, 32 (27): 1902039.

**95** Koepfli, S.M., Baumann, M., Koyaz, Y., Gadola, R., Güngör, A., Keller, K., Horst, Y., Nashashibi, S., Schwanninger, R., Doderer, M., Passerini, E., Fedoryshyn, Y., Leuthold, J. Metamaterial graphene photodetector with bandwidth exceeding 500 Gigahertz. *Science* 2023, 380 (6650): 1169–1174.

**96** Landy, N.I., Sajuyigbe, S., Mock, J.J., Smith, D.R., Padilla, W.J. Perfect metamaterial absorber. *Phys. Rev. Lett.* 2008, 100 (20): 207402.

**97** Xia, F., Mueller, T., Lin, Y., Valdes-Garcia, A., Avouris, P. Ultrafast graphene photodetector. *Nat. Nanotechnol.* 2009, 4 (12): 839–843.

**98** Sagade, A.A., Neumaier, D., Schall, D., Otto, M., Pesquera, A., Centeno, A., Elorza, A. Z., Kurz, H. Highly air stable passivation of graphene based field effect devices. *Nanoscale* 2015, 7 (8): 3558–3564.

**99** Tian, P., Tang, L., Teng, K.S., Xiang, J., Lau, S.P. Recent advances in graphene homogeneous p–n junction for optoelectronics. *Adv. Mater. Technol.* 2019, 4 (7): 1900007.

**100** Mylnikov, D.A., Titova, E.I., Kashchenko, M.A., Safonov, I.V., Zhukov, S.S., Semkin, V.A., Novoselov, K.S., Bandurin, D.A., Svintsov, D.A. Terahertz photoconductivity in bilayer graphene transistors: evidence for tunneling at gate-induced junctions. *Nano Lett.* 2023, 23 (1): 220–226.

**101** Titova, E., Mylnikov, D., Kashchenko, M., Safonov, I., Zhukov, S., Dzhikirba, K., Novoselov, K.S., Bandurin, D.A., Alymov, G., Svintsov, D. Ultralow-noise terahertz detection by p–n junctions in gapped bilayer graphene. *ACS Nano* 2023, 17 (9): 8223–8232.

**102** Gayduchenko, I., Xu, S.G., Alymov, G., Moskotin, M., Tretyakov, I., Taniguchi, T., Watanabe, K., Goltsman, G., Geim, A.K., Fedorov, G., Svintsov, D., Bandurin, D.A. Tunnel field-effect transistors for sensitive terahertz detection. *Nat. Commun.* 2021, 12 (1): 543.

**103** Icking, E., Banszerus, L., Wörtche, F., Volmer, F., Schmidt, P., Steiner, C., Engels, S., Hesselmann, J., Goldsche, M., Watanabe, K., Taniguchi, T., Volk, C., Beschoten, B., Stampfer, C. Transport spectroscopy of ultraclean tunable band gaps in bilayer graphene. *Adv. Electron. Mater.* 2022, 8 (11): 2200510.

**104** Brihuega, I., Mallet, P., González-Herrero, H., Trambly De Laissardière, G., Ugeda, M.M., Magaud, L., Gómez-Rodríguez, J.M., Ynduráin, F., Veuillen, J.-Y. Unraveling the intrinsic and robust nature of van Hove singularities in twisted bilayer graphene by scanning tunneling microscopy and theoretical analysis. *Phys. Rev. Lett.* 2012, 109 (19): 196802.

**105** Jain, S.K., Juričić, V., Barkema, G.T. Structure of twisted and buckled bilayer graphene. *2D Mater.* 2016, 4 (1): 015018.

**106** Wu, Q., Qian, J., Wang, Y., Xing, L., Wei, Z., Gao, X., Li, Y., Liu, Z., Liu, H., Shu, H., Yin, J., Wang, X., Peng, H. Waveguide-integrated twisted bilayer graphene photodetectors. *Nat. Commun.* 2024, 15 (1): 3688.

**107** Novoselov, K.S., Geim, A.K., Morozov, S.V., Jiang, D., Katsnelson, M.I., Grigorieva, I.V., Dubonos, S.V., Firsov, A.A. Two-dimensional gas of massless Dirac fermions in graphene. *Nature* 2005, 438 (7065): 197–200.

**108** Zhang, G., Wu, H., Yang, L., Jin, W., Zhang, W., Chang, H. Graphene-based spintronics. *Appl. Phys. Rev.* 2024, 11 (2): 021308.

**109** Sha, Y., Zheng, J., Liu, K., Du, H., Watanabe, K., Taniguchi, T., Jia, J., Shi, Z., Zhong, R., Chen, G. Observation of a Chern insulator in crystalline ABCA-tetralayer graphene with spin-orbit coupling. *Science* 2024, 384 (6694): 414–419.

**110** Cao, Y., Fatemi, V., Demir, A., Fang, S., Tomarken, S.L., Luo, J.Y., Sanchez-Yamagishi, J.D., Watanabe, K., Taniguchi, T., Kaxiras, E., Ashoori, R.C., Jarillo-Herrero, P. Correlated insulator behaviour at half-filling in magic-angle graphene superlattices. *Nature* 2018, 556 (7699): 80–84.

**111** Cao, Y., Fatemi, V., Fang, S., Watanabe, K., Taniguchi, T., Kaxiras, E., Jarillo-Herrero, P. Unconventional superconductivity in magic-angle graphene superlattices. *Nature* 2018, 556 (7699): 43–50.

**112** Bistritzer, R., MacDonald, A.H. Moiré bands in twisted double-layer graphene. *Proc. Natl. Acad. Sci.* 2011, 108 (30): 12233–12237.

**113** Chen, G., Sharpe, A.L., Gallagher, P., Rosen, I.T., Fox, E.J., Jiang, L., Lyu, B., Li, H., Watanabe, K., Taniguchi, T., Jung, J., Shi, Z., Goldhaber-Gordon, D., Zhang, Y., Wang, F. Signatures of tunable superconductivity in a trilayer graphene Moiré superlattice. *Nature* 2019, 572 (7768): 215–219.

**114** Klein, D.R., Xia, L.-Q., MacNeill, D., Watanabe, K., Taniguchi, T., Jarillo-Herrero, P. Electrical switching of a bistable Moiré superconductor. *Nat. Nanotechnol.* 2023, 18 (4): 331–335.

**115** Zheng, Z., Ma, Q., Bi, Z., De La Barrera, S., Liu, M.-H., Mao, N., Zhang, Y., Kiper, N., Watanabe, K., Taniguchi, T., Kong, J., Tisdale, W.A., Ashoori, R., Gedik, N., Fu, L., Xu, S.-Y., Jarillo-Herrero, P. Unconventional ferroelectricity in Moiré heterostructures. *Nature* 2020, 588 (7836): 71–76.

**116** Tang, D.-M., Cretu, O., Ishihara, S., Zheng, Y., Otsuka, K., Xiang, R., Maruyama, S., Cheng, H.-M., Liu, C., Golberg, D. Chirality engineering for carbon nanotube electronics. *Nat. Rev. Electr. Eng.* 2024, 1 (3): 149–162.

**117** Samsonidze, G.G., Saito, R., Kobayashi, N., Grüneis, A., Jiang, J., Jorio, A., Chou, S.G., Dresselhaus, G., Dresselhaus, M.S. Family behavior of the optical transition energies in single-wall carbon nanotubes of smaller diameters. *Appl. Phys. Lett.* 2004, 85 (23): 5703–5705.

**118** Su, W., Li, X., Li, L., Yang, D., Wang, F., Wei, X., Zhou, W., Kataura, H., Xie, S., Liu, H. Chirality-dependent electrical transport properties of carbon nanotubes obtained by experimental measurement. *Nat. Commun.* 2023, 14 (1): 1672.

**119** Yang, D., Li, L., Li, X., Xi, W., Zhang, Y., Liu, Y., Wei, X., Zhou, W., Wei, F., Xie, S., Liu, H. Preparing high-concentration individualized carbon nanotubes for industrial separation of multiple single-chirality species. *Nat. Commun.* 2023, 14 (1): 2491.

**120** Lin, Y., Cao, Y., Ding, S., Zhang, P., Xu, L., Liu, C., Hu, Q., Jin, C., Peng, L.-M., Zhang, Z. Scaling aligned carbon nanotube transistors to a sub-10 nm node. *Nat. Electron.* 2023, 6 (7): 506–515.

**121** Auth, C., Aliyarukunju, A., Asoro, M., Bergstrom, D., Bhagwat, V., Birdsall, J., Bisnik, N., Buehler, M., Chikarmane, V., Ding, G., Fu, Q., Gomez, H., Han, W., Hanken, D., Haran, M., Hattendorf, M., Heussner, R., Hiramatsu, H., Ho, B., Jaloviar, S., Jin, I., Joshi, S., Kirby, S., Kosaraju, S., Kothari, H., Leatherman, G., Lee, K., Leib, J., Madhavan, A., Maria, K., Meyer, H., Mule, T., Parker, C., Parthasarathy, S., Pelto, C., Pipes, L., Post, I., Prince, M., Rahman, A., Rajamani, S., Saha, A., Santos, J.D., Sharma, M., Sharma, V., Shin, J., Sinha, R., Smith, R., Sprinkle, M., St Amour, A., Staus, C., Suri, R., Towner, D., Tripathi, A., Tura, A., Ward, C., Yeoh, A. A 10nm high performance and low-power CMOS technology featuring 3rd generation FinFET transistors, self-aligned quad patterning, contact over active gate and cobalt local interconnects. In: *2017 IEEE International Electron Devices Meeting (IEDM)*, IEEE, 2017.

**122** Wu, W., Ma, H., Cai, X., Han, B., Li, Y., Xu, K., Lin, H., Zhang, F., Chen, Z., Zhang, Z., Peng, L.-M., Wang, S. High-speed carbon nanotube photodetectors for 2 μm communications. *ACS Nano* 2023, 17 (15): 15155–15164.

**123** Charlier, J.-C., Blase, X., Roche, S. Electronic and transport properties of nanotubes. *Rev. Mod. Phys.* 2007, 79 (2): 677–732.

**124** Yang, L., Wang, S., Zeng, Q., Zhang, Z., Peng, L. Carbon nanotube photoelectronic and photovoltaic devices and their applications in infrared detection. *Small* 2013, 9 (8): 1225–1236.

**125** Avouris, P., Freitag, M., Perebeinos, V. Carbon-nanotube photonics and optoelectronics. *Nat. Photonics* 2008, 2 (6): 341–350.

**126** Fang, Y., Liu, Y., Qi, L., Xue, Y., Li, Y. 2D graphdiyne: an emerging carbon material. *Chem. Soc. Rev.* 2022, 51 (7): 2681–2709.

**127** Sang, P., Ma, X., Wang, Q., Wei, W., Wang, F., Wu, J., Zhan, X., Li, Y., Chen, J. Toward high-performance monolayer graphdiyne transistor: strain engineering matters. *Appl. Surf. Sci.* 2021, 536: 147836.

**128** Zhang, Y., Huang, P., Guo, J., Shi, R., Huang, W., Shi, Z., Wu, L., Zhang, F., Gao, L., Li, C., Zhang, X., Xu, J., Zhang, H. Graphdiyne-based flexible photodetectors with high responsivity and detectivity. *Adv. Mater.* 2020, 32 (23): 2001082.

**129** He, J., Wang, N., Cui, Z., Du, H., Fu, L., Huang, C., Yang, Z., Shen, X., Yi, Y., Tu, Z., Li, Y. Hydrogen substituted graphdiyne as carbon-rich flexible electrode for lithium and sodium ion batteries. *Nat. Commun.* 2017, 8 (1): 1172.

# II

# Silicon

*Yi Shi*

## 2.1 Introduction

Nanowires (NWs) typically refer to tiny structures with diameters of less than 100 nm and aspect ratios typically reaching 1 : 1000 or higher. Silicon nanowires (Si NWs) have unique features different from bulk silicon, including a high specific surface area that makes chemical modification convenient and effective, unique quasi-one-dimensional (1D) electron transport, tunable energy band shifts, enhanced Raman scattering, and more. This makes Si NWs promising for applications in energy, electronics, and sensing. On-chip integration of Si NW-based devices is also attractive because of the large reserves of silicon materials and high complementary metal oxide semiconductor (CMOS) compatibility. The preparation of high-quality and characteristically controllable NWs has received a lot of attention from the industry. This chapter provides an overview of methods for the preparation of Si NWs, which include vapor–liquid–solid (VLS), vapor–solid–solid (VSS), Oriented Attachment (OA), VS, dissolution methods, and metal-assisted chemical etching (MACE). Among others, the growth behavior of VLS and its control are discussed in this chapter, and the chapter also discusses the distinction between VLS and OA and possible correlations. Subsequently, this chapter describes how the grown NW arrays can be assembled and focuses on a method for in situ lateral growth of NW arrays without the need for an assembly process, the in-plane solid–liquid–solid (IPSLS) method, which uses a B-type catalyst with a low surface tension, such as In or Bi, to direct the front end of the Si NWs to roll in the plane of the substrate.

Applications of Si NWs in photovoltaics, lithium, thermoelectrics, and electronics are summarized, with electronics of Si NWs being the focus as more attention and expectations. NW technology is expected to become a mainstay in silicon ICs to achieve scaling beyond Moore's Law, and this boost is reflected in the following aspects: lithography-free definition of very small diameter channels, availability of doping engineering, interface engineering and passivation, improved electrostatic integrity and short-channel effects (SCE) resistance, and nontraditional transport. Finally, this chapter describes the realization and progress of two novel architectures, vertical Si NW transistors and vertically stacked horizontal Si NW transistors, which, by liberating the z-axis degrees of freedom, continue to increase integration density despite scaling toward the limit in the horizontal direction.

*Single Element Semiconductors: Properties and Devices*, First Edition. Yi Shi and Shancheng Yan.

## 2.2 Synthesis of Si NWs

The preparation of NWs is first categorized into bottom-up growth methods and top-down etching methods. For the bottom-up method, the VLS method is the oldest and most central method since 1964. At the growth temperature, a metal catalyst forms eutectic droplets with Si, absorbing Si from the gas-phase Si-containing precursor, reaching supersaturation, and precipitating Si crystals underneath the droplets. VLS has a long history and a great deal of research, so we will focus on VLS methods in which the kinetics and crystallography of Si NW growth are revealed. The VSS method is the closest to the VLS method, with the difference being mainly that the growth occurs at different points on the metal-Si phase diagram, leading to the replacement of eutectic droplets by metal-catalyst-Si clusters in the solid phase. The solid–liquid–solid (SLS) method differs in how the Si precursors are supplied, which may be laser-ablated or thermally evaporated Si-$SiO_2$ particles and pretreated Si substrates. Oxide-assisted growth (OAG) has been reported to be possible in conjunction with VLS or SLS growth, with the respective dominant growth of Si oxide and metal presenting different NW morphologies, with OAG NWs having an intrinsic oxide shell, with outward migration of O elements from the active $SiO_x$ clusters during the growth process. Therefore, OAG synthesizes core–shell NWs with advantages of stable diameters and low metal contamination. In contrast, oxygen-free processes, such as silane chemical vapor deposition (CVD) or tetrachlorosilane CVD, are usually chosen for pure VLS to avoid the interference of oxides and obtain high-quality NWs without oxide sheaths.

The IPSLS method is another, in which eutectic catalyst droplets roll along the substrate surface, converting amorphous silicon at the front into crystalline silicon at the back. The difference in free energy in the process provides the driving force for the rolling behavior, while the involvement of friction and surface tension leads to a unique growth behavior.

NW synthesis in solution has also been investigated. The initial program used silicon-containing organics, such as monophenylsilane and diphenylsilane (DPS), as precursors. As for solution growth in inorganics, now that silanes and disilanes remain in vapor phase at room temperature, using trisilanes has become the preferred choice. What happens in solution is similar to VLS growth, where a metallic droplet dominates the growth in 1D, and this is called the supercritical fluid–liquid–solid (SFLS) method. In addition, hot organic solutions at atmospheric pressure can be used as long as their boiling points are above the eutectic temperature and the temperature of precursor cleavage.

Etching methods are categorized into dry etching and wet etching. Dry etching is based on photolithography and has mature processes and mechanisms. MACE is expected to achieve higher yields and lower costs. The catalysts are metal nanoparticles (NPs) suspended in solution or pre-patterned on the substrate's surface. Due to the electronegativity of the particles, etching occurs only below the particles. It is worth mentioning that catalyzed by gold filtered at the atomic level through polymer globular membranes, MACE yielded the smallest diameter NWs to date, as fine as 0.9 nm. Small-diameter catalysts in VLS are unstable due to factors such as Oswald maturation of hot droplets and hydrodynamic stresses due to Si incidence. On the contrary, MACE occurring in solution can be a room temperature process that makes maintaining minimal diameters possible.

It is important to note that most methods (VLS, VSS, etching) only produce NW arrays standing perpendicular to the substrate and that the orientation of OAG and solution methods is usually wandering because of scattered and immobile seeds. For applications requiring horizontal NWs, such as field effect transistors (FETs), an assembly of the NW arrays is required. In-plane in situ growth (template-constrained VLS, IPSLS), on the other hand, can skip assembly.

Different growth modes lead to different morphologies: length, diameter, and crystallinity, as well as growth direction (kinks), cross section, sidewall morphology, dislocations, and stacking faults. The study of these morphologies has been progressively deepened with the development of characterization techniques, such as scanning electron microscopy (SEM), transmission electron microscopy (TEM), and energy dispersive spectrometer (EDS), which are crucial for NWs, and in particular, the development of in situ NW growth techniques under the microscope. The quantitative study of various growth processes and growth effects and the revelation of growth mechanisms are the basis for precise control of the growth process and the realization of high-quality device applications.

### 2.2.1 Fundamental Aspects and Morphology Analysis of VLS Growth

The first to use the VLS method to grow Si NWs was Wagner and Ellis at Bell Labs [1] in 1964. In the 1970s, Givargizov enriched the VLS theory for wire growth [2]. In the 1990s, actual nanoscale diameters were realized based on silane CVD [3] and laser ablation [4, 5] precursor supply methods, and the theory of controlled diameter was proposed [6] of controlling the diameter was proposed.

A typical VLS (or VSS) process is depicted in Figure 2.1a (or b). First, metal NPs are deposited on a bare Si substrate. Subsequently, the substrate enters the reaction zone, which is heated to the growth temperature and energized with a gas-phase Si-containing precursor. The Si in the precursor preferentially nucleates at the metal particles. The metal-Si eutectic droplet continues to absorb Si from the gas-phase precursor, which is supersaturated in the

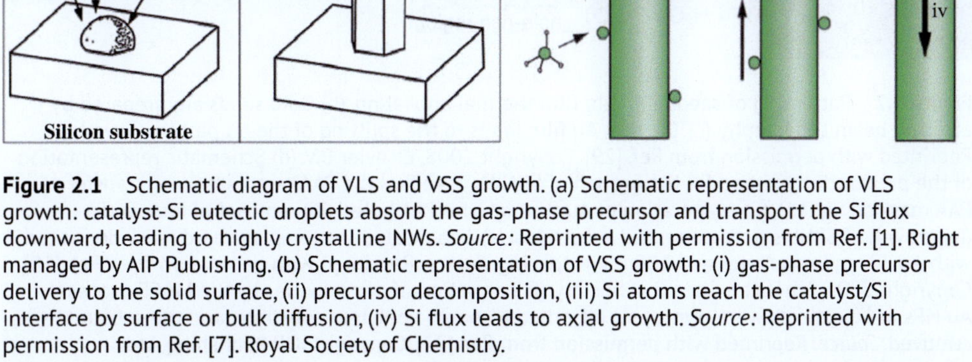


**Figure 2.1** Schematic diagram of VLS and VSS growth. (a) Schematic representation of VLS growth: catalyst-Si eutectic droplets absorb the gas-phase precursor and transport the Si flux downward, leading to highly crystalline NWs. *Source:* Reprinted with permission from Ref. [1]. Right managed by AIP Publishing. (b) Schematic representation of VSS growth: (i) gas-phase precursor delivery to the solid surface, (ii) precursor decomposition, (iii) Si atoms reach the catalyst/Si interface by surface or bulk diffusion, (iv) Si flux leads to axial growth. *Source:* Reprinted with permission from Ref. [7]. Royal Society of Chemistry.

droplet and precipitates downward, resulting in continued vertical growth of NWs below the droplet. In contrast, the eutectic droplet is stable during this process, always at the top of the newly grown Si NWs.

For all types of NW growth, the heart of the matter lies in controlling the wire to grow at a high rate in one particular direction and at a much smaller rate in other directions. This depends on the activation energy. In the case of Au-catalyzed Si NW growth, for example, the activation energy for VLS growth is about half that of Si film growth [8]: 19 kcal/mol (catalyzed by droplets) and 35 kcal/mol (film deposition) for silane and 12 kcal/mol (catalyzed by droplets) and 28 kcal/mol (film deposition) for disilane. This approximately twofold energy difference results in weak Si deposition on the sidewalls of the NWs; i.e. the Au-catalyzed NWs show only slight needle-like growth [9].

Based on this traditional VLS model, three steps are considered and enriched in the process: (i) the patterning of the metal NPs before growth, (ii) the supply mode of the precursor during growth, and (iii) the morphology of the Si-catalyst complexes (eutectic droplets for the VLS and solid-phase mixtures for the VSS) during growth.

#### 2.2.1.1 The Process of Patterning Metal NPs

Different fabrication processes, such as the deposition of porous anodic alumina (PAA) masks [10], deposition on pre-patterned silicon and silicon oxide substrates [11, 12], nanosphere lithography [13], contact printing [14], and electroplating displacement [15], were employed to form size-controllable seeds since the diameter of the NWs is essentially comparable to that of the catalyst droplets (Figure 2.2). As a concise and generalized method, deposition on Si substrates [16, 17] or magnetron sputtering of ultrathin metal layers (several nanometers or tens of nanometers) followed by the [18, 19] thermal annealing has also

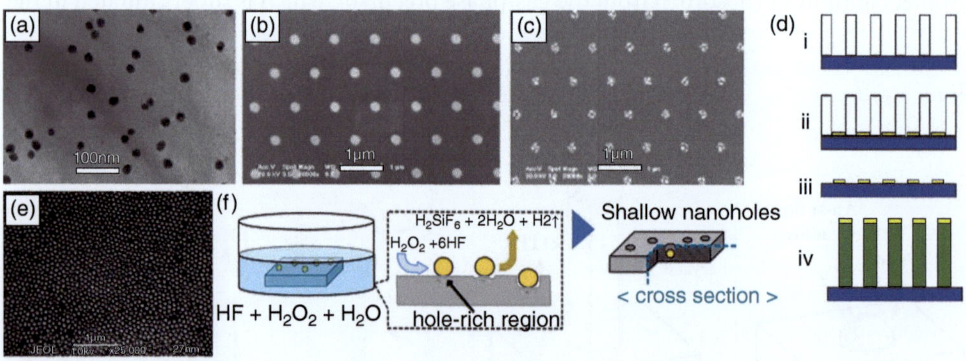


**Figure 2.2** Patterning of seeds. (a) Thin film thermal annealing. (b) Gold seeds are prepared by electron beam lithography. (c) Too thin Au film leads to the splitting of the Au pattern. *Source:* Reprinted with permission from Ref. [29]. Copyright 2008, Elsevier B.V. (d) Schematic representation of the preparation method for the growth of Si NWs based on the PAA template for (i) transfer of PAA mask on Si(111), (ii) evaporation of Au through the PAA holes, (iii) removal of PAA, and (iv) growth of Si NWs from patterned substrates. (e) Anodic alumina template hard mask deposited with Au, 53nm in diameter and 100nm in spacing. *Source:* Reprinted with permission from Ref. [10]. Copyright 2006, American Chemical Society. (f) Formation of nanocavities by the MACE process. Au NPs in Au colloids were previously adsorbed through a polymer substrate, and the substrate was removed. *Source:* Reprinted with permission from Ref. [22]. Copyright IOP Publishing, Ltd.

been reported to form nano-sized metal-Si eutectic droplets, with the melted metal tending to fragment into nanodroplets as a result of Oswald maturation, which makes for a much smaller morphology on the surface. A self-assembled layer exfoliation patterning method based on the inclusion of metal NPs has also been reported [20], which can be used to implant NPs in complex structures, such as on the sidewalls of microgrooves, for achieving in situ growth of NWs.

Briefly, catalysts formed using pre-patterned layers (substrates, oxide templates, or metal layers) have more controllable dimensions, wherein the oxide layers have a barrier effect that prevents diffusion of the metal in the Si. This pre-patterning is limited by lithography.

Relatively, schemes based on nonphotolithographic physical processes, such as thermal film fragmentation and colloidal adsorption, provide less stable control of diameter. The degree of control of the diameter by the NP self-assembly process depends on the NPs used.

A droplet in the plane may catalyze growth in all equivalent directions, e.g. the equivalent 4 <111> directions [21]. It is reported that close to 100% perpendicularity is realized based on the liquid phase epitaxy mechanism [16]. Another attempt to improve the uniformity of the growth orientation was to place Au NPs in shallow nanopits formed by MACE [20, 22, 23]. Au was successfully used as a catalyst for the MACE and VLS steps. Considering only how the NWs are created, the method is still a bottom-up VLS approach.

#### 2.2.1.2 The Process of Silicon Transportation

In the growth of Si NWs, silicon gas-phase precursors are transported to the top of the NW by CVD or molecular beam epitaxy (MBE) (Figure 2.2). CVD is the most versatile method with the broadest range of options for precursors and catalysts. It has been shown that for most VLS catalysts except Cu, Ni, and Fe, even slight aerobic environments are detrimental, exhibiting strong oxidation of the metal droplets and growth instabilities [24]. Therefore, VLS employs oxygen-free silicides including $SiCl_4$, $SiH_2Cl_2$, $SiH_4$, and $Si_2H_6$. The significant difference between these precursors is that chlorinated silanes are more stable and cleave at higher and lower temperatures, leading to different growth temperature intervals, where the decomposition temperature of $SiCl_4$ is higher than 900°C and is affected by partial pressure [10, 11, 25]. $SiH_2Cl_2$ cleaves at 700–800°C to produce a $SiCl_2$-intermediate [26]. $SiH_4$ decomposes at ~350°C and is commonly used as a precursor in low-temperature CVD. It is generally believed that $Si_2H_6$ has high reactivity and decomposes at 400–800°C to produce highly reactive multiple radicals. It is generally believed that $Si_2H_6$ has high reactivity and decomposes at 400–800°C to produce highly active radicals. The cleavage temperature of the precursor, together with the triple phase point of the metal, determines the different types of catalyst metals available.

Another critical effect of chlorine-containing precursors is that chlorinated silanes in the $H_2$ atmosphere cause HCl etching [9]. Some growths utilize this etching to eliminate oxide interference. The natural oxides of Si are easy to remove before growth, but for some types of metal catalysts (Al, Sn, Zn), oxidation causes severe disturbances.

As for the growth rate, the dependence of the partial pressure of silane was investigated by Jew et al. Meanwhile, the dependence of temperature was investigated by Schmid et al. According to them [9], the growth rate can be estimated by the following equation:

$$\nu \approx 1.7 \times 10^6 \frac{\text{nm}}{\text{s mbar}} \text{p} \exp\left(\frac{-9500\text{K}}{\text{T}}\right) \tag{2.1}$$

High-temperature CVDs have fast growth rates but multiple defects and rough walls. Oswald maturation [27], i.e. the exchange of material between catalysts that leads to larger droplets' growth and smaller droplets' decay during the growth duration, prevents high-temperature CVD from growing NWs with clearly defined diameters. In addition, diffusion of the metal catalyst is another consequence of high growth temperatures. Low-temperature CVD has a slow growth rate and high crystal quality. In low-temperature CVD, plasma irradiation has been shown to reduce the activation energy and increase the growth rate [28].

Of course, these silicides could also be supplied using MBE. The first realization of gold-silane MBE in 2004 [30]. After that, MBE has been studied far less than those of CVD. One problem is that Si is deposited not only on the NWs but also partially on the substrate, as shown in Figure 2.3. The existence of a competing growth mechanism has resulted in the inability to grow NWs below a diameter of 35 nm [31]. The growth of NWs below 35 nm in diameter is impossible due to this competing growth mechanism. In addition, the MBE growth rate is only nm/min, while the growth rate of low-temperature CVD is also on the order of nm/s. Although it has been argued [31] that MBE allows for more precise doping by controlling the incident particle flux, doping based on shallow energy metal catalysts or gas-phase doping is also available.

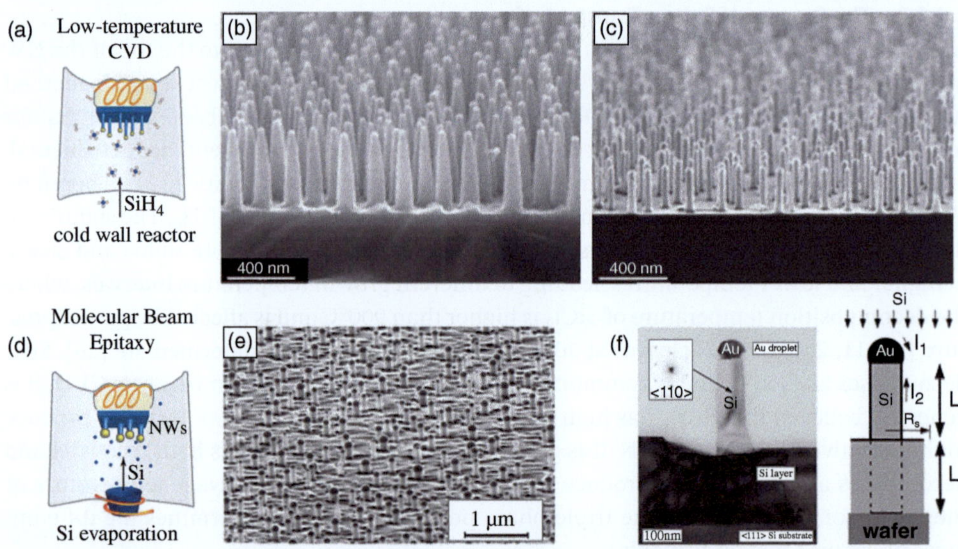


**Figure 2.3** Transportation of precursors. (a) Schematic of low-temperature CVD. (b, c) SEM cross-sectional images (tilt 10°) of Si NWs grown at different temperatures: 490°C (b) and 430°C (c). *Source:* Reprinted with permission from Ref. [32]. Copyright 2006, Springer Nature Limited. (d) Schematic of MBE. (d) SEM overview of Si filaments grown on <111> Si substrate at 525°C for 120min. (e) Left: TEM cross section of a Si filament with a length of about 300nm and a diameter of 95nm on a <111> Si substrate. Right: schematic of MBE wire growth. $I_1$ and $I_2$ are fluxes of Si-adsorbed atoms pointing toward the gold cap. *Source:* Reprinted with permission from Ref. [31]. Right managed by AIP Publishing.

#### 2.2.1.3 Catalysts

Numerous types of metallic materials are helpful for the growth of Si NWs, as shown in Figure 2.4g. Among them, Au is an ideal material for laboratory growth of Si NWs: gold has nonoxidizing chemical properties. The fact that Au is nontoxic makes its safety constraints less stringent. The Au-Si binary phase diagram (Figure 2.4a) shows the Au-Si eutectic point is as low as 363°C, where the percentage of Si atoms is as high as 19%. There is a steep liquid-phase line slope near the eutectic point, properties that imply fast growth at low temperatures and a smaller energetic cost for increasing the Si concentration beyond its equilibrium value.

Unfortunately, Au is well known to be a deep energy level impurity in Si, leading to the incompatibility of gold with CMOS production standards. It has been shown that Au diffuses inside NWs with a thermodynamic equilibrium concentration of $\sim 10^{16}\ cm^{-3}$ [33, 34], accompanied by the aggregation of Au clusters on the sidewalls. Au will aggregate in the direction of the lowest energy of the Au/Si interface, for example [35], 20 nm <111>. Si NWs with six {112} surfaces, the nanoclusters exhibit a triple symmetric distribution on the discontinuous sidewalls corresponding to the (11-2), (1-21), and (-211) planes in the six available crystals, forming clusters on a 4-nm scale. In contrast, the other three surfaces exhibit a smooth and gold-free morphology. In conclusion, the diffusion of Au in Si makes it challenging to remove, and the Au-catalyzed NWs make it difficult to eliminate the toxicity of Au on the carrier properties. Figure 2.4h illustrates the doping energy levels of each catalyst element in silicon, where diffusion of gold in silicon will introduce unexpected deep impurity energy levels. In contrast, shallow energy level doping is valuable for applications such as Bi and In.

There has been a long search for catalysts to replace gold. However, despite the emergence of many promising materials, none can completely replace Au in terms of growth quality and convenience: many process improvements and device applications still use Au as a catalyst. At the same time, most non-Au catalysts are being worked on to discuss growth mechanisms and improve growth quality, as described in the following section. Before this [36], metal catalysts are classified according to binary phase diagram characteristics and chemistry: type A: phase diagram dominated by eutectic sites with >10% silicon concentration; no metal-silicide phases are present. Type B: phase diagram dominated by eutectic sites with <1% silicon concentration; no metal-silicide phases are present. Type C: phase diagram with one or more metal-silicide phases; eutectic sites at temperatures above 800°C. Type C, phase diagram with one or more metal-silicide phases; eutectic sites located at temperatures above 800°C. Type D: phase diagrams dominated by eutectic sites with <1% silicon concentration; no metal-silicide phase is present.

A-type catalysts are gold-like metals. Their phase diagram is simple eutectic; a single eutectic site dominates it. This eutectic site is located at more than 10 atoms % Si. In addition, A-type catalysts do not have any metal-silicide phases. There are only three A-type metals: Al, Ag, and Au.

B-type catalysts are low-silica-solubility metals. Their phase diagrams also show a central eutectic point but no silicide phase. The Si concentration at the eutectic point is much lower than that of the A-type catalysts, below 1 atom % Si.

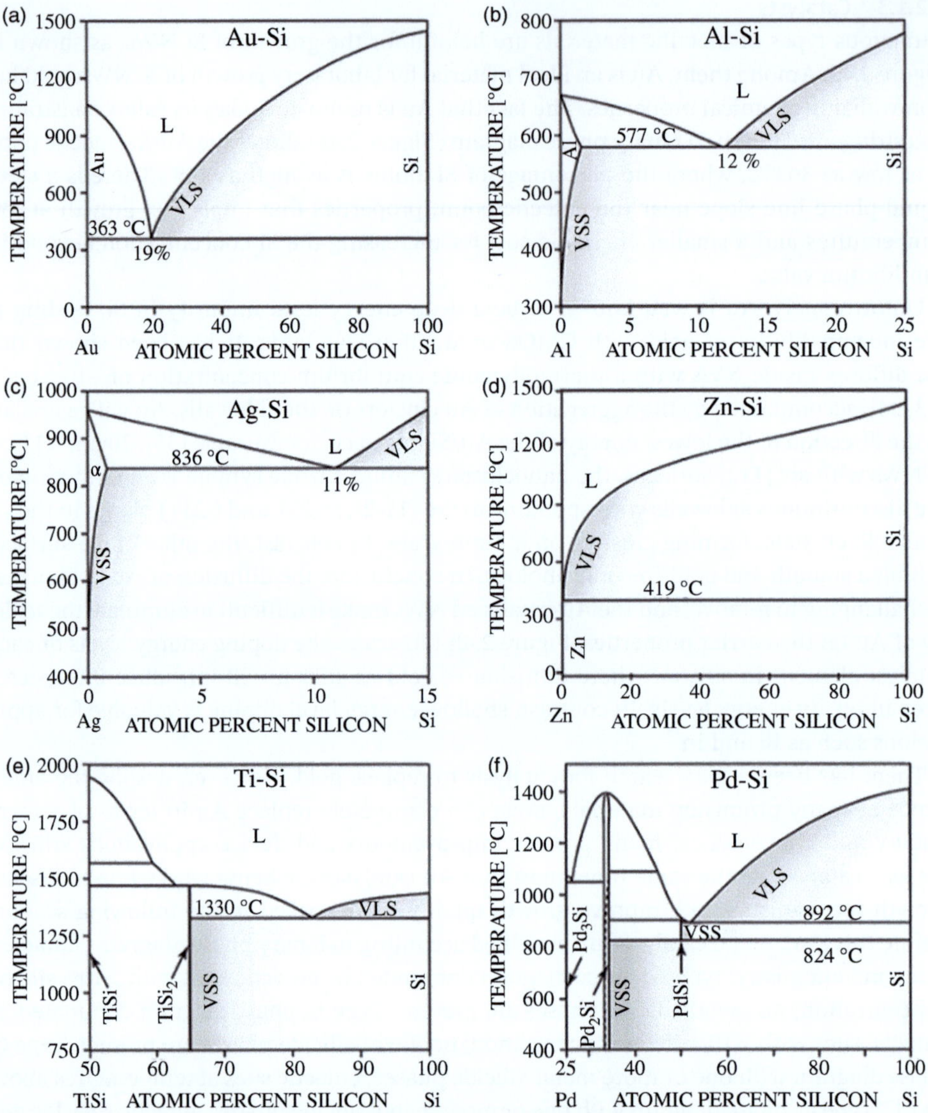


**Figure 2.4** (a–f) Schematic phase diagrams of different metal-Si systems. (a) Au-Si, (b) Al-Si, (c) Ag-Si, (d) Zn-Si, (e) Ti-Si, (f) Pd-Si, (g) Periodic table of potential catalyst metals classified according to the phase diagrams. Type A: phase diagram dominated by eutectic sites with >10% silicon concentration; no metal-silicide phases are present. Type B: phase diagram dominated by eutectic sites with <1% silicon concentration; no metal-silicide phases are present. Type C: phase diagram with one or more metal-silicide phases; eutectic sites at temperatures above 800°C. Type C: phase diagram with one or more metal-silicide phases; eutectic sites located at temperatures above 800°C. Type D: phase diagrams dominated by eutectic sites with <1% silicon concentration; no metal-silicide phase is present. (h) The required minimum temperatures of certain metals used for VLS growth of Si NWs are plotted versus their respective impurity level energies. These impurity level energies are given at the right-hand side relative to the middle of the bandgap (assuming a bandgap of 1.12 eV for Si). Color codes refer to the catalyst classification of Figure 2.10: red, type A; orange, type B; blue, type C. *Source:* Reprinted with permission from Ref. [9]. Copyright 2006, American Chemical Society.

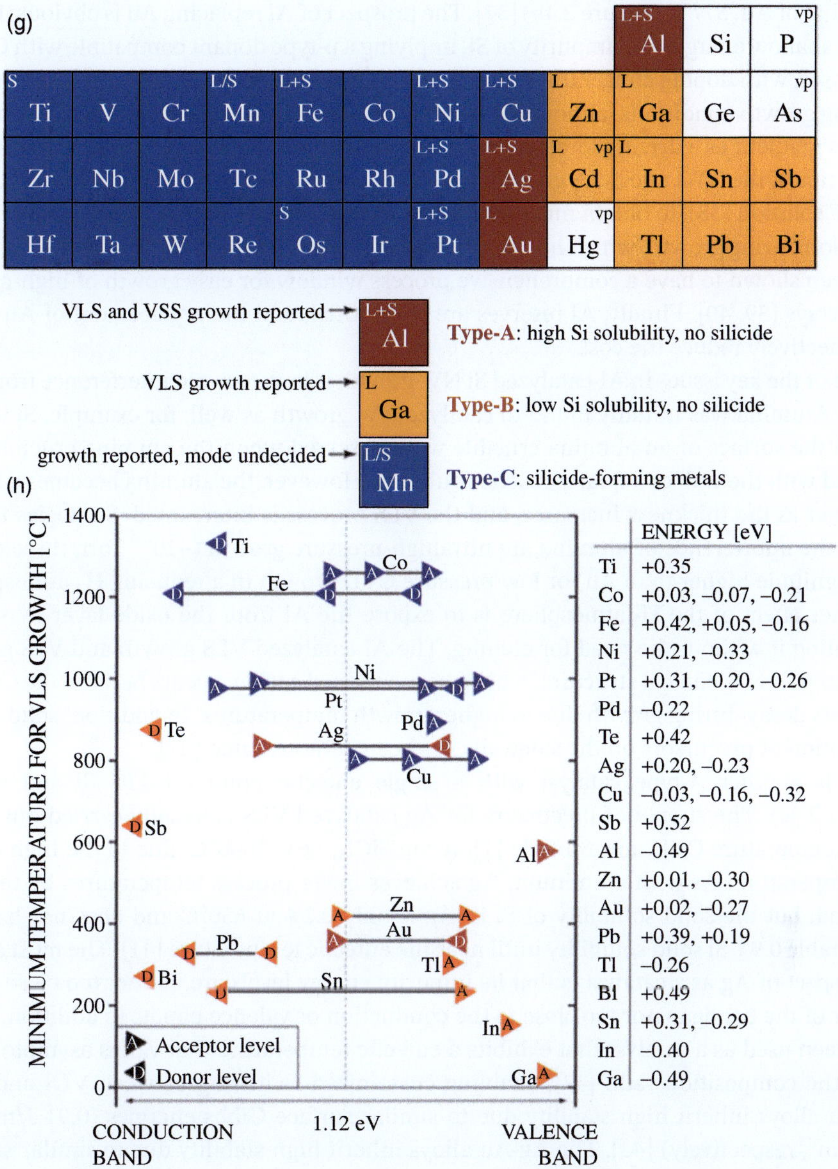


**Figure 2.4 (Cont'd)**

C-type catalysts are silicide-forming metals. Their phase diagrams indicate the presence of one or more silicide phases. In addition, the minimum eutectic temperature is higher than 800°C. Typical C-type catalysts are Cu, Pt, or Ti.

#### *2.2.1.3.1 Type A Gold Catalysts: Al and Ag*

Al is a well-known gold-like catalyst with a clean phase diagram similar to that of gold, where the eutectic fraction of Si is as high as 12.6%, albeit at a slightly higher temperature than that of Au, 577°C (Figure 2.4b) [37]. The prospect of Al replacing Au is obvious: firstly, Al is a shallow energy level impurity of Si, implying a p-type dopant compatible with CMOS processes, with doping concentrations reported to be in the range of $10^{18}$–$10^{20}$ $cm^{-3}$, under varying growth conditions, although it is difficult to avoid the presence of an axial concentration gradient as a driving force for aluminum transport [37, 38]; secondly, Al aggregates at the tips of the NWs after VLS growth and can be removed simply from the Al, e.g. using a 5% HF solution [38], to obtain metal-free wires of interest for electronics, in contrast to Au diffusion during growth, which has long been a nuisance. Thirdly, Al-catalyzed NW growth has been shown to have a comprehensive process window for easy growth of high-quality NW arrays [39, 40]. Finally, Al reserves are substantial, and the replacement of Au by Al can effectively reduce the cost.

One of the key issues in Al-catalyzed Si NW growth is how to avoid interference from alumina. Alumina was initially found to catalyze NW growth as well; for example, Si whiskers on the surface of an alumina crucible were observed when the alumina crucible wall reacted with the wall of a quartz reaction tube [1]. However, the alumina becomes a blocking layer as the thickness increases, and the VLS process is interrupted. Strategies to preclude the interference of alumina are ultrahigh-pressure growth (~$10^{-9}$ Torr, three orders of magnitude higher than Au) or low-pressure CVD growth in a reducing $H_2$ atmosphere. Another effect of the $H_2$ atmosphere is to expose the Al from the oxide layer by simply annealing it without the need for etching. The Al-catalyzed VLS growth and VSS growth generally show a conical structure, which, as mentioned earlier, occurs because the catalyst droplets decay during growth due to higher growth temperatures. In addition, solid-phase deposition of precursors on the sidewalls has been demonstrated [37].

Ag is another A-type catalyst with a single eutectic point (at 11% Si and 836°C, Figure 2.4c). The supply of precursors for Ag-catalyzed VLS is usually carried out using high-temperature CVD, for example [1], using $SiCl_4$, near 1000°C, due to the high eutectic temperature. As with aluminum, Ag achieves lower process temperatures by the VSS method, but the solid solubility of Si in Ag is only 0.2% at 650°C and does not have an acceptable 0.9% Si solid solubility until near the eutectic temperature [41]. The most attractive aspect of Ag as a catalyst is that its impurity energy levels are neither too close to the center of the bandgap nor too close to the conduction or valence bands. In addition, it has also been used as a catalyst that exhibits a eutectic temperature that varies asymptotically with the composition ratio [42], enabling customized switching between VLS and VSS. Ag-Au alloys inherit high stability due to similar surface Gibbs energies (0.91 $J/m^2$ and 0.9 $J/m^2$, respectively) [43]. The Ag-Au alloys inherit high stability due to similar surface Gibbs energies (0.91 J/m and 0.9 J/m, respectively), low critical polymerization temperatures ranging from 120 to 164°C, and readily achieved diffusion [19, 41], the same lattice structure (face-centered cubic lattice), and no compound formation [19].

#### 2.2.1.3.2 Type B Low Si Solubility Catalysts: Zn, Sn, Bi, In, and Ga

Zn has a single eutectic point (at 0.02% Si atomic fraction and 420°C), as shown in Figure 2.4d. Zn is significantly cheaper than other B-type catalysts, about eight times cheaper than Sn, and 40 times more abundant [44]. The cost of Zn is significantly lower than other B-type catalysts, being about eight times cheaper and 40 times more abundant than Sn. Zn also has the highest surface tension of all the B-type catalysts, making it easier to form high-contact catalysts [45]. The difficulty in applying Zn lies in its volatile and oxidizing nature. Fortunately, the $H_2$ atmosphere of silane CVD can alleviate this problem. 440°C silane CVD can yield intrinsically doped NWs with diameters ranging from 15 to 35 nm [46]. Pulsed plasma-enhanced chemical vapor deposition (PPECVD) has also been applied [47, 48], where precursors are pre-decomposed into active reactants, allowing the growth temperature to be reduced to 400°C. The compatibility of the electroplating process [49] allows for better cost-effectiveness and higher throughput for the pre-preparation of Zn catalysts.

Sn appears to be an attractive catalyst when considering thermal budget cuts and low-temperature growth, because of its Sn-Si eutectic point as low as 232°C. In practice, however, the growth temperature of VLS cannot currently be much lower than Au's due to the need for silanes to decompose at temperatures higher than 400°C. The suggested temperature for Sn-catalyzed VLS is ~350°C, in contrast to poor growth at 250°C [18]. The diameter of Sn-catalyzed NWs is significantly larger than that of the catalyst droplets due to the deposition of Si films on the sidewalls [50]. Sn is a catalyst species of interest for solar cells [17, 50, 51]. The catalysts for such applications usually do not require thin and precise NWs.

The difficulty with Bi is its very small surface tension (~0.36 $J/m^2$), which leads to a puzzling growth process, in addition to the evaporative nature of Bi, which makes high growth temperatures unavailable. The earliest catalytic growth of Bi was reported in 2012 by Kumar et al. [52] at 280°C, in the vicinity of the Bi-Si eutectic temperature. Benabderrahmane Zaghouani et al. [53] conducted a study for the formation of Bi NPs by a fast thermal annealing process, and his optimized process conditions produced quasi-spherical Bi-NPs. However, whether according to Kumar or Benabderrahmane Zaghouani's method, the final Si NWs obtained were kinked, bent, and randomly oriented. An alternative approach is the fragmentation of Bi films in $H_2$-plasma at lower temperatures (e.g. ~300°C), which helps to obtain more reliable catalyst droplets. The advantage of Bi is the introduction of a shallower donor level in c-Si, located 69 meV below the conduction band edge, as shown in Figure 11d, which makes in situ n-type doped cores feasible, implying that no gas-phase dopants are required for the radial n-i-p junctions that can be constructed, which has been successfully applied to photovoltaic NWs [54]. Meanwhile, the demonstration of p-type in situ doping of photovoltaic NWs using Ga has also been reported [55].

Another creative approach based on these properties of Bi is to allow Bi droplets to roll in-plane for lateral SLS growth, which we will discuss specifically in Chapter 1.2.2. In is another catalyst with similar properties.

#### *2.2.1.3.3 Type C Silicide-forming Catalysts*

C-type catalysts are silicide-forming catalyst metals, like Figure 2.4e–f. The phase diagram of C-type catalysts is usually quite complex, exhibiting several silicide phases and a variety of eutectic sites. Due to the presence of silicide phases, C-type catalysts can be used not only for VLS but also for VSS wire growth through silicide particles.

### 2.2.2 Morphology Analysis for VLS or VSS Methods

In the last section, we discussed why it is enabled to grow Si NWs by the VLS or VSS method, discussing the physical phenomena during the growth, which are seed formation, precursor transport, and control of catalyst eutectic droplets. In this section, we will discuss how well it grows. Besides the control of crystal orientation and crystallinity. It is a matter of keeping the control diameter as constant as possible during growth.

#### 2.2.2.1 Control of Diameter and Growth Direction

The diameter of Si is thought to determine the direction of Si growth: it is generally believed that for larger diameters (30–100 nm), NW growth is oriented in the <111> direction, whereas for smaller diameters, the NW growth is oriented in the <110> direction. Conversion occurs near 20 nm, with a narrow growth window in the <112> direction [56]. This can be understood in terms of growth kinetics: the growth direction depends on the direction that minimizes the total energy. For large diameters, the catalyst-Si interfacial energy determines the system's energy, and the sidewalls' surface energy gradually dominates as the diameter decreases and the specific surface area increases [57]. For the vicinity of the critical diameter, kinking is observed by changing the growth parameters [37], i.e. an abrupt change in the growth direction.

In addition, the reduction in diameter also leads to slower growth, with the diameter limit of VLS-grown Si NWs being near 2 nm, limited by the nearly halting growth rate. As early as 2001, VLS made obtaining NWs with smaller scales easier than the lithographic node [6]. Recently, gold-catalyzed VLS reported the synthesis of NWs with diameters smaller than 5 nm, a scale already smaller than the block's [58]. This scale is already smaller than the Bohr radius of free excitons in bulk silicon, although the simultaneous generation of coarse NWs is still not suppressed.

#### 2.2.2.2 Diameter Stability

It is generally accepted that eutectic droplets provide good soft catalysis and soft confinement during growth, contributing to the formation of high-quality nanostructures. The ideal scenario is for the eutectic droplets to act as catalysts to maintain a stable composition, leading to cylindrical NWs. In practice, however, this vision could not be realized even under conditions where all experimental parameters (surface structure, gas cleanliness, and background contaminants) were tightly controlled [59].

Diameter variations during VLS growth need to be explained by the dynamic behavior of eutectic droplets [60]: the diameter inhomogeneities of NW growth are grouped into the nine phenomena shown in Figure 2.5, where S4–S9 are combinations of S1–S3. The first phenomenon during the VLS process is the catalyst's absorption of the gas-phase precursor, which generates droplet seeds on the substrate surface. As the growth time accumulates, the droplets may experience the so-called overgrowth phenomenon, i.e. the droplets absorb vapor components, increasing in diameter, corresponding to the conical structure shown in S1. In addition, scattering of catalyst particles in the gas phase, transfer of catalyst elements, and autocatalytic effects can lead to droplet overgrowth. Conversely, the droplets may decay, corresponding to progressively smaller diameters and needle-like structures S3. The causes of droplet decay include: (i) droplet fragmentation caused by lateral temperature gradients or sudden temperature changes [57]; (ii) droplet oscillations caused by oversupply of precursors [51]; (iii) hydrodynamic stress due to precursor incidence, which alters the surface energy of the droplet [57]; (iv) droplet accumulation of charge due to frictional charging, etc., which decays under electrostatic force [61]. The hydrodynamic stress

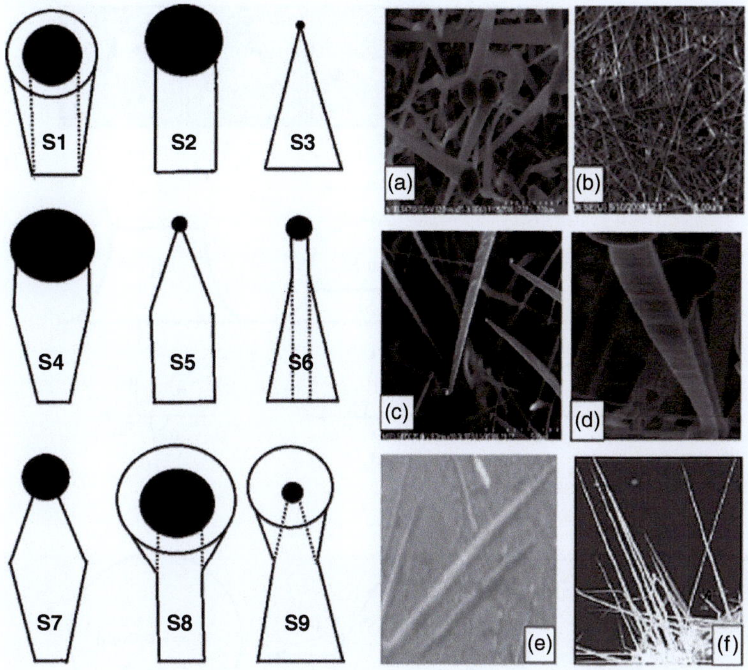


**Figure 2.5** Left: theoretically, there exist nine variations of diameters. Right: examples of grown NWs with (a) conical structure S1, (b) homogeneous structure S2, (c) needle-like structure S3, (d) hybridized structure S4 (e.g. a combination of the S1 and S2 structures), and (e) hybridized structure S7 (e.g. a combination of the S1 and S3 structures). *Source:* Reprinted with permission from Ref. [62]. Copyright 2008, American Chemical Society.

changes the surface energy of the droplet. Small particles may be scattered when the alloy droplet ruptures, forming a secondary catalytic effect, and many new small NWs will grow. As time passes, more and more NWs will gradually aggregate [60].

Based on growth stability considerations, growth temperatures too high above the eutectic temperature are not recommended, and precursor pressures should not be too high. However, this is beneficial for growth rate enhancement; e.g. according to the conclusion in the review [9], the contribution of the growth temperature to the growth rate is exponential.

From the force point of view, the diameter stability control of VSL requires the analysis of surface tension, thermal stress, fluid dynamic stress, and electrostatic force, which is currently based on theories and theories. From the energy point of view, monitoring and stabilizing the surface energy of the droplet is the general route to achieve stable growth.

#### 2.2.2.3 Contact Angle, Infiltration, and Base Exbandation

Suppose the droplets are in force equilibrium during growth. In that case, a generalized phenomenon of base expansion occurs, as shown in Figure 2.6, where wires growing on a substrate usually show diameter expansion at their base caused by changes in droplet shape. Considering the equilibrium balance of surface tension and the corresponding droplet shape, a model for the swelling at the conductor can be derived. The equilibrium contact angle of the droplet depends on the surface energy and geometry [63].

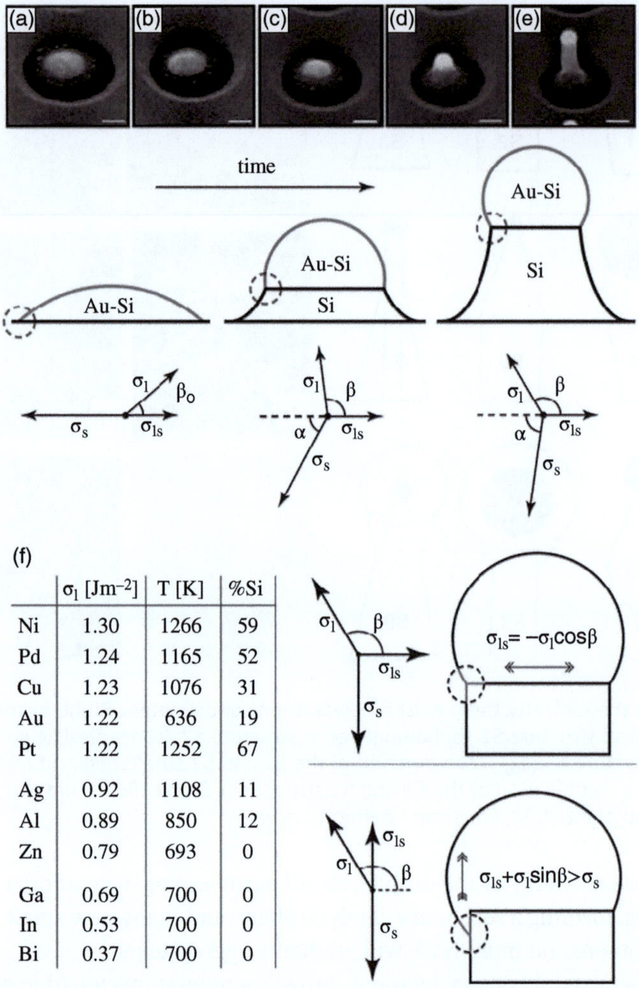


| | $\sigma_l$ [Jm$^{-2}$] | T [K] | %Si |
|---|---|---|---|
| Ni | 1.30 | 1266 | 59 |
| Pd | 1.24 | 1165 | 52 |
| Cu | 1.23 | 1076 | 31 |
| Au | 1.22 | 636 | 19 |
| Pt | 1.22 | 1252 | 67 |
| Ag | 0.92 | 1108 | 11 |
| Al | 0.89 | 850 | 12 |
| Zn | 0.79 | 693 | 0 |
| Ga | 0.69 | 700 | 0 |
| In | 0.53 | 700 | 0 |
| Bi | 0.37 | 700 | 0 |

**Figure 2.6** Analysis of contact angle and surface tension. (a–e) Scanning electron micrographs showing the formation of the droplet shape at the early stage of growth. *Source:* Reprinted with permission from Ref. [62]. Right managed by AIP Publishing. Lower panel: schematic of the droplet and linear shape at the early growth stage, with the corresponding equilibrium balance of surface forces at the left edge of the droplet (dashed circle). Note that the horizontal force components sum to zero. (f) Equilibrium shape of the droplet during growth; the horizontal force components must cancel. (b) In a nonequilibrium position, net vertical force is needed to re-establish equilibrium after some perturbations. (c) Different metal-silicon alloy catalyst systems' surface tension is calculated from the temperatures and compositions in two neighboring columns. *Source:* Reprinted with permission from Ref. [9]. Copyright 2006, American Chemical Society.

As shown in Figure 2.6f, the force equilibrium becomes difficult for a surface tension that is too small, and the critical value is generally considered to be around 0.7 J/m$^2$. However, growths smaller than this value have also been reported, mainly for several B-type catalysts, but the growth behavior becomes complicated.

### 2.2.3 Oxide-assisted Growth

This section discusses another time-honored method, OAG. The method occurs within a quartz or alumina tube and employs a solid-phase target. Although the same method is referred to differently, including, among these, OAG has wide recognition. Considering the similarity of the growth environments and the historical origins, the VLS or SLS processes occurring in heat pipes are also presented in this chapter, as long as their Si source is a solid powder.

#### 2.2.3.1 Development of OAG Theory

In 1998, Morales and Lieber [4] reported a method for Si NW growth in hot quartz tubes: a FeSi target (10% Fe, 90% Si) was laser ablated at 1200°C and entered the quartz tube as Si and Fe vapors, which formed nanoclusters that the Si then absorbed on the tube wall, and they interpreted this growth as Fe-Si eutectic droplet-mediated VLS growth. In the same year, Lee et al. [5] subsequently pointed out that [64, 65] that the involvement of $SiO_2$ particles contributes significantly to the reaction, as evidenced visually by increasing the growth rate. At the same time, pure Si powder at 1200°C leads to almost no growth. As shown in Figure 2.7b–e, two NWs with completely different properties were obtained by varying the temperature in the same device, proving the independence and coexistence of the OAG and VLS mechanisms [66]. During the OAG growth process [66], oxygen atoms in the silicon oxide clusters are expelled by silicon atoms and diffuse to the edges, forming chemically inert silicon oxide sheaths. Thus, the highly active $SiO_x(1<x<2)$ at the NW's tip is a vaporized silicon oxide collector. At the same time, the outer $SiO_2$ layer of the Si NWs prevents the diameter growth of the NW.

In the laser ablation method, the reaction is not under thermodynamic equilibrium conditions. This may explain that VLS growth simultaneously yields very small diameters (~10 nm) and <111> orientations [30]. In contrast, OAG growth is oriented in the <112> and <110> directions, which are dominated by the minimized surface energy associated with the diameter [67], which is similar to the previously mentioned CVD-VLS growth of NWs with similar diameters [8] similar to the previously mentioned CVD-VLS-grown NWs of similar diameters.

In summary, despite having overlapping process windows, OAG forms unique products compared to VLS: (i) the absence of metal particles at the tips of the NWs; (ii) the core–shell structure of silicon oxide; and (iii) a different growth direction than VLS, as well as smooth steering instead of kinking.

Questioning, based on an investigation of the composition of the tube wall [68], is believed that the Al and Cu retained by the quartz tube wall reached the surface by annealing and guided the NW growth. In the case of the alumina tube wall, it was reduced to Al in the $H_2$ atmosphere and then melted to form eutectic droplets.

#### 2.2.3.2 VLS or SLS: A Discussion of Precursor States in the Heat Pipes

The nature of Si-bearing precursors in quartz or alumina tubes has also been described in different ways: the review [31] summarizes laser ablation and thermal evaporation methods as SLS methods, emphasizing the difference between solid-phase powder particles containing Si and gas-phase silicides such as silanes: as an example, commercially available powder

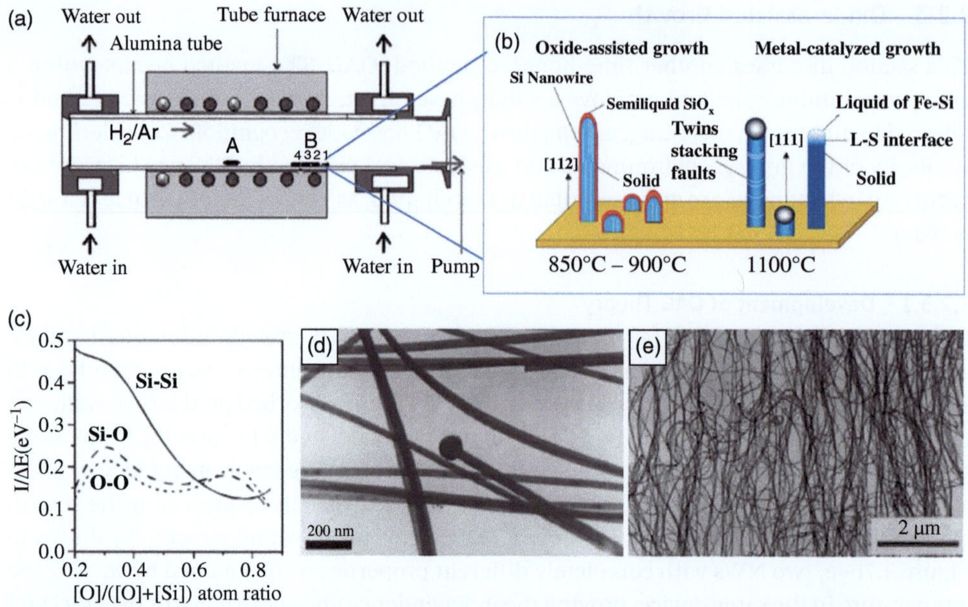


**Figure 2.7** (a) Schematic diagram of the thermal evaporation growth device. A temperature gradient of 900–1350°C is generated in the tube furnace. Unlike laser ablation, growth begins on the placed substrate. An alumina tube wall and $H_2$ atmosphere usually accompany it. (b) Different NWs appear in different temperature zones on the substrate, showing OAG and VLS mechanisms, respectively (c). The inverse of the energy difference hE = LUMO ± HOMO (electron acceptor ± electron donor) and thus the reactivity (proportional to the inverse of the energy difference) for the formation of a Si ± Si bond, a Si ± O bond, or an O ± O bond between two silicon oxide clusters as a function of Si/O ratio. (d) Morphology of NWs grown in the high-temperature region on the substrate. (e) Morphology of NWs grown in the low-temperature region on the substrate. *Source:* Reprinted with permission from Ref. [66]. Copyright 2003, WILEY-VCH Verlag GmbH & Co. KGaA, Weinheim.

particles of SiO [68] (diameters of about $4.5 \times 10^4$ nm and $1.02 \times 10^{-7}$ g) are typically about 16 orders of magnitude heavier and five orders of magnitude larger than the latter. The large mean free range of the particles also better explains that growth occurs farther from the nozzle [31]. However, for proponents of VLS, "vapor containing Fe and Si" is used to describe the evaporation products of FeSi targets [4]. For evaporation products of mixtures of Si and $SiO_2$ particles, it is expressed as SiO gas, or $SiO_x$ ($1<x<2$) [66, 67]. These two descriptions may point to different reaction stages: the production of SiO gas followed by the formation of small silicon oxide clusters $Si_nO_m$ in the growth zone at lower temperatures, which are amenable to Si-Si bonding [30]. These clusters are easy for Si-Si bonding. Clusters with $n = 3 - 8$ are considered favorable for nucleation, as shown in Figure 2.7. Some studies have examined the reaction parameters required for the evolution of SiO [69], suggesting that different crystallinity of silica in the feedstock promotes the formation of clusters with different properties.

$$\mathrm{Si(solid)} + \mathrm{SiO_2(solid)} \xrightarrow{\text{high temperature}} \mathrm{2SiO(gas)} \tag{2.2}$$

$$2\text{SiO(gas)} \xrightarrow{\text{low temperature}} \text{Si(solid)} + \text{SiO}_2\text{(solid)} \tag{2.3}$$

$$\text{Si}_x\text{O(gas)} \xrightarrow{\text{low temperature}} \text{Si}_{x-1}\text{O(solid)} + \text{Si(solid)}(x > 1) \tag{2.4}$$

In summary, there is still controversy about what happens inside the quartz (or aluminum oxide) tube [31]. The controversy centers first and foremost on the growth mechanism and the type of catalyst. At a temperature high enough to potentially cause surface rearrangement, a subtle line of logic lies in the fact that proponents of VLS cannot rule out the influence of oxides in the tube wall, even with an oxygen-free FeSi hybrid target; similarly, proponents of OAG cannot rule out the possible interference of metals due to less than 0.1% impurities in the tube wall: at least it modifies the substrate, and at best, it completely alters or even dominates the catalytic mechanism. Secondly, the unknown lies like the precursors and intermediates. In conclusion, there is still more to be explored about the growth mechanism of this method.

#### 2.2.3.3 Characterization of OAG NWs

The hard confinement of the silicon oxide sheath ensures diameter stability. NWs grown by the OAG mechanism have a uniform diameter of about 20 nm, even though the NWs are several hundred micrometers in length. OAG is more suitable for producing small-diameter NWs, reportedly as thin as 1 nm [70]. In contrast, the flanks of VLS are unconstrained, relying solely on the stability of the droplets themselves, which is fragile at high temperatures, small droplet sizes, or any of the low surface tension crystal species. However, the silicon oxide sheath hampers its optical and electrical characterization; for example, the carrier trapping effect of defect centers present at the Si-$SiO_x$ interface has been reported to lead to the splitting of bands for low-temperature photoluminescence [71].

OAG-based NWs exhibit smooth steering. The growing NWs are neither perpendicular to the wall of the tube nor can they grow in-plane; they are haphazard. This is a disadvantage for applications. In addition, metal impurities were found to be trapped inside the NWs [72]. Although this approach is almost "metal-free" compared to VLS, it exhibits convincing electrical properties. Finally, an advantage worth considering is that solid particle precursors are much safer than silane or tetrachlorosilane, which spontaneously combust in air.

Similarly, sulfide-assisted growth has been reported in tube furnaces at growth temperatures above 1000°C with SiO as the precursor. Silicon sulfide (melting point 1090°C) as the precursor and a tiny liquid catalyst dominated the VLS growth, yielding <111> oriented, oxygen-rich, core–shell, needle-like NWs [73].

#### 2.2.3.4 OAG Growth on Substrate

A method based on catalytic nucleation on pretreated silicon substrates was also investigated and described from the perspective of oxide-dominated growth and metal-dominated growth. Hydrogen-encapsulated silicon wafers were etched with 100°C ultrapure water to produce reactive silicon oxide-rich ($SiO_x$) layers [77]. Selective NW growth in the $SiO_x$ layer region was observed [74] using silane CVD (Figure 2.8a–c). Based on these, the VS theory was proposed, suggesting that $SiO_x$ active sites on the substrate capture silicon atoms in the

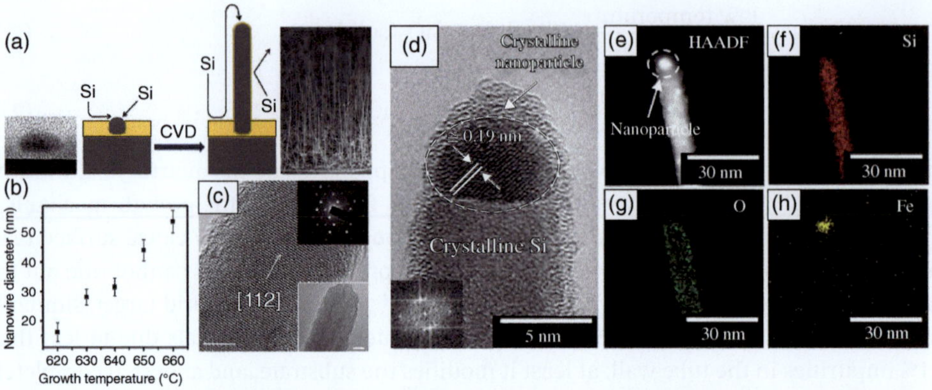


**Figure 2.8** OAG based on pretreated silicon surface. (a) Schematic diagram of VS growth. In silane CVD, Si is captured by $SiO_x$ active sites, and the NWs show vertical growth. (b) Temperature dependence of NW diameter. The wires were grown at 620–660°C under the additional flow of HCl gas to avoid nonspecific deposition of amorphous silicon during growth. (c) Lattice-resolved TEM images were taken at the middle and end of a single-crystal Si NW (lower inset). Arrows highlight the NW axis, corresponding to the <112> direction; scale bar = 5 nm. (Upper inset) Corresponding diffractogram *Source:* Reprinted with permission from Ref. [74]. Copyright 2009, American Chemical Society was recorded along the [75] region axis. (d) TEM image of the tip region showing crystalline NPs at a crystal lattice plane distance of about 0.19nm. The inset shows the corresponding FFT image. (e) HAADF-STEM image of Si NW. (f–h) Energy dispersive X-ray spectroscopy elemental mapping. Showing Si atoms present in the Si NW stem and tip (red), oxygen that should come from the silicon oxide Si NW overlay (green), and different concentrations of iron atoms in the tip region of the Si NW (yellow). *Source:* Reprinted with permission from Ref. [76]. Copyright 2021, The Authors. Advanced Materials Interfaces published by Wiley-VCH GmbH.

gas without any metal droplet catalyst. In addition, the method exhibits good control of the diameter by temperature and the etching of the oxide shell layer in an atmosphere of HCl.

However, similar to the OAG, growth happening on substrate could be triggered by metal impurities in the substrate [78]. As shown in Figure 2.8d–h, the growth after KOH solution treatment revealed 0.19-nm-diameter Fe NPs at the tip, and the KOH solution may contain a very small amount of Fe impurities due to its production process. However, the growth temperature was only 680°C, well below the Fe-Si eutectic point of 1207°C, and the growth showed a classical oxide sheath. Although the growth mechanism has not yet been thoroughly investigated, this growth method is expected to lead to shallow metal content and perpendicular-to-substrate NW arrays at a much lower temperature.

### 2.2.4 Thermal Annealing of Silicon Substrates

A growth method that directly heats the Si substrate has also been reported. This is the undisputed SLS approach, where the Si for growing NWs comes from the rearrangement of the substrate material. As shown in Figure 2.9, in the case of Fe catalysts, the Fe film is broken by heat to form droplets. Subsequently, the Fe-Si eutectic droplets are oxidized, the oxygen element is crowded to the outer end of the droplets, and larger droplets consisting of the Fe-Si core and the Fe-Si-O shell layer produce $Si/SiO_x$ core–shell lines. As a result,

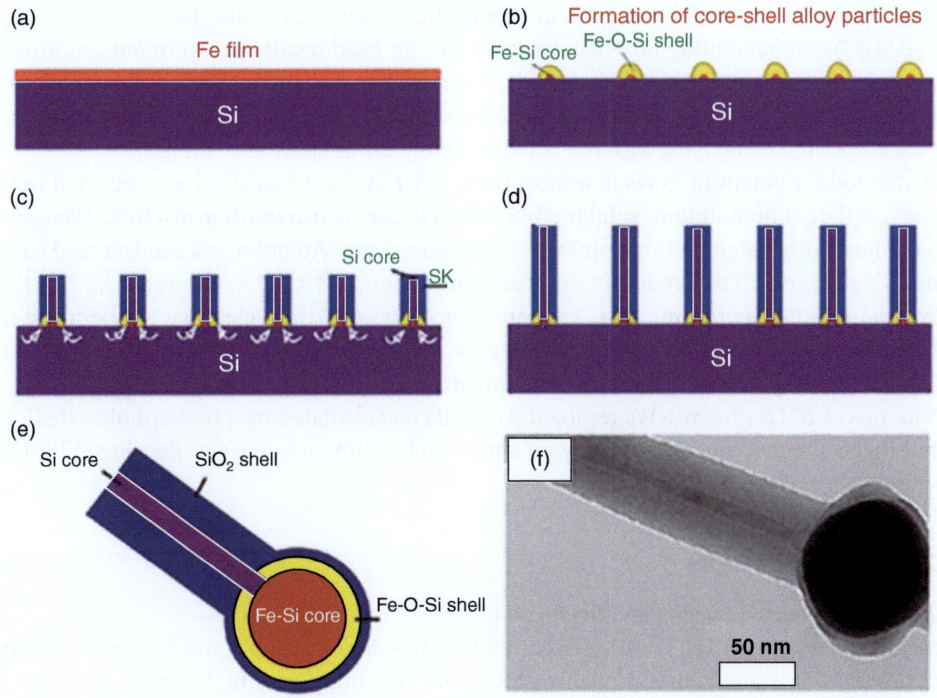


**Figure 2.9** Schematic representation of the $Si/SiO_x$ core–shell line SLS mechanism. (a) Deposition of a thin layer of Fe on a Si substrate. (b) Formation of Fe-$FeO_x$ by Si core–shell alloy droplets. (c) Continuous diffusion of Si atoms into the droplets through the substrate/droplet interface and growth of $Si/SiO_x$ core–shell wires through the catalyst-wire interface. (d) The final state of Si wire growth. The smooth surface of the pristine substrate becomes rough at the end of Si wire growth. (e) Schematic structure of the core–shell wires grown from the catalyst. (f) TEM image of a core–shell wire terminated with a core–shell catalyst. *Source:* Reprinted with permission from Ref. [31]. Right managed by AIP Publishing.

the eutectic droplets remain on the substrate at all times, directing the silicon flux from the substrate into the NWs. Oxidation accompanies the growth in parallel, yielding a rich morphology, such as coaxial silicon wires outside, biaxial, tri-concentric, diameter-oscillating, Ni or Au trapping, and transition morphologies [31].

### 2.2.5 Dissolution-based Methods

The SFLS method was first reported in 2000 [79]: Using alkanethiol-coated gold nanocrystals (25 nm in diameter) as crystal seeds, 1D Si crystallization was directed in a solvent heated and pressurized above the critical point, with the Si coming from DPS in solution. Other silicon-containing organics, such as monophenylsilane [80], have also been reported. The inorganic silicon precursor, trisilane, has also been employed to avoid carbonaceous shells. The high reactivity of trisilanes leads to fast synthesis rates, e.g. nearly 80 mg of Si NW product from a 10-mL reactor volume in less than 2 min [81]. The high reactivity of trisilane leads to fast synthesis rates, e.g. nearly 80 mg of Si NW product from a 10-mL reactor volume in less than 2 min. Sn-catalyzed NWs in supercritical toluene have also been reported [76]: Sn $(HMDS)_2$ for generating Sn NPs. A higher Si : Sn ratio results in a pronounced amorphous silicon shell, which arises from the growth of Sn seeds etched by hydrogen from the decomposition of trisilane. As with VLS NW growth, the diameter of the NWs can be tuned by the size of the metal NPs, with the smallest diameters as short as 4 nm [82].

A method for obtaining NWs in atmospheric environments has also been reported [83], known as the solution–liquid–solid method. Growth occurs in a solution at 410°C. Eicosane or squalane with a high boiling point was used as a solvent, Au or Sn as a catalyst, and trisilane as a precursor. The predominant growth direction was <111>, but about 20% of the NWs had <110>growth direction. Trisilanes decompose at their cleavage temperature to form various reaction intermediates, which are eventually reduced to silanes through the 1,2-hydrogen displacement reaction mechanism and decompose into silicon atoms [84].

The tips of SFLS-grown NWs retained Au seeds encapsulated by a hydrophobic shell. A rapid thermal quenching followed by an aqua regia etching process was developed [85] to eliminate 95% of the Au.

## 2.2.6 Etching Methods

### 2.2.6.1 Reactive Ion Etching and Dry Etching

Reactive ion etching (RIE) is a CMOS-compatible top-down method, and deep reactive ion etching (DRIE) is an evolution of this process for obtaining very high aspect ratio structures. In this process, the cycle of RIE etching is followed by the passivation of the sample, and both steps are repeated several times. In this way, the influence of the ionic direction plays a more significant role than in standard etching, obtaining higher anisotropic etching and smoother walls. However, this method makes it challenging to achieve aspect ratios above 50 : 1 [86].

E-beam lithography (EBL) is the most commonly used method [87] in dry etching because of its flexibility. In the EBL process, electrons are used instead of photons, which has the advantage of shorter wavelengths (with a relevant $\lambda$ ranging from ~12 to 0.04 nm, respectively). EBL with DRIE has been reported to produce Si NWs as thin as 50 nm and, in specific contexts, as thin as 10 nm [88] and even 5 nm [89, 90]. However, expensive special equipment is required.

### 2.2.6.2 Metal-assisted Chemical Etching

In 2002, Peng et al. [91] achieved dense Si NWs aligned perpendicular to the MACE by using a 5.0 : 0.02 M HF : $AgNO_3$ solution, which was made into a silver salt scheme: the $AgNO_3$ acted both as an oxidizing agent and a metal source, and the electronegativity of the Ag NPs guided the preferential oxidation of Si by nitrate ions underneath the Ag and the etching by the HF. Ag dendrites were concomitantly produced to form on the Si NWs a dense network several micrometers thick (~40 μm), as shown in the SEM cross section (Figure 2.8d). Finally, the excess silver dendrites were selectively etched with nitric acid (Figure 2.8c), and vertical arrays of Si NWs were obtained as shown in the SEM cross section (Figure 2.8e) at the end of the removal by the nitric acid solution. In addition to the silver salt method, $KAuCl_4^-$ [91], Fe $(NO_3)_3^-$ [92], $KMnO_4$ [93], $KBrO_3^-$ [94], and $K_2CrO_{27}^-$ [92] have also been demonstrated to drive the etching of Si by similar principles.

The advantages are room temperature, fast reaction, and a simple process flow. However, a problem is the generation of NW arrays. The metal concentration determines the diameter, density, and spacing of NWs, and metal particles are reduced in the solution. Noting that metal particles dominate anisotropic etching, etching methods based on pre-patterned metal particles have been developed to achieve more precise control. Patterning can be based on UV lithography [95], colloidal lithography [96], anodized aluminum masks [97], etc. Ag and Au are commonly used metal types for MACE, where Au has high solution stability and adjustable etching rates. Ti and AZO are often used as adhesives for Au. The metal film determines the average diameter of the NW and is usually around 50 nm [97, 98], but suitable sizes up to a few nanometers and quantum confinement are also possible [99]. Sub-nanometer diameters (minimum 0.9 nm) were first obtained by the MAEC method [100], which fills the gap of atomic-scale Si NWs. The key idea is the combination of ultrafine filtration and MAEC, e.g. catalyst downscaling strategies using polymeric spherical membrane filtration.

## 2.3 Horizontal Si NWs: Self-assembly or In-plane Epitaxy

As mentioned previously, multiple methods (VLS, substrate thermal annealing, etching) can efficiently yield arrays of NWs standing perpendicular to the substrate with controllable diameters and reliable quality. Based on these growths, collection and assembly strategies are widely used to produce devices composed of horizontal NWs [101, 102].

### 2.3.1 Growth-assembly Methods

#### 2.3.1.1 Fluid Alignment Strategy

The assembly of NWs is carried out by using fluid motion to generate shear forces on the solid boundary, and the orientation of the NWs suspended in solution will be realigned to the direction of fluid flow to minimize the fluid resistance. Periodic alignment of NW arrays is demonstrated by combining surface modification and flow-assisted assembly techniques. Crossed NW arrays can be easily realized by two fluid assembly processes in perpendicular flow directions, and the junction devices can be generated on the crossed nodes to form a network [101, 102].

#### 2.3.1.2 Langmuir–Blodgett Assembly Strategy

In the Langmuir–Blodgett assembly method, surface-activated NWs float on the surface of an organic solvent to form an NW-surfactant monolayer. A barrier at one end of the sink compresses the NW monolayer at an appropriate rate, and based on the rule of surface energy minimization, the NWs present a parallel arrangement, and finally, the wafers immersed in the solvent carrying the NW arrays are pulled out [103, 104].

#### 2.3.1.3 Dielectrophoresis of NWs

Si NWs can be assembled in large flux self-limiting by dielectrophoresis (DEP): a colloidal suspension containing Si NWs flows through a substrate prefabricated with DEP electrodes. It is confined at the electrodes under the combined control of dielectrophoretic, hydrodynamic, and surface forces. The coupling of capillary forces with DEP has also been studied [105], which is used to realize heterogeneous NW devices. In addition, when the quality and characteristics of NW growth in the fluid vary, the DEP method exhibits strong collection selectivity, which can significantly improve device homogeneity and quality [106].

#### 2.3.1.4 Printing

Transfer printing (TP), or contact printing, is a dry assembly strategy that utilizes tangential forces on the stamp to align standing or randomly oriented NWs. A simple contact printing process demonstrates the wafer-level assembly of highly ordered, dense, and regular NW arrays with a high degree of uniformity and reproducibility, where the surface chemistry of the receiver substrate [107] regulates the NW density. A differential web printing method was developed [75] that transfers semiconductor NWs directly from a cylindrical donor substrate to a patterned receiver substrate, where the transferred NWs are ordered and aligned. The scalability of the TP process is discussed [108]. Dual transfer printing with the stepwise transfer of NWs and electrodes and multiple transfer printing based on thermal release tape is proposed. Three-dimensional (3D) NW circuits can be fabricated by iterating the contact printing, device fabrication, and separation layer deposition steps multiple times [109].

#### 2.3.1.5 Blown Bubble Film Method

The blown bubble film (BBF) method is a versatile and scalable assembly method with potential for large-area NW array applications. Uniformly aligned and density-controlled NW films are achieved by controlled bubble expansion of a homogeneous polymer suspension containing NW materials. The alignment is attributed to the shear stresses present in the epoxy fluid during the bubble expansion process [110]. Electrical measurements show that large-area NW-based FET arrays can be efficiently fabricated on a wafer scale [111].

In summary, numerous methods can transform Si NWs from a vertical or fragmented state to high-quality horizontal arrays. However, most of the methods are solution-based. Surfactants usually functionalize the NWs to form stable suspensions in organic solvents, and DEP methods require patterned electrodes. BBF methods are considered difficult to achieve higher density NW integration, and, in general, each method has its own shortage [112].

### 2.3.2 In Situ Growth of Lateral NWs

A novel growth method, the IPSLS method, was proposed in 2009 [112, 113], which is used to skip the post-growth self-assembly process and grow NWs directly along the horizontal direction of the substrate. This method is fundamentally different from previous SLS: while the classical SLS process forms solid-phase precursors by laser ablation, for example, in contrast, in the IPSLS growth process, an a-Si thin film as a substrate is used as a precursor layer, which is absorbed by the nanodroplets, moves horizontally, and produces a continuous succession of well-defined NWs with controllable size, morphology, and even growth paths. For IPSLS, In becomes an ideal catalyst because of the slow diffusion process of In in the a-Si:H matrix [114]. Another reason is the enormous surface tension of In [115]. One of the problems with using In catalysts is the disabling of the gas-phase n-type dopant P.

Still, solutions have been proposed: Bi catalysts [116] or P doping in solid-phase amorphous silicon layers [117] in the solid-phase amorphous silicon layer.

The growth picture of IPSLS is unique and interesting. The front and back ends of the catalyst-Si eutectic droplet form solid–liquid interfaces with a-Si and c-Si, respectively. For Si atoms, the two solid–liquid interfaces have a Gibbs free energy difference of $\Delta E = 0.12$ eV, which leads to Si crossing the droplet from the front to the back, i.e. generating Si flux. The droplet is subjected to both the thrust from the rear-generating NWs and the drag force absorbed by the Si in the front, and, at the same time, the friction during the travel needs to be overcome. When the thrust from the rear becomes dominant, the NWs have a straighter surface morphology. Due to the dynamic equilibrium of the forces, the droplet itself thus does not remain spherical as in a vertical VLS. Instead, it exhibits a cycle of deformation accompanied by fluctuations in silicon fluxes [115]. This cycle is evidenced by the observed surface ripples [118]. At the end of growth, this pulling causes the droplet to flatten until it collapses and growth terminates [116]. The growth is terminated.

Two growth modes were observed [108]: the grounded growth (GG) mode and the suspended growth (SG) mode. It is well understood that the GG mode grows slower and is limited by the nucleation rate at the Si-In interface. In the SG mode, the newly generated NWs at the back end no longer adhere to the substrate. Still, they are suspended, and the front and back interfaces of the catalyst droplets are decoupled, with the droplets having rolling degrees of freedom and exhibiting higher tolerance to fluctuations in silicon flux. On the contrary, in the GG mode, the velocity difference at the interface is presented by a periodic change in the direction of droplet travel, leading to a sinusoidal morphology. This phenomenon can be used to define silicon nanosprings [119].

The most characteristic and attractive feature of the IPSLS method is NWs' in situ lateral growth. Second, the solid-phase Si supply, as well as the higher $\Delta E$, lead to high growth rates – $10^2$ nm/s, two orders of magnitude higher than In-catalyzed vertical VSL growth – and, again, the avoidance of the use of flammable and explosive Si gas-phase precursors, as pursued by other SLS methods. Finally, possessing the promise of pre-patterning processes, including the recently demonstrated vertical stacked Si NW arrays grown and integrated on parallel sidewall recesses for 3D integration [120] that utilize a-Si layer modification [121] and lithography [122, 123] to define the NW morphology precisely. However, IPSLS has a growth mechanism different from conventional vertical growth. Despite the group's long-term work on this, there is still less research on the growth mechanism and parameter window of IPSLS.

## 2.4 Applications of Si NWs

### 2.4.1 Si NW Solar Cells

Silicon is the classical material used for solar cells. The bandgap width of silicon can match the natural light spectrum and realize a high photovoltaic response. Therefore, silicon solar cells have always had the advantages of cost and power conversion efficiency. After the first generation of solar cells, represented by planar silicon solar cells, Si NW solar cells are considered the second generation [124], moving toward the third generation. On the one hand, these cells require less silicon to produce the same power. In addition, these cells can be manufactured with low-grade silicon while having the same output performance, and ultimately, these advantages can significantly reduce the cost of cell production [124].

On the other hand, the advantages of NW materials, such as reduced reflection, extreme light trapping, improved bandgap tuning, and simple strain relaxation, are introduced [125]. Finally, simulation work, including analysis and optimization of parameters such as NW length, diameter, array, and interfacial properties, has led to a new understanding of Si NW arrays for photovoltaic applications. Advances in Si NW fabrication processes have provided flexibility in fabricating these products. In summary, Si NW solar cells are a technological extension of planar silicon solar cells, demonstrating the potential for large-scale photovoltaic applications. This chapter will present the advances and challenges of Si NW solar cells, starting with their structure and principles.

#### 2.4.1.1 Principle and Structure of Si NW Solar Cells

In general, the operation of a photovoltaic device involves three important steps: photon absorption, charge separation, and charge collection. Incident above-bandgap photons are converted into electron–hole pairs in the base region. The photoinduced charge carriers are then dissociated and swept to the front and back electrodes through the p-n junction between the emitter and base regions.

##### *2.4.1.1.1 Si NW Arrays for Optical Trapping*

Properly designed Si NW arrays can achieve good light trapping, which requires avoiding light reflection and penetration. As shown in Figure 2.10a, when the light wave wavelength λ is much longer than the array periodicity P, the incident light easily passes through the NW array. For the case where λ tends to be close to P, significant light scattering leads to an extended optical range length and improved light absorption. When λ < < P, there is strong light loss due to reflection and transmission [126]. Therefore, photovoltaic devices based on high-performance silicon micro/NW arrays should have optimal array periodicity matching the wavelength region with a relatively large photon number. For crystalline Si structures with a bandgap of 1.12 eV, the optimal P is about 600 nm. Otherwise, the enhanced light scattering redshifts or blueshifts out of the energy region with a large photon number, leading to a degradation of the solar energy collection quality.

The principles of optical trapping achieved by NW arrays are multiple. Optical antenna effects are believed to dominate the light absorption of NW arrays. In addition, studies have reported fluctuations in absorption efficiency due to mode resonance effects, as shown in Figure 2.10c, where a comparison of the absorption spectra of the two samples (cf. right inset of Figure 2.10b) indicates that the guided resonance formed in the latter array produces several sharp absorption peaks, which leads to an increase of the optical trapping in a localized range. However, the overall trend of the curves still reflects that light scattering is the dominant mechanism for enhancing light trapping rather than guided resonance.

In many studies, arrays of Si NWs can achieve extremely low reflectivity, e.g. around 1% [129, 131]. The disordered NW arrays exhibit enhanced light absorption due to enhanced optical scattering [132].

Optical capture optimization for solar cells is significant in terms of the increased number of photons, improved photocurrent, and increased open-circuit voltage ($V_{OC}$) through

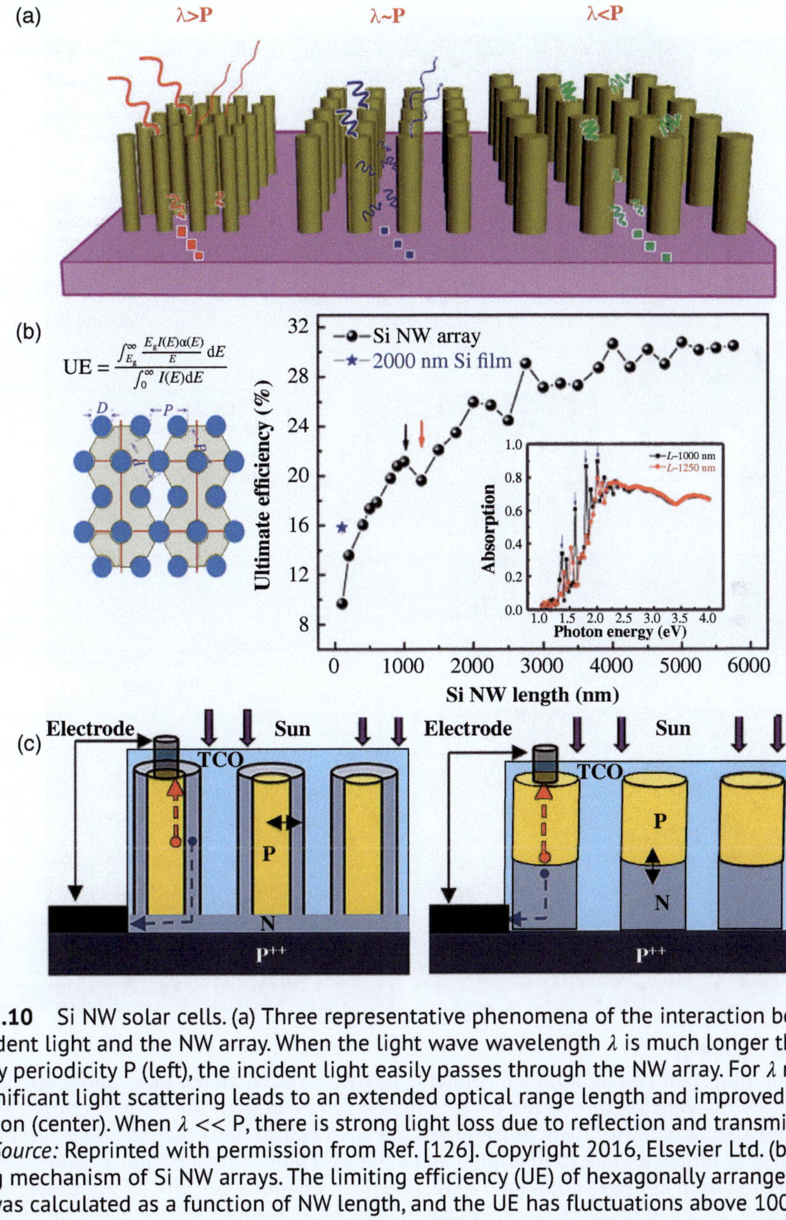


$$UE = \frac{\int_{E_g}^{\infty} \frac{E_g I(E)\alpha(E)}{E} dE}{\int_0^{\infty} I(E) dE}$$

**Figure 2.10** Si NW solar cells. (a) Three representative phenomena of the interaction between the incident light and the NW array. When the light wave wavelength $\lambda$ is much longer than the array periodicity P (left), the incident light easily passes through the NW array. For $\lambda$ near to P, significant light scattering leads to an extended optical range length and improved light absorption (center). When $\lambda << P$, there is strong light loss due to reflection and transmission (right). *Source:* Reprinted with permission from Ref. [126]. Copyright 2016, Elsevier Ltd. (b) Light trapping mechanism of Si NW arrays. The limiting efficiency (UE) of hexagonally arranged Si NW arrays was calculated as a function of NW length, and the UE has fluctuations above 1000nm due to enhanced light absorption induced by waveguide modes. Array periodicity P = 600, NW diameter D = 500nm (left). *Source:* Reprinted with permission from Ref. [127]. Copyright IOP Publishing, Ltd. The scattering-based light trapping effect is considered to be dominant (right). *Source:* Reprinted with permission from Ref. [128]. Copyright 2018, the Chinese Chemical Society, publication and hosting by Elsevier B.V. (c) There are two architectures of Si NW solar cells: left, radial, and right, axial. The arrows depict the separation and collection of photogenerated carriers. *Source:* Reprinted with permission from Ref. [128]. Copyright 2018, the Chinese Chemical Society, publication and hosting by Elsevier B.V. (d) Schematic of the multilayer structure of a radial

*(Continued)*

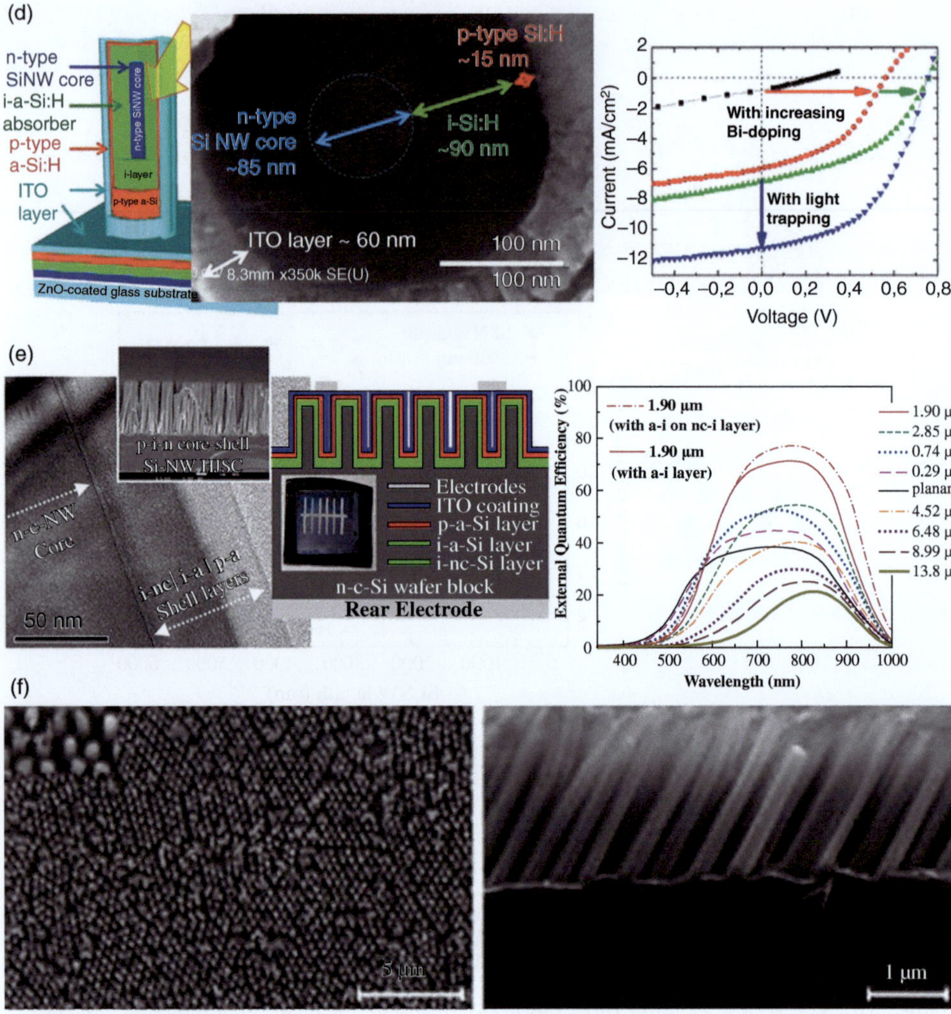


**Figure 2.10 (Cont'd)**
junction thin film solar cell built on a bicatalytic Si NW on a ZnO-coated glass substrate (left). The coaxial multilayer structure in the radial junction can be seen in the cross-sectional SEM image (center). Voltammetric characterization of the solar cell, as shown by the blue curve, with improved device efficiency due to Bi doping and enhanced light trapping (right). *Source:* Reprinted with permission from Ref. [59]. Copyright 2012, American Chemical Society. (e) Multilayer p-a-Si/i-a-Si/n-c-Si-NW solar cell produced by chemical etching of an n-type (100) Si wafer and PECVD-grown shell layer (left). The maximum external quantum efficiency (EQE) was obtained at an array length of 1.9 μm, and the curves show the variation of EQE in the visible band for NW array devices with different lengths versus planar devices (right). *Source:* Reprinted with permission from Ref. [129]. Copyright 2023, American Chemical Society. (f) Top view (left) of a large-area ordered Si NW generated on an n-type Si(113) substrate. The cross-sectional SEM image shows the n-type Si(113) substrate (right). *Source:* Reprinted with permission from Ref. [130]. Copyright 2013, American Chemical Society.

a change in photon entropy [133]. The optical properties of NW arrays have made them popular for photovoltaic devices, and planar solar cells have tended to use NW arrays and other surface nanostructures to enhance optical trapping [134]. However, because the junction region remains planar, they are unable to obtain the other performance advantages of NW solar cells, including radial charge separation.

#### *2.4.1.1.2 Radial Junctions in Si NWs*

Two typical structures of Si NW solar cells are shown in Figure 2.10c, where the NWs are encapsulated by transparent conductive oxides (TCOs), with indium tin oxide being the most common TCO material. The core of the NW is usually n-type or p-type monocrystalline silicon, and due to the photovoltaic effect, the electron–hole pairs are first separated and then transported along the NW and eventually into the collector electrode, as shown by the direction of the arrow in Figure 2.10c.

Radial junctions are a widely preferred configuration to most thoroughly exploit the performance advantages of Si NW solar cells. Due to the larger junction area, radial junctions have better carrier collection than axial junctions and are more tolerant to defects. The radial distribution of the doping distribution leads to an asymmetric distribution of the space charge density and electric field strength, which is more likely to completely deplete the dopant layer, especially in the case of thin cores. Thus, the doping concentration and thickness in the p and n regions must be carefully tuned. Systematic studies and recommendations for doping concentration and material defect density for radial junction devices are provided by the review [135]. Photoelectrochemical studies have shown that charge separation and collection occur only in the space charge region of the p-n junction, so the diameter of the Si NW should be smaller than the width of the space charge region [136]. Based on radial junctions, p-i-n structures [59, 129] have also been proposed, as shown in Figure 2.10d and e, where hydrogen-containing intrinsic Si single crystals at the interface are used for enhanced light trapping in the presence of defects. An intrinsic nanocrystalline silicon (i-nc-Si) layer at the interface minimizes the substantial charge complex loss due to significant lattice mismatch at the i-a-Si/n-c-Si-NW heterojunction.

### 2.4.1.2 Si NW Solar Cell Performance Improvements

For photovoltaic devices, a typical voltammetric curve is shown in Figure 2.11a, and the power of the solar cell is represented by the shaded area, which depends on the actual operating point. Thus, the solar cell characteristics depend on the combination of short-circuit current ($J_{SC}$), $V_{OC}$, and fill factor (FF).

$$\eta = \frac{V_{oc} \times I_{sc} \times FF}{P_{in}} \tag{2.5}$$

where the $J_{SC}$ depends on processes that depend on photon absorption, carrier collection, and separation, which are typically characterized using internal quantum efficiency (IQE) and EQE, and so that

$$J_{sc} = q \int_{\Delta\lambda} \frac{\lambda}{hc} \Phi(\lambda).EQE(\lambda)d\lambda \tag{2.6}$$

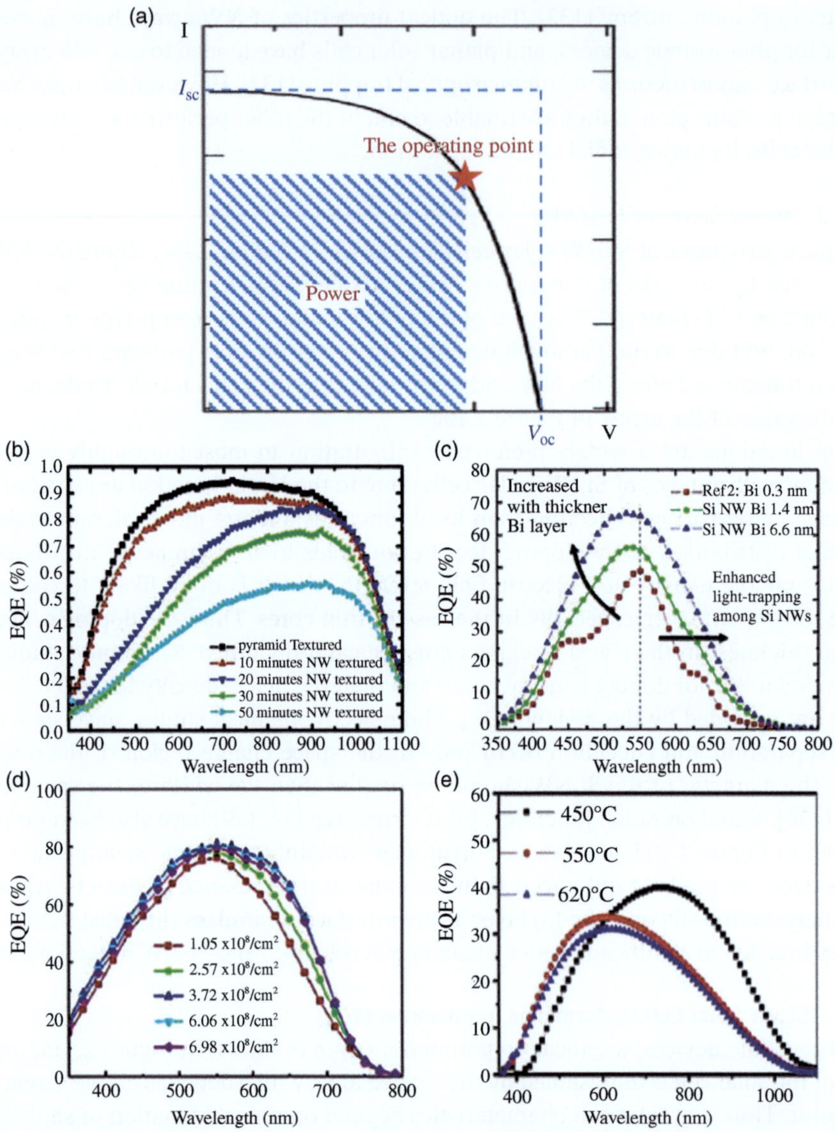


**Figure 2.11** (a) Schematic of the volt-ampere characteristic curve of a solar cell, which qualitatively shows the performance parameters. (b–e) Variation of EQE with the parameters of the NW (b) length (controlled by etching time). (c) The thickness of the Bi layer in the Bi-catalytic NW. (d) Density of the NW array. (e) Deposition temperature. *Source:* Reprinted with permission from Ref. [59]. Copyright 2012, American Chemical Society.

Therefore, the effort to increase the total power can be done by the improvement of EQE. For example, as shown in Figure 2.11b–e. The effects of length, passivation layer, density of the array, and process temperature of the passivation layer on the EQE were investigated, and the optimized process parameters were selected.

$V_{OC}$ equals the radiative limit at the Shockley–Queisser UE at AM1.5G conditions.

$$V_{oc} = \frac{K_B T}{q} \ln\left(\frac{R_{sun}}{R_{blackbody}}\right) \tag{2.7}$$

where $K_B$ is the Boltzmann constant, T is the temperature, q is the electron charge, and $V_{OC}$ depends only on the solar power generation rate ($R_{sun}$) and the blackbody emissivity ($R_{blackbody}$).

The FF describes the ratio of the actual converted power at the operating point of the battery as a percentage of $J_{SC} \times V_{OC}$, i.e. the ratio of the two rectangular areas in Figure 2.11a, and it depends on the structure and operating conditions of the battery itself. The FF can be optimized under some conditions [133].

#### 2.4.1.3 Passivation of Si NW Solar Cells

Ideally, a long NW length reduces the blackbody emissivity, which helps to obtain a large $V_{OC}$. In contrast, the photocurrent generated by the Si NW first grows and then saturates as the length increases. The efficiency of the solar cell increases as the NW length grows. However, longer length also means that more of the composite centers generated during fabrication, such as surface dangling bonds and defects, are located in the charge transport route, which prevents obtaining a saturated photocurrent [131, 137]. Therefore, efforts to reduce carrier complex centers are ongoing. This effort is mainly carried out through surface passivation. Furthermore, according to Hong et al.'s first-principles calculations [138], the surface dangling bonds of Si nanostructures are a fundamental barrier to nano-doping, and surface passivation is also very important to prevent dopant deactivation.

The classical materials to passivate photovoltaic NWs include alumina (Figure 2.12a), silicon nitride (Figure 2.12b), and amorphous silicon hydride (a-Si:H) (Figure 2.12c). Kato et al. [139] reported a 15.9 μs effective carrier lifetime in $Al_2O_3$-passivated polysilicon NWs. Kim et al. [140] prepared Si nanopillar array structures by MACE. They demonstrated that $Al_2O_3$ passivation exhibited the highest effective carrier lifetime of 31.9 μs, which is higher than that of intrinsic a-Si:H, thermal oxides, wet oxides, and organic layer passivation in the same process. Dan et al. [141] deposited a 10-nm-thick shell of intrinsic a-Si:H by VLS and reported that the shell layer resulted in two orders of magnitude carrier lifetime growth and significant optical absorption growth. As shown in Figure 2.11a, Jia et al. [142] reported a multicore shell TCO/$Al_2O_3$/a-Si:H(p+i)/c-Si NW structure that requires only a very thin $Al_2O_3$ passivation layer to be effective, as the mechanism is field-effect passivation, whereby the negatively charged $Al_2O_3$ efficiently reduces the complexation in the a-Si:H layer by repelling minority carriers from the surface. This field effect of the passivation layer also modulates the bandgap of the TCO, thus changing the optical transparency [142]. An effective carrier lifetime of up to 1.7 ms was reported by Chen et al. [143]. Achieved by $SiN_x$-passivated c-Si NW arrays, an a-Si:H galvanic isolation layer was inserted between Si NWs and c-Si wafers to improve the carrier lifetime. Otherwise, the effective carrier lifetime was only 50 μs. Jura et al. [144] reported an effective carrier lifetime of 530 μs for Si NWs passivated in silicon nitride, which also demonstrated high-performance Si NW solar cells with efficiencies of up to 18% on polysilicon and 19% on monocrystalline silicon, respectively. Lin demonstrated an effective and industrially compatible dielectric passivation technique that has achieved an area of 125 × 125 $mm^2$ for 17.11% efficiency of

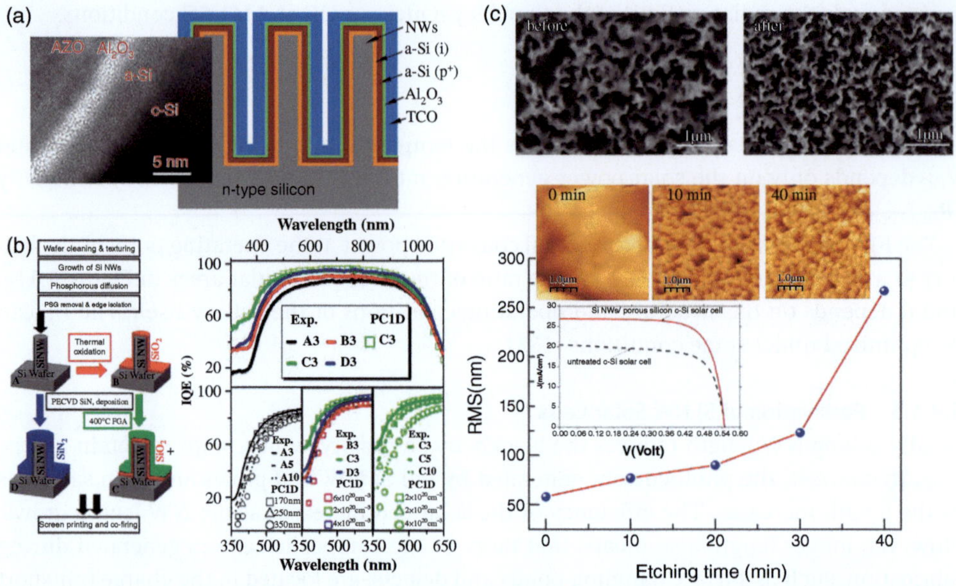


**Figure 2.12** Passivation of Si NW solar cells. (a) Schematic of the multicore-shell TCO/$Al_2O_3$/a-Si:H(p+i)/c-Si NW structure, left-metered inset: a plasmon energy loss-filtered (23 eV) TEM image, showing the thin $Al_2O_3$ layer. *Source:* Reprinted with permission from Ref. [142]. Copyright 2013, American Chemical Society. (b) Left: schematic fabrication process diagrams of Si NW-based solar cells with different passivation. A (black, bare Si NW without dielectric shell); B (red, thermally oxidized layer of ~10-nm $SiO_2$); C (green, thermally oxidized ≈10-nm $SiO_2$ layer combined with 60-nm $SiN_x$); and D (blue, direct deposition of 70-nm $SiN_x$). Right: experimental average IQE spectra in the experimentally averaged IQE spectra in the wavelength range 300–1100 nm. (c) SEM micrograph images of Si nanostructures before and after dye etching treatment (top). Atomic force microscopy scanning (middle). Root mean square variation with etching duration (bottom) J-V characteristics of Si NWs/porous Si solar cells at AM 1.5 (inset), with red curves depicting the enhanced photocurrent and energy conversion efficiency. *Source:* Reprinted with permission from Ref. [147]. Copyright 2017, Elsevier B.V.

the solar cell, 70 nm of $SiN_x$ By PECVD deposition as shown in Figure 2.11d, the passivation exhibits good defect state density reduction and RUSH composite suppression due to surface passivation and hydrogen diffusion-induced bulk passivation [145]. Atomic layer deposition (ALD) is a reliable means of passivation layer deposition even in nanostructures with high aspect ratios [146]. PECVD has been used to deposit hydrogen-containing amorphous silicon and silicon nitride, as these passivation layers are typically ~10 nm thick. In addition, Ben Rabha et al. [146] obtained Si NWs/porous Si-based solar cells with an effective carrier lifetime of ~10 μs and increased power conversion efficiency from 8 to 14.2% by MACE and acid dye etching. This method relies only on the wet realization of the passivation, with potential advantages in yield and cost.

In addition, a high-quality super-$SiO_2$/Si interface with compositional mutations, consisting of ultrathin $SiO_2$ layers on Si(111) wafers, was successfully defined using plasma oxidation of neutral thermal oxygen atoms, which is expected to enable the intentional tuning of the absorption spectra of Si NW solar cells to improve the overlap between the absorption spectra and the solar spectrum [147] and thus increase the utilization rate.

It is important to note that surface passivation affects multiple aspects of a PV device simultaneously rather than just improving charge transfer. For example [141], surface passivation of porous silicon based on acid dye etching treatment resulted in a 36% decrease in light reflection. Thick silicon nitride passivation layers also exhibit enhanced anti-reflective properties. As reported by Lin et al., the passivation layer dominated the length-dependent light trapping effect of Si NW within their experiments [148].

Si NW arrays are not necessarily perpendicular to the Si substrate. In the anisotropic etching method, the Si NWs generated by the Si(100) and Si(111) substrates are perpendicular to the Si surface. In contrast, the Si NWs generated by the Si(110) and Si(113) substrates are tilted to the substrate surface [130]. The tilted-arrayed Si NWs [149] solar cells exhibit the highest power conversion efficiency of 11.37%, and the special tilt-aligned structure achieves excellent electrode contact.

The CVD (VLS) method can prepare photovoltaic NW arrays with high crystallinity and perpendicularity. However, a lower PCE [150]. Avoiding Au catalysts is a strategy because Au diffuses on the NW surface and introduces complex centers, e.g. Sn-catalyzed silane CVD [17]. Slight differences in the CVD process impact the performance of the devices, e.g. vertical NW arrays only appear at the appropriate precursor partial pressure and growth temperature [151]. The synthesis conditions also affect the reflectivity. Synthesis conditions also affect the reflectivity [55]. Ag-catalyzed MACE preparation follows the spirit of the current pursuit of simple low-temperature solar cell fabrication routes, with less dangling bond generation in solution environments, and, in addition, avoids the problems of metal diffusion in thermal CVD; in short, the MACE approach is the preferred choice for producing high-performance Si NW solar cells [152]. However, one of the challenges lies in the instability of silver grain and silver dendrite formation in solution. Although photovoltaic Si NW arrays accommodate an appropriate degree of structural randomness, further process optimization is necessary.

### 2.4.2 Si NW-based Anode Materials for Lithium Batteries

The original negative electrode material was lithium metal. Lithium electrodes and polar nonprotonic electrolyte solvents create a dense surface film that prevents adequate passivation. As the battery was charged and discharged, severe lithium dendrites were created, causing safety issues and significant lithium consumption. The currently used anode material is mainly graphite, which has a low specific capacity and cannot meet the market demand for high-performance lithium batteries [153].

Figure 2.13a shows how a silicon-based lithium battery works, including two processes: silicon on the electrodes absorbs electrons and lithium ions during charging, and silicon-lithium compounds on the electrodes release electrons and lithium ions during discharging. The reaction mechanism between Si-based materials and lithium ions is an alloying reaction mechanism [154]. Both produce a series of alloying compounds ($Li_X$ Si), as shown in Figure 2.13c. Silicon-based materials are highly desirable because the specific capacity is more than 10 times that of commercial graphite anodes: 3580 mAh/g for $Li_{15}Si_4$ and 4200 mAh/g for $Li_{22}Si_4$. As an alloying reactive material, the biggest challenge for silicon-based anodes is the large swelling effect [155]. The molar volume of the intermetallic compounds of Li/Si is larger than that of the nanostructured Si phase, and thus, the failure of the Si anode is mainly caused by the larger volume expansion/contraction during lithium insertion/extraction. The large stresses caused by repetitive volume expansion and contraction lead to cracking and pulverization of the Si anode, which results in loss of electrical contact and eventual capacity degradation. Figure 2.13b depicts three mechanisms of

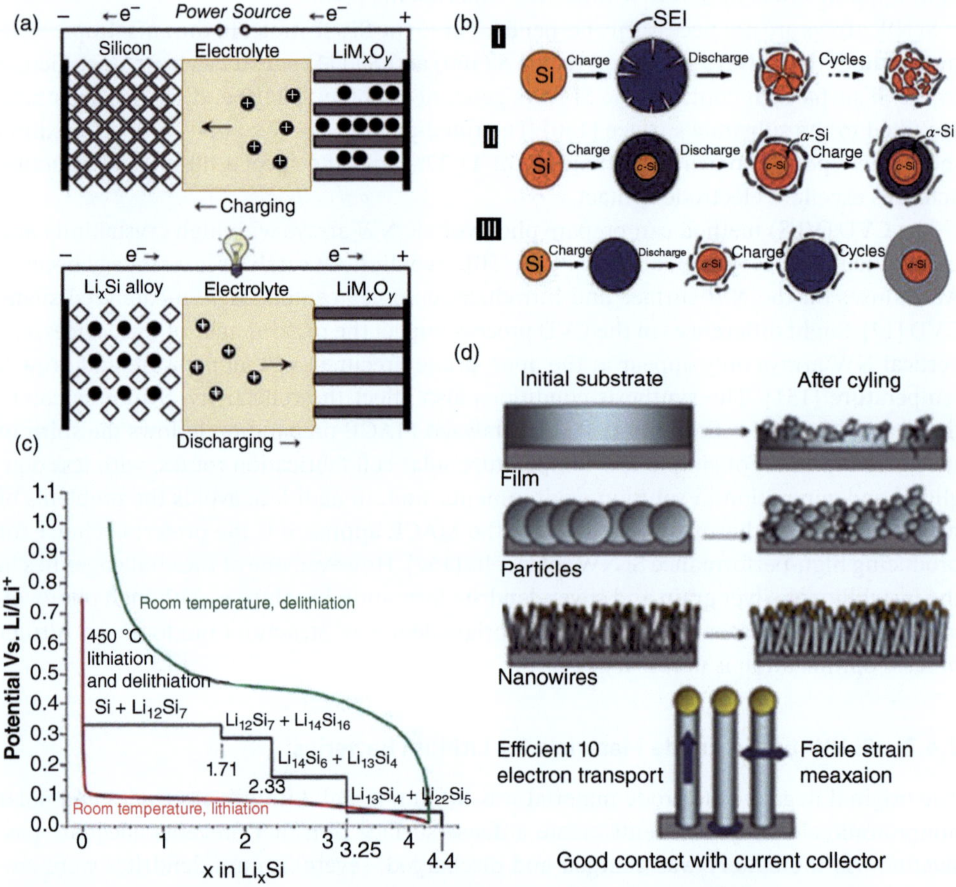


**Figure 2.13** Basic understanding of silicon-based lithium batteries and electrode material failure. (a) Schematic diagram of charging and discharging of a lithium battery with silicon-based electrodes. *Source:* Reprinted with permission from Ref. [155]. Copyright 2008, Elsevier B.V. (b) Degradation mechanisms of large volume change anodes in LIB. (i) Fracture and pulverization of high-capacity electrode materials. (ii) Delayed lithiation due to compressive stress reduces rate performance and decreases effective capacity. (iii) Larger volume changes induce the growth of unstable solid electrolyte interfaces (SEI) on the electrode surface. *Source:* Reprinted with permission from Ref. [157]. Copyright 2008, Elsevier B.V. (c) Electrochemical lithiation and delithiation curves for Si at room temperature. Black line: theoretical voltage profile at 450°C. The red and green lines indicate the lithiation and delithiation of crystalline Si at room temperature. The red and green lines indicate the lithiation and delithiation of crystalline Si at room temperature, respectively. (d) Schematic of charge/discharge cycling of electrode materials with three structures: thin film, particles, and NWs, explaining why NWs have higher resistance to cycling. *Source:* Reprinted with permission from Ref. [158]. Copyright 2018, Wiley-VCH Verlag GmbH & Co. KGaA, Weinheim.

electrode failure due to large electrode material strains. The study [156] showed that the volume expansion of silicon-based anode materials after lithium insertion can reach up to 320%. Si NWs are considered promising in solving this problem because small-diameter Si NWs can easily adapt to the large volume change caused by the alloying reaction, as shown in Figure 2.13d. Compared to thin-film or particulate materials, NW materials have the advantage of ordered arrays after many cycles and the perpendicularity of the direction of strain to the direction of charge transport. In addition, the Si NWs are directly connected to the NWs. In addition, Si NWs are directly connected to the collector, providing a direct 1D electron path for fast charge transport. Si NWs have a high surface-area ratio, which can provide a fast charging and discharging rate [155].

The electrode material prefers highly parallel NW arrays rather than scattered-oriented NWs. As we have discussed in Section 2.3.2 on growth orientation control, in summary, crystalline Si NWs for lithium-ion battery anodes were fabricated by the VLS method [152], OAG method [159], VS method [158], and etching methods [160, 161] obtained with high parallelism and crystallinity, combined with large fluxes. Si NWs can be synthesized directly on the collector material without any bonding process. The core–shell structure defined by VS or OAG growth improves the resilience to swelling effects. As shown in Figure 2.14, a crystalline-amorphous core–shell NW of silicon [158] is designed to use only amorphous shell materials for lithium storage while retaining the crystalline core as a mechanical support and efficient electron transport pathway based on the difference in lithium potentials between the two. When Si-$SiO_x$ NWs are used as anodes [159], amorphous $SiO_x$ can also accommodate large volume changes of Si during cycling.

Etching methods have realized devices with high performance and high cycling stability, e.g. [160]. In addition, magnetron sputtering deposition combined with metal-catalyzed chemical etching has prepared Si NW films that can be used as anode materials for lithium-ion batteries, with an initial capacity of more than 3000 mAh/g and maintaining 90% of the capacity after 100 cycles. The MACE methodology allows simple access to porous Si NWs. Oxide-free porous silicon [161] and highly porous structures with thin silicon walls [162] improve the lithiation process by several mechanisms, reducing the specific capacity and suppressing the volume expansion. Surface modifications based on coating techniques have also been used; for example [163], Sn coatings can enhance the cyclic coulombic efficiency by twice that of bare NW.

### 2.4.3 Thermoelectric Generators

Thermoelectric generators (TEGs) convert heat to electricity by inducing the movement of charge carriers through the Seebeck effect. TEGs can be used to generate electricity from waste heat, and other places of application include watches, automobiles, biometric sensors, and cooling chambers [152]. A detailed theoretical study of the thermoelectric effect of NWs can be found in the review [164]. Briefly, the dimensionless quality factor of TEG is as follows:

$$ZT = S^2\sigma T/\kappa \tag{2.8}$$

where S-Seebeck coefficient ($\mu V\ K^{-1}$), $\sigma$-conductivity (S/m), T-ambient temperature (K), and $\kappa$-thermal conductivity (W/m-K). The strategies to enhance the value of ZT are

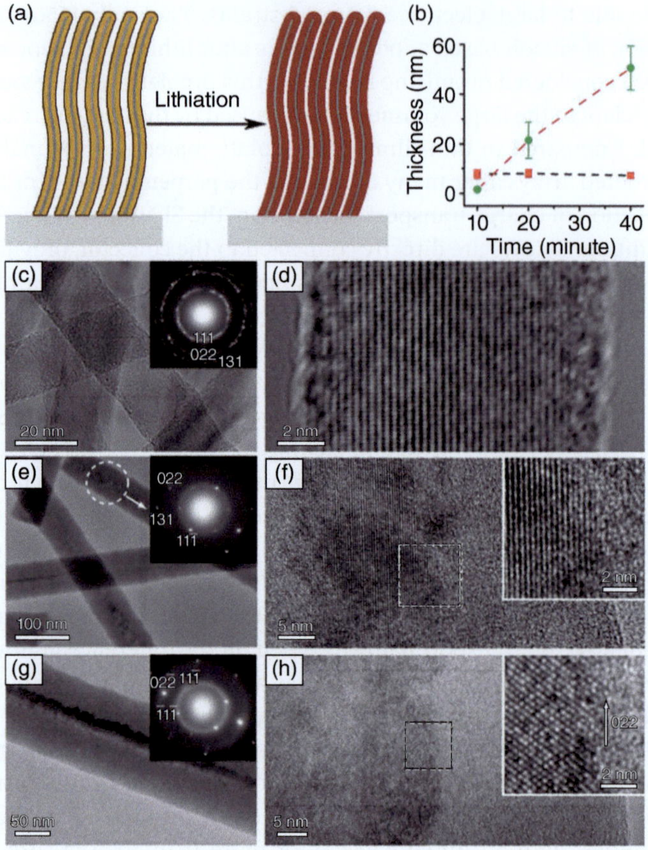


**Figure 2.14** (a) Schematic of Si c-a core–shell NWs lithiation grown directly on stainless steel collectors. (b) Statistical data of core radius (red) and shell thickness (green) versus growth time. The growth temperature is 485°C. The data show that the core radius remains constant, and the shell thickness increases steadily with growth time. (c) TEM and selected area electron diffraction (SAED) images of NW grown for 10 min. (d) HRTEM image of NW grown for 10 min. (e) TEM and SAED images of NW grown for 20 min. (f) HRTEM image of NW grown for 20 min. (g) TEM and SAED images of NW grown for 40 min. (h) HRTEM image of NW grown for 40 min. *Source:* Reprinted with permission from Ref. [158]. Copyright 2009, American Chemical Society.

enhancement of the Seebeck coefficient S, enhancement of the electrical conductivity κ, and reduction of the thermal conductivity κ.

### 2.4.3.1 Si NW-based TEGs

TEG is considered Silicon due to its high storage capacity and nontoxicity. Silicon nanostructures are prioritized over bulk silicon. The Bessel factor of Si NWs is three times higher than that of bulk silicon [165]. Si NWs can realize highly efficient TEG devices (ZT ~0.1) [152]. For the Bessel factor, alumina-Si core–shell NWs [23] were shown to improve the Bessel factor, and porous Si NWs with large surface areas [166] also exhibit higher Seebeck coefficients. Figure 2.15a shows the process flow. Increasing the surface roughness, increasing the doping concentration, and decreasing the diameter of the NWs are a few ways to

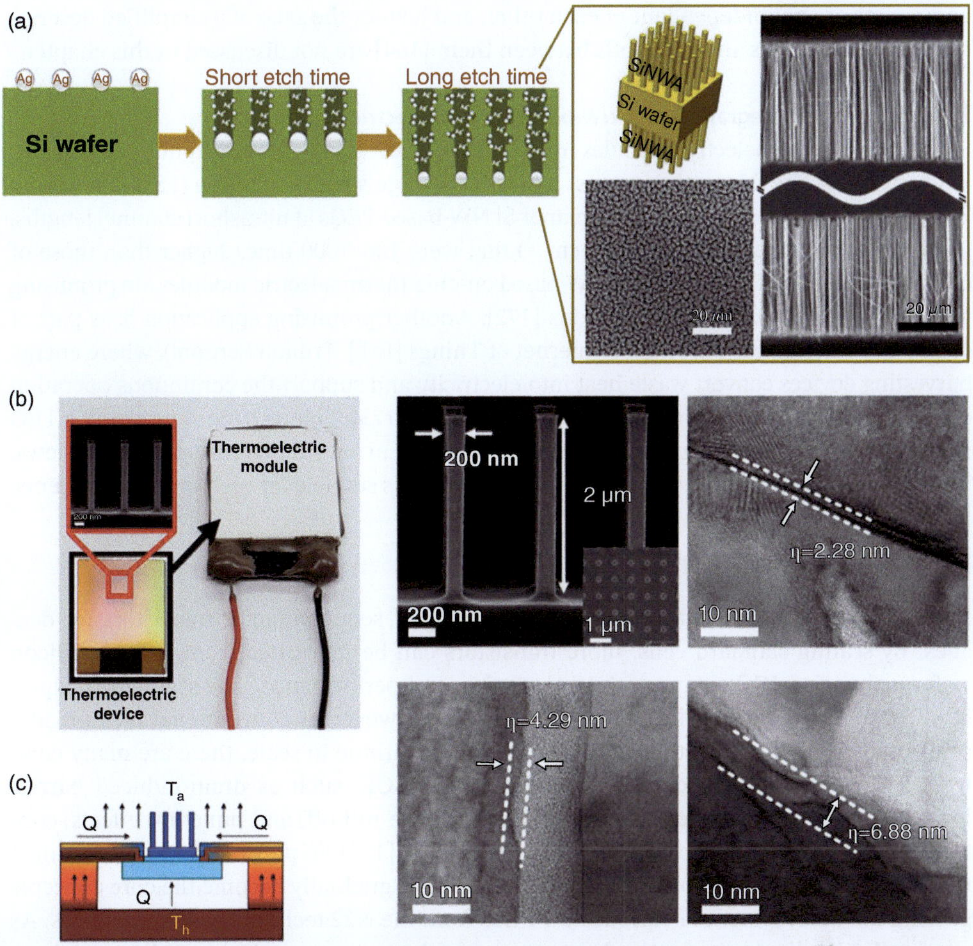


**Figure 2.15** Fabrication and on-chip integration of Si NW TEGs. (a) Porous Si NW arrays prepared by a longer (240 min) MACE. *Source:* Reprinted with permission from Ref. [166]. Copyright 2015, Elsevier Ltd. (b) Photographs of fabricated thermoelectric device chips and modules with a size of 1 cm × 1 cm (left). Structural characterization of vertical Si NWs with rough surfaces. Cross-sectional view SEM image of a vertical Si NW array with a diameter of 200 nm and a height of 2 μm. The inset shows the top view SEM image. High-resolution transmission electron microscopy (HRTEM) images (surface roughness n = 2.28, 4.29, and 6.88 nm) of the vertical Si NW (right panel). *Source:* Reprinted with permission from Ref. [167]. Copyright 2019, American Chemical Society. (c) Si NW-based transverse micro-TEG integrated in a micromechanical device. *Source:* Reprinted with permission from Ref. [170]. Copyright 2016, IOP Publishing Ltd.

reduce the thermal conductivity [167]. Figure 2.15b shows Si NWs with increased surface roughness. Width-modulated NW structures have high thermoelectric efficiency because width modulation enhances boundary scattering and reduces thermal conductivity [168]. Relatively long Si NW arrays have better ZT since thermal resistance increases with length. For improving conductivity, the spin-on doping (SOD) technique is used to prevent surface depletion [169], which improves the conductivity. It is essential to point out that these parameters are not independent of each other, and just for the sake of a simplified description, the correlations and trade-offs between them [164] are not discussed in this chapter.

#### 2.4.3.2 On-chip Integration of Si NW-based Thermoelectric Modules

Si NW-based thermoelectric modules have also been developed, demonstrating the compatibility of the Si NW fabrication process with CMOS large-scale integration (LSI) fabrication. Zhang H et al. reported miniaturized planar Si NW-based TEGs at ultrashort channel lengths, generating power densities (4.2 mW $cm^{-2}$) that were 100–1000 times higher than those of conventional planar TEGs [171]. Si NW-based on-chip thermoelectric modules are promising for on-chip hotspot cooling applications [172]. Another promising application is as part of distributed intelligence paradigms (Internet of Things [IoT], Trillion Sensors), where energy harvesting devices convert waste heat into electricity and support the continuous operation of many miniature sensors connected to the IoT [170, 173]. Figure 2.15c shows how a TEG is involved in on-chip thermal management, with red arrows representing the heat flows. The vertical scalability of multistage TE modules [174] is suitable for high thermal gradients.

### 2.4.4 Electronics of Si NWs

Moore's Law has predicted the continuous scaling of semiconductor transistors for decades. By scaling standard cells, more transistors can be fabricated from a single silicon wafer, enhancing ICs' compactness, economics, and performance. The scaling principles of effect transistors are reviewed in the literature. However, as conventional metal-oxide-semiconductor field effect transistors (MOSFETs) continue to scale, there are many challenges: first, the leakage current mechanism and SCE, such as drain-induced barrier lowering (DIBL), hot-carrier effects, threshold voltage roll-off, and puncture effects, continue to increase, leading to performance degradation [175]. In particular, the fight against the SCE and the suppression of leakage currents have gradually become the core concepts guiding the development of various new FETs from the N22 technology node onwards. As we specifically discuss in the third paragraph of this section, transistor architectures have undergone several leaps, from planar transistors to Fin field-effect transistor (FinFETs) and multigate FETs, then to gate-all-around FET (GAA FET), and finally, to 3D-stacked FETs. Second, the study reveals that the carrier density of silicon material channels decreases at the nanoscale, for different scales, due to, respectively, quantum confinement effects [176], impurity surface polarization [177], and impurity ionization energy enhancement (or dielectric mismatch) [178, 179]. A popular and promising direction is the introduction of other low-dimensional materials with high mobility, such as channel materials. For silicon, a deeper understanding of carrier behavior at small scales and developing novel carrier transport modes are necessary. Finally, at nanoscale scales, doping becomes difficult and even dopant failure [180]. Advanced doping engineering and silicide intrusion contacts are two widely adopted strategies.

As a 1D or quasi-1D material, Si NWs have novel electrical and physical properties that are given expectations by extended Moore's law. We will start with the material properties of Si NWs and then review the device-level advances realized in Si NWs. These devices include, first of all, MOSFETs and unconventional transistors such as Schottky barrier field effect transistors (SBFETs) and junctionless field effect transistors.

#### 2.4.4.1 Properties of Si NWs

##### *2.4.4.1.1 Bandgap Shift in Si NWs*

Si NWs with diameters of several nanometers have shown promise since they [181] are stably prepared. The electronics of Si NWs have shown great promise. Based on the quantum confinement effect, the energy bands of NWs deviate from bulk silicon at small enough diameters, e.g. [70], 1.1 eV at 7 nm diameter, increasing to 3.5 eV at 1.3 nm diameter. Figure 2.16a illustrates the prediction of the bandgap as a function of diameter. The high surface area ratio of the structure allows for the richness of electronic structure engineering based on surface effects; e.g. hybridization effects introduced by chemical passivation of the NW surfaces by different chemical functional groups can provide bandgap shifts, and the core–shell interface leads to similar shifts [71]. Strain can change a 1D lattice's lattice constant and provide robust energy band shifts and modulation.

##### *2.4.4.1.2 Electron Transport in Si NWs*

Si NWs are applied as 1D or quasi-1D channel structures for FETs to counteract the SCE, reduce the leakage current, and ensure the efficient conduction of the silicon channel. The electronic behavior in Si NWs exhibits a high degree of anisotropy due to confinement in two directions. The length-axis crystallographic orientation determines the carrier transport mechanism in Si NWs, the wire diameter and cross-sectional shape, and the channel length. For NW channels with lengths much larger than the carrier mean free range, scattering effects will determine the mobility and carrier transport. Due to the NW's very high aspect ratios, the mobility is strongly influenced by the orientation and roughness of the semiconductor surface and the interface with the surrounding oxide [182]. The mobility of NWs is affected by interband scattering effects, and Das et al. [183] found that the mobility fluctuates with increasing diameter due to the emergence of new channels for interband scattering. The mobility is also affected by the cross section. Patterned and etched NWs typically have rectangular, quadratic, or trapezoidal cross sections.

In contrast, synthetic NWs exhibit a multifaceted cross section with a predominantly hexagonal arrangement, and the NWs can be altered to change their surfaces through machining and heat treatment. To counteract dielectric mismatch, cylindrical NWs and NWs with elliptical cross sections are sometimes preferred in electronic devices [184] (Figure 2.16b,c). According to experimental and simulation studies, Si NW electron tubes with triangular cross sections have higher on/off ratios and better subthreshold characteristics, and such NWs are fabricated by anisotropic wet etching in the {111} plane [185, 186]. Gunawan et al. [187] report the first direct capacitance measurements of Si NWs and the extraction of carrier mobility in undoped channel Si NW FETs at room temperature, demonstrating comparable mobility to low-index-oriented bulk silicon.

As early as 2000, controlled doping of Si NWs has demonstrated the technical potential of Si NWs to realize FETs [189]. As shown in Figure 2.17, NW bottom-gate transistors are the most classical type, similar to planar silicon devices. Based on this architecture, Cui [190] and Lieber [191] et al. demonstrated the first n-doped and p-doped accumulation mode transistors made from bottom-up-grown Si NWs, achieving a mobility of ~100 $cm^2/V\text{-}s$. It is shown that source–drain contact thermal annealing and surface passivation significantly

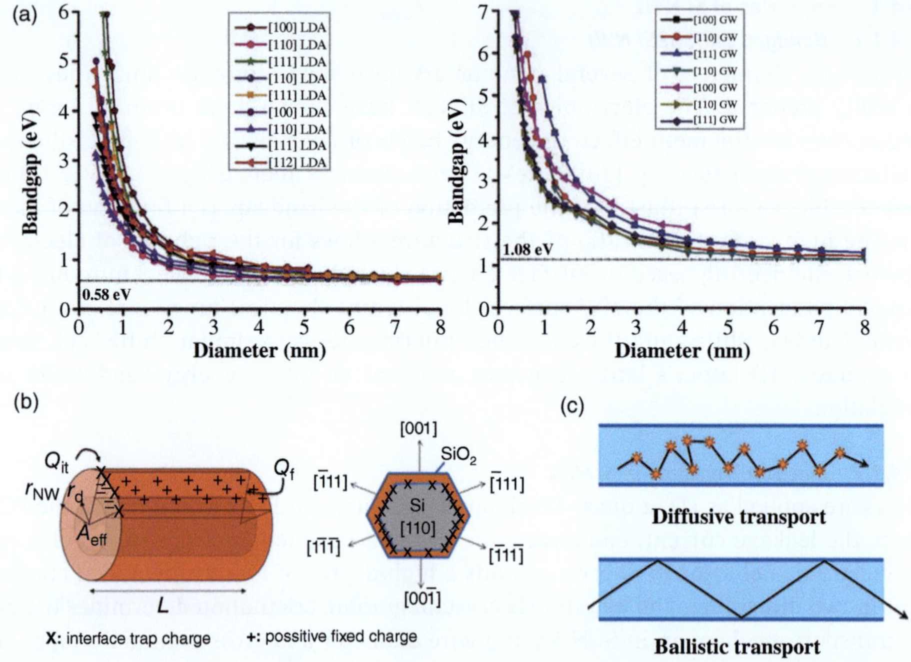


**Figure 2.16** Electrical properties of Si NWs. (a) Calculated results: the bandgap of Si NWs with different crystallographic orientations varies with the diameter. LDA: load-density approximation; GW: green function (G) and screen coulomb interaction (W). *Source:* Reprinted with permission from Ref. [128]. Copyright 2018, the Chinese Chemical Society, publication and hosting by Elsevier B.V. (b) Effect of surface charge on NWs. Schematic representation of the effect of surface and interfacial charges on the conducting area of p-doped NWs. A positive surface charge depletes the surface-bounded volume in the NW, leaving an inner cylinder as the active conducting volume (left panel). Oriented Si NWs with a hexagonal <110> cross section are typically obtained in catalyst-assisted growth (right panel). *Source:* Reprinted with permission from Ref. [184]. Copyright 2019, American Chemical Society. (c) Both drift-diffusion transport and ballistic transport are likely to occur in NW channels [188]. OA.

improve the performance of the FETs. In addition, strain doping is necessary for the mobility enhancement of Si NWs [184].

In 1D semiconductors and metals, electron transport may be ballistic; i.e. when the mean free range of the charge carriers is greater than the length of the wire, scattering effects are negligible. The average carrier velocity in ballistic transport is independent of the local electric field [186], in which case mobility is not a meaningful parameter for evaluating ballistic transistors. The conductivity depends on the contacts and connections in the 1D quantum structure, as described by the following equation:

$$G = 2e^2/h\ T/R \tag{2.9}$$

where T and R are the quantum mechanical transmission and reflection coefficients at the potential barrier, respectively, satisfying T + R = 1.

The ballistic transport mechanism is expected to observe quantum conductivity effects in NWs when the separation of the quantized subbands of a 1D system is larger than

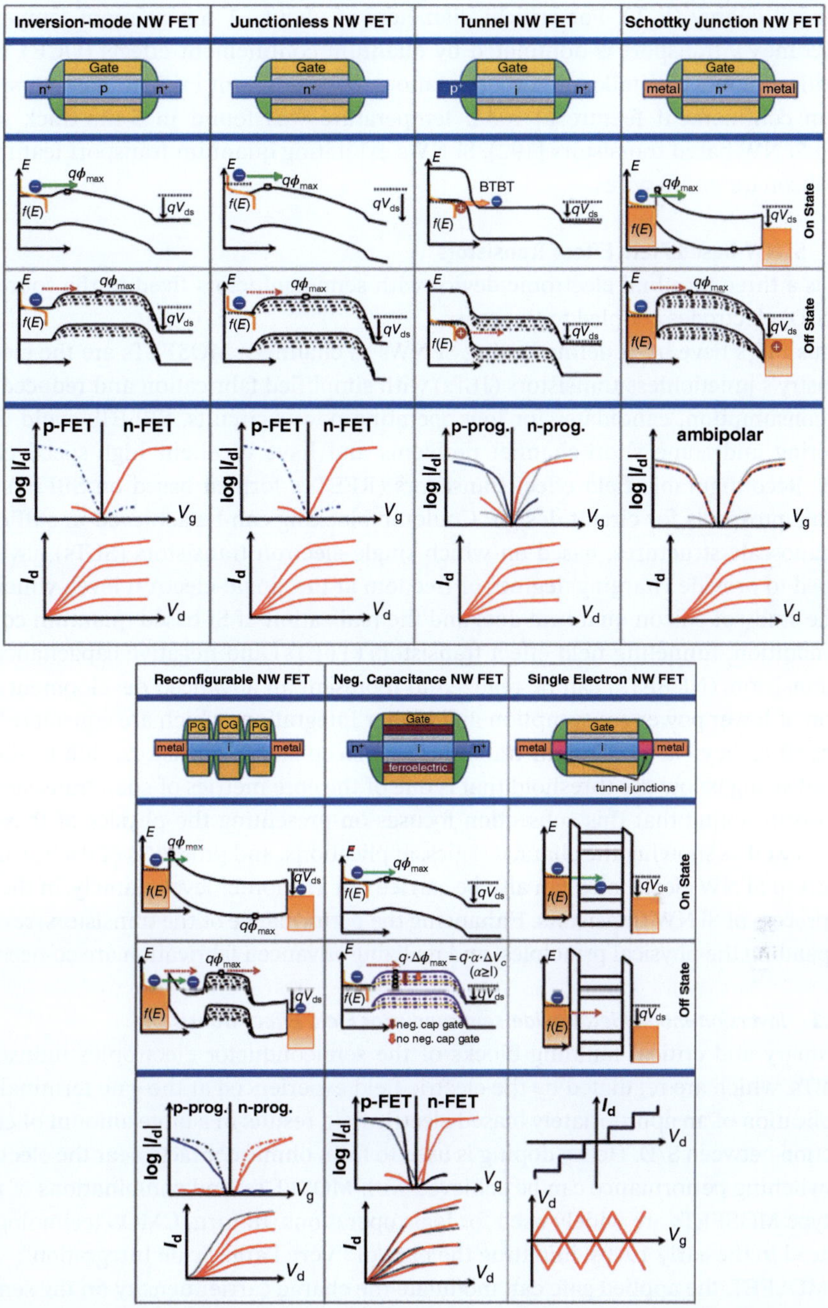


**Figure 2.17** Summary of transistor types that can be realized with Si NWs. Comparison of inversion-mode FETs, junctionless FETs, interband TFETs, SB NW FETs, RFETs, NCFETs, and single-electron FETs. The first row shows a schematic cross section of an NW FET with wrap-around gating. The second row shows a schematic of the band diagram of an NW FET operating in the on-state and off-state. The third row shows the typical subthreshold transmission and output characteristics of the FETs [188]. OA.

the thermal energy ($k_BT$). For critical diameters of about ~1 nm at room temperature ($k_BT$ ~25 meV), transport is dominated by quantum confinement effects (QCE), and Si NWs with different crystallographic orientations have different critical diameters; e.g. a quantum confinement feature at room temperature was found in 5-nm-thick <110> p-doped Si NW-based transistors [192]. Si NWs exhibiting quantum transport features are called silicon quantum wires.

#### 2.4.4.2 Si NW-based Field Effect Transistors

A FET is a three-terminal electronic device with semiconductors fixed to the source and drain (S/D) electrodes coupled to the gate.

Various FETs have been defined using Si NWs as channels. MOSFETs are the electronics industry's junctionless transistors (JLTs) with simplified fabrication and reduced static power consumption, candidates for low operating power circuits. SBFETs avoid doping engineering and some short-channel problems and have excellent high-speed characteristics. Reconfigurable field effect transistors (RFETs), formed based on SBFETs, offer additional prospects for circuit design. Coulomb blocking can be achieved in sufficiently small nanoscale structures, based on which single-electron transistors (SETs) have been developed to provide charging degrees of freedom at the single-electron level, which supports the study of silicon quantum dots and the realization of Si-based quantum computing. In addition, tunneling field effect transistors (TFETs) and negative capacitance field effect transistors (NCFETs) can be realized to represent an advanced development in the direction of lower power consumption and higher integration, which are considered to be categorized as beyond Boltzmann transistors, or steep-slope transistors, due to the subthreshold swing below the threshold that is one of the core metrics of such transistors.

It is worth noting that this subsection focuses on presenting the physics of these constructs, as well as showing the characteristics, applications, and prospects of each structure. Advances in Si NW electronics can also be carried out at another level, namely, in the fabrication process of Si NW transistors. Enhancing the performance of the transistors, revealing and expanding the physical principles, and realizing advanced fabrication are co-occurring.

##### *2.4.4.2.1 Inversion-mode Metal-oxide-semiconductor Field Effect Transistors*

The primary and critical building blocks of the semiconductor electronics industry are MOSFETs, which are regulated by the electric field experienced at the gate terminals, and the application of an appropriately biased electric field results in a large amount of current conduction between S/D. Heavy doping is used to form ohmic contacts near the electrodes. Good switching performance can be achieved with MOSFETs, and combinations of n-type and p-type MOSFETs are widely used for logic operations. In turn, CMOS technology was introduced in the early 1970s, initiating the pace of "Very Large Scale Integration".

In a MOSFET, the applied gate can modulate the charge carrier density on the semiconductor channel surface by accumulation, depletion, or inversion. Strong inversion occurs when the concentration of inverted charge carriers exceeds the charge density produced by doping. This mode of inversion and accumulation in transistors has been widely used by the microelectronics industry, from lumped MOSFETs to silicon-on-insulator (SOI) transistors and 3D multigate FinFETs. The NW MOS transistor is a promising prediction that is in line with this development trend.

##### 2.4.4.2.2 Junctionless Field Effect Transistor

JLTs, or JL-FETs, also known as gated NW resistors, and junctionless accumulation mode field effect transistors (JAM FETs), were proposed by Colinge and Lee et al. in 2009–2010 [193, 194]. The Si-based junctionless nanowire transistor (JNT) is a Si NW transistor that exhibits improved SCE resistance and good gating and is an up-and-coming candidate for further scaling up of MOSFET applications [195, 196].

Unlike conventional FETs, JLTs consist of a single, homogeneous, highly doped (typically $10^{19}$ $cm^{-3}$) semiconductor region that houses the charge transport channel, source, and drain regions. In a junctionless structure, current conduction does not require an inversion mode. When the gate voltage is less than the threshold voltage, all carriers are depleted, and the gate terminal voltage is increased from the threshold voltage; the transistor is in partial depletion mode, and hence, conduction starts. Unlike surface conduction in conventional MOSFETs, intrinsic conduction occurs in JLTs. Despite the different modes of transport, the $I_d$-$V_g$ transfer and $I_d$-$V_d$ output characteristics are similar to those of an inverting mode MOSFET, which has led to the JLT being categorized as a conventional transistor [184].

Junctionless nanowire FETs have several advantages over conventional nanowire FETs [188]. The charge carriers in a junctionless nanowire FET are distributed in the center of the NW in both the off and on states, which reduces the interaction between the charge carriers and the dielectric layer, especially in the off state. This favors long-term reliability. During junctionless FET operation, the gate electric field falling through the dielectric layer is low, allowing gating to be performed using an ultrathin and efficient dielectric layer. Since homogeneous doped NWs without S/D injectors are used in junctionless FETs, device fabrication constraints can be relaxed, and the variability of electrical characteristics in the device is expected to be mitigated for small devices.

JNTs are compatible with CMOS processes and can mimic conventional inversion-mode FETs to build functional circuits [188]. The NW transistor is compatible with the CMOS process. Although commercialization has not yet been completed, numerous technology-enabled [197] have been made to combine with JLT technology, achieving circuit-level demonstrations, such as a 6T static random access memory (SRAM) memory circuit based on nanoscale JLTs with high-κ dielectrics [198] with enhanced electron tunneling resistance and improved read and write noise margins. A novel hybrid C-FinFET common-source amplifier (H-CA) using charge-coupled n-FinFETs and JL p-FinFETs [199]. The gain and gain bandwidth product of the H-CA are increased by ~40%, respectively, compared to the corresponding values of the JL-CA of the same size. Vacuum dielectric structures for source or drain stacks [200] are proposed for suppressing hot-carrier degradation and SCE. Vertically integrated junctionless FET [201] Consisting of vertically stacked multiple Si NWs with a full wrap-around gate (GAA) structure, the low fabrication complexity and high-performance nonvolatile flash memory are demonstrated by utilizing the JLT's features of no need for fabrication of S/D and bulk charge transfer.

As an industry attempt, IMEC proposes a novel SRAM cell design with two vertically stacked junctionless vertical NWFETs that can reduce SRAM area per bit by 39%. According to IMEC, JLT may also be a viable analog/RF applications [197] option. JL-FETs are also candidates in quantum computing because they operate even at very low temperatures. However, researchers continue to address several challenges, including lower ON currents, temperature sensitivity, and scalability limited by junction-free energy ($E_J$).

#### *2.4.4.2.3 Schottky Barrier Field Effect Transistors*

The SBFET is an interesting alternative to the classical MOSFET. Structurally, the main difference is that metal electrodes replace the simply doped S/D regions. The transport characteristics of SBFETs are considerably affected by the Schottky barrier height (SBH). However, the prediction and control of the SBH is a challenging task, as Tung in his review [202] described in the review that electrostatics, solid-state quantum theory, interface chemistry, and quantum transport each play a role in the SBH, and this review publishes systematic suggestions on how to define and correct the SBH. Specifically, in the case of silicon channels, the interface forms a stable metal-Si compound [203].

The metal silicide/Si interface can also realize ohmic contacts with MOSFET-like behavior at shallow barrier heights. For example [204], PtSi/Si/PtSi heterostructure NWs prepared by utilizing the diffusion process of Pt in silicon have nanoscale junction regions, presenting high crystallinity and sharp interfaces, and the resistivity of PtSi is as low as ~28.6 μΩ·cm. Compared to conventional electrodes, this structure presents attractive high-speed performance, e.g. a transmission frequency of 280 GHz [205]. In addition, applications of Ni-Si and Al-Si systems for reconfigurable transistors are described in the following section. A distinctive feature of NW SBFETs, compared to thin films or bulk materials, is the sharp interfaces with atomic-level clarity, which are thought to evade some of the SCE due to bypassing dopant engineering [184].

#### *2.4.4.2.4 Reconfigurable NW FETs*

Based on low-dimensional semiconductors, Schottky field effect transistors can be realized with polarity-controllable or reconfigurable NW FETs. RFETs are novel multifunctional or functionally enhanced devices that only provide n- and p-FET characteristics through electrical signals. These four-terminal NW devices rely on the unique properties of nanoscale Schottky junctions in the NW to regulate the polarity and concentration of charge carriers. Depending on the difference in polarity control methods, reconfigurable devices can be categorized into three types [206]:

**i)** Selective charge carrier injection and concentration control at contacts.
**ii)** Contact polarity selection and control of charge concentration in the channel.
**iii)** Bipolar operation with charge carrier filtering in the channel.

Weber et al. added to this by suggesting that the polarity control can also combine the first and second in the case of having more than three independent gates [207].

The development history of Si NW RFETs is summarized in the review [208]. In 2008, bipolar operation based on Schottky contacts and Si NWs was by Weber et al. [209]. In 2012, the name "reconfigurable field effect transistor" was first introduced in 2012 by Heinzig et al. [210]. A year later, devices with improved symmetry were reported by the team [211]. A year later, the team reported that selective tunability of charge-carrier transport was crucial for symmetry, which was achieved by combining the nanoscale dimensions of the junction with the application of radial compressive strain. RFET-based logic gates were developed during this period, and more complex reconfigurable nanowire transistors were demonstrated. In these studies, RFETs could enhance the circuit's implicit reconfigurability [212] and explicit reconfigurability [213] to realize more functions per computing unit, and explicit reconfigurable circuits can switch logic functions when receiving the corresponding signals, for example [213]. An RFET circuit with both NAND and NOR functions is presented. Explicit reconfigurable logic gates have a higher expressive power of the truth table, and an example of an application consists of a bipolar Si NW FET transistor with XOR gate functions [212]. Since 2014, RFET research has entered a new period, with a proliferation of research on new channel materials and structures that enable RFETs to realize more functionality [208]. For Si NW-based RFETs, one direction is to realize the simplification from dual gate to single gate [214].

On the other hand, trigate RFETs realize dual-threshold voltage operation for reduced standby power consumption [215], which reduces leakage power consumption by 54% compared to a FinFET counterpart with comparable performance. In the case of three independent gates, polarity control is extended to mode control [216]. A Si NW FinFET-based study [217] exploits the weak ionization shock-induced positive feedback effect and the dynamic modulation of SBs to achieve an ultra-steep subthreshold swing (SS) down five orders of magnitude, as low as 6.0 mV/dec, with a switching ratio of $10^7$ on average. Structures with four or more gates on a single Si NW channel have also been proposed [218] called multi-independent gate RFETs, which provide programmable multiplexing and can be used as transmission gates in multiplexers. In addition to Ni-Si material systems, Al-Si material systems have been developed due to the relatively symmetric barrier heights for electrons and holes. The Al-Si-Al RFET platform is proposed to demonstrate a multi-CG RFET similar to a three-input wired AND [219].

In summary, RFETs achieve circuit-level benefits: dynamic reconfiguration, intrinsic heterodyne and wired AND functionality, control of threshold voltages, and suppression of parasitic charge-sharing effects in dynamic logic gates. As an outlook for future applications [208], the polymorphic nature of RFET circuits provides new approaches to hardware security solutions, including logic locking, masquerading, physically unclonable functions, or chip authentication. Reconfigurable features also hold many prospects for analog and mixed-signal designs. RFETs enable XOR and XNOR, potentially driving their use in binary neuronal networks. RFETs offer more flexibility and functionality for LSIs as a competition and complement to CMOS. In 2018, the first complete physical synthesis flow for emerging reconfigurable nanotechnology using open-source tools [220] was introduced, which undoubtedly encouraged the adoption of RFETs at the industrial level. However, compared to the more mature CMOS industry, RFETs require further model development and standardization efforts.

#### *2.4.4.2.5 Si NW Single-electron Field Effect Transistors and Si NW Quantum Dots*

In Si NWs, the charging energy of a single electron is described by the following equation:

$$E_c = e^2/2C_i \tag{2.10}$$

It is clear that if the NW has a sufficiently small capacitance ($C_i$) coupled to the electrode through an insulating tunnel junction, then $E_c$ will be higher than the thermal energy of the electrons, kT, and the control of the charging behavior through the channel, in this case, will be at the one-electron level. Transistors operating on this principle are called SETs. When a bias voltage is applied, electrons in the source can cross the first thin insulating junction into the semiconductor channel and then follow the channel through the second junction to the drain, thus enabling current to flow from the source to the drain. If this process is fast enough and the electrons charge the channel one at a time, the additional charge on the channel is a discrete multiple of the charge of a single electron.

Voisin et al. investigated edge states' formation and electronic properties in Trigate NW transistors in the one-electron limit [221]. At low temperatures, 1D edge modes localize into quantum dots with a spacing between the first two single-particle energy levels of about 1 meV due to the attractive confinement potential generated by the positive charge traps in the gate dielectric layer. Lavieville et al. reported a combination of a narrow (3.4 nm) diameter of MOS NWs and thick gate oxides (7 nm), coupled with designed electrode lengths, to achieve coulombic solid interactions and high charging energy (more than 200 meV), demonstrating a room-temperature artificial atom transistor platform based on deep ultraviolet lithography NWs [222]. Duan et al. reported bipolar gate-defined quantum dots in SOI NWs fabricated using custom CMOS, demonstrated reconfigurable single and double quantum dots in the n-type and p-type ranges by gate-based dispersive readout, and achieved a minimum integration time of 160 (100) μs for electrons (holes) with a signal-to-noise ratio of 1 for the detection of the jumps [223]. Phonon-mediated quantum transport is a byproduct of the Coulomb-blocked electron spectra of semiconductor quantum dots at low temperatures, and single-phonon-mediated quantum transport can be observed through the characterization of electronic noise in quantum dots in silicon SETs [224].

Physical structures similar to SETs are also described in some literature as quantum dots in Si NWs due to different contexts and purposes, with silicon quantum dots acting as the core of quantum transport of carriers in the former and being used for quantum computation in the latter based on realizing the charging flexibility of single electrons. For quantum computing, silicon is an attractive material due to the long expected electron–spin coherence time of quantum dots in silicon, which is a consequence of the small spin–orbit coupling and the predominantly spin-zero nuclear background in silicon [225]. As mentioned earlier, in both top-down and bottom-up approaches, Si NW technology allows the definition of nanoscale channels with diameters smaller than 10 nm in high yields, with strong confinement of lateral transport and easy formation of silicon quantum dots. In 2009, measurements of quantum dots in NiSi-Si-NiSi NWs were reported with a clarity that allows a complete understanding of the spin of the first four-hole states [226]. The promise of metal silicide contacts for silicon quantum devices was dialectically analyzed [227]. The benefits are that flexible contact interfaces and intentionally introduced localized states provide the design space to increase the low-temperature conductance, e.g. it is possible to increase the low-temperature conductance of a device by more than an order of magnitude by implementing contacts with meandering shapes. The challenge is to concentrate the magnetic field in the semiconductor near the silicide, thereby diminishing the SB width while staying below any breakdown field of the material of interest (e.g. the gate oxide).

#### *2.4.4.2.6 NW Transistors Beyond Boltzmann*

Suman et al. [228] suggest that four directions will work together to drive electronics into the era of hyper-scaling to meet the demands of data-intensive workloads: transistors beyond Boltzmann, embedded high-performance memories beyond SRAM and dynamic random access memory (DRAM); monolithic 3D integration of logic, memory, analog, and input/output transistors; and heterogeneous integration of functionally diverse ICs with monolithic-like performance. Given that the subject of this chapter, Si NWs, is at the level of the material, we offer an outlook for only two directions at the device level, and they are the first two of these: beyond Boltzmann transistors and high-performance memories.

Let's go back to the basics of transistors to understand subthreshold swing (SS, or S for short): As the working principle of transistors, to turn off the transistor, the gate introduces an energy barrier in the channel to limit the pure diffusion of the drain current $I_{diff}$ to charge carriers with energies E exceeding the height of the energy barrier. S is defined as the change in gate voltage needed to describe the change in source-drain current of an order of magnitude as the FET moves from the off state to the on state.

$$S = \frac{dV_g}{d(\log_{10} I_{diff})} \cong \left(1 + \frac{C_{dm}}{C_{ox}}\right) \mathrm{In}10 \frac{kT}{q} \tag{2.11}$$

where k is the Boltzmann constant, $C_{ox}$ is the coupling capacitance between the gate voltage and the semiconductor surface potential, and $C_{dm}$ is the depletion layer capacitance. In the ideal case, i.e. when the oxide gate capacitance dominates, S depends only on the temperature:

$$S \cong \mathrm{In}10 \frac{kT}{q} \approx 60 \frac{mV}{decade_{Id}} \mid \mathrm{for} T = 300\mathrm{K} \tag{2.12}$$

S is one of the key measures of the energy efficiency of a transistor, and a smaller S value means that a smaller voltage change is required during switching to achieve fast switching of the current, which reduces power consumption.

#### *2.4.4.2.7 Memory*

##### 2.4.4.2.7.1 *Si NW-implemented SRAM and DRAM*

In recent years, structures based on Si NWs have been demonstrated for storage. These devices typically use silicon nitride as the charge storage layer. The combination of Si NW MOS structures with the positive feedback effect of weak shock ionization (FBFET) has made progress for memory use. A single-gate p-n-i-n Si-NW channel full-gate (GAA) feedback FET [229] SRAM was realized, with storage characteristics that can be generated by weak shock ionization, with 0.76 V VTH and 1 μs for programming/erasing operations as nonvolatile memory (NVM), and 10 ns read/write speeds and 6.3 μA current-detection margins as volatile memory. Bendable FBFETs with n+ -p-n+ Si NW channels exhibit SRAM characteristics [230]: 3600-second hold time, $10^4$ PE cycles of endurance, and high memory margins of over 15 A. Bistable Resistors with n+ -p-n+ Si NW Channels Exhibit Lossless Readout Memory Characteristics [231]. While enabling memory operation, these devices also feature the steep subthreshold swing and excellent switching ratio of FBFETs, as well as the nanoscale features of Si NWs and CMOS compatibility. Si NW FETs can replace conventional six-transistor SRAMs to enhance circuit compactness and optimize performance [230]. In addition, the individually addressable, reconfigurable, nonvolatile transistors of Si NW RFETs were successfully demonstrated for the first time in 2018, showing the potential of combining storage functionality with reconfigurable operation [232].

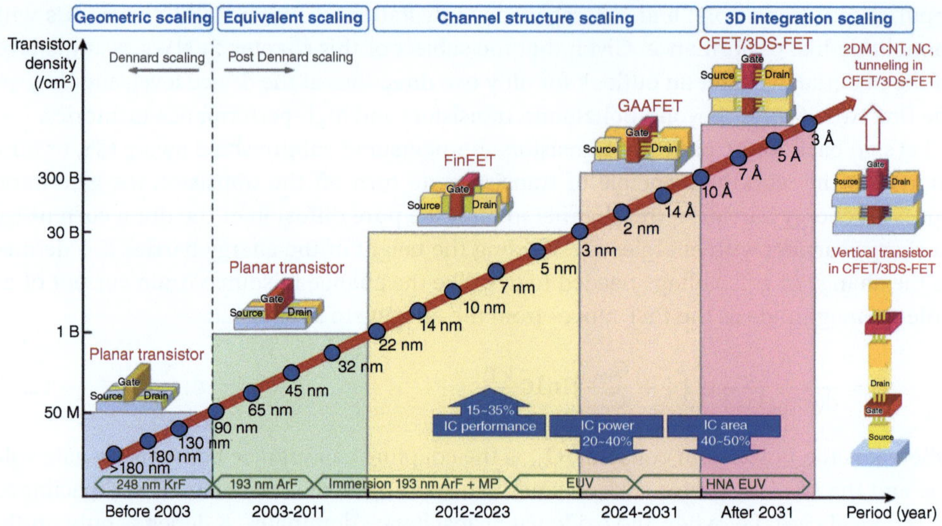


**Figure 2.18** Development history and emerging trends of MOSFETs with different transistor structures and scaling stages. *Source:* Reprinted with permission from Ref. [233]. Copyright 2024, Qingzhu Zhang, Yongkui Zhang, Yanna Luo, Huaxiang Yin 2024. Published by Oxford University Press on behalf of China Science Publishing & Media Ltd.

### 2.4.4.3 Fabrication of Si NW Transistors

The previous paragraph reviewed and introduced the electronics of Si NWs. While numerous physical architectures based on NWs have been proposed and demonstrated, advances in fabrication methods are equally critical to achieving high-quality electronics. This chapter provides a noncomprehensive collection of some of the fabrication strategies for Si NWs that cover the three stages of transistor development, presenting as clearly as possible the involvement and contribution of Si NW materials in the CMOS development journey.

#### *2.4.4.3.1 Si NW Plainar Transistor*

The self-alignment process is reported to produce Si NW FETs [234], where 20-nm-diameter NW self-patterned Au layers were grown in-plane by silane CVD, followed by processing of the electrodes. A demonstration of the direct selection of high-quality Si NWs in heterogeneously grown SFLS NWs with different conductivities and crystalline qualities is reported. The assembly of NWs is based on a dielectrophoretic process. It incorporates impedance spectroscopy analysis of real-time NW acquisition rates at different signal frequencies, which produces “aligned monolayers” of NWs with controlled densities, as shown in Figure 2.19b [106]. This approach promises low cost, high throughput, industrial scalability, and the potential for heterogeneous integration, although the subthreshold swing is high [235]. Low cost and high CMOS compatibility can be achieved by top-down etching [236] (Figure 2.19a). The silicon nitride layer is used as a mask and removed after etching. A self-limiting oxidation process is used to produce the gate oxide. High etch depth control can be obtained using self-stopping etching on (111) wafers.

In summary, in-plane FET implementations have proven to be highly reproducible and flexible, whether based on a top-down or bottom-up approach. These in-plane FET devices

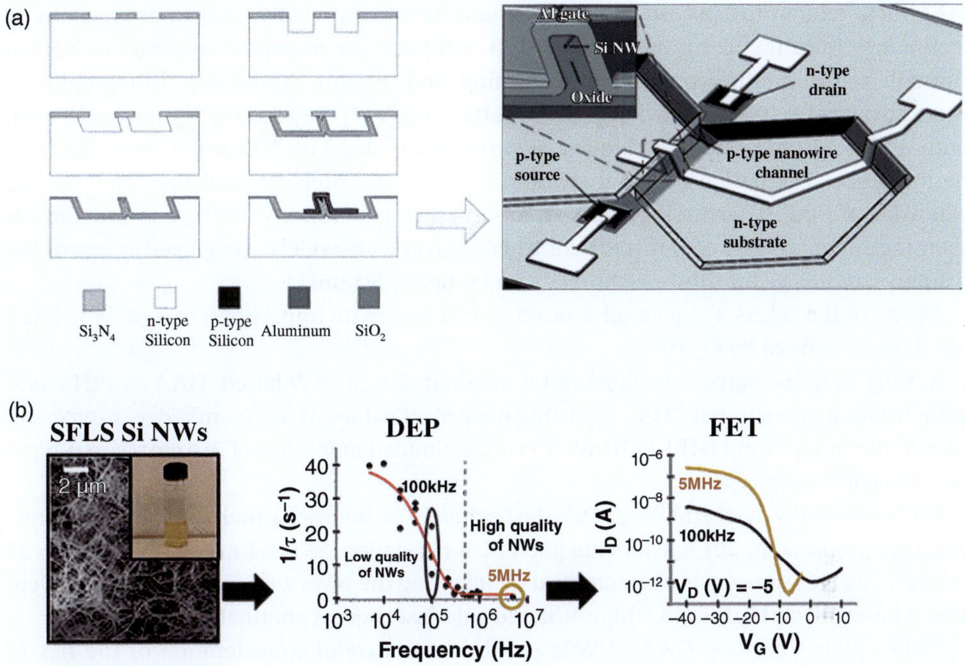


**Figure 2.19** Si NW planar FET process. (a) Etching and self-limiting oxidation. *Source:* Reprinted with permission from Ref. [236]. Copyright 2013, WILEY-VCH Verlag GmbH & Co. KGaA, Weinheim. (b) DEP and impedance spectroscopy. *Source:* Reprinted with permission from Ref. [106]. Copyright 2016, American Chemical Society.

demonstrate the process availability and performance benefits of implementing FETs from Si NWs and are the basis for advanced FinFETs based on Si NWs.

#### 2.4.4.3.2 Gate-all-around Nanowire Transistor

The transistor roadmap is shown in Figure 2.18. Simple scaling fails in nodes beyond N130 nm, and since the N22 nm node, countering various SCE and resisting degradation of electrostatic control while continuing to scale has increasingly become a core requirement for transistor architecture change. Transistor electrostatic integrity is quantified using two key metrics: subthreshold slope (SS) and DIBL.

The transistor architecture underwent its first major change, evolving from single-gate planar to three-gate 3D fins to obtain more robust electrostatic channel control. With the conversion of the channel profile from planar Si to 3D Si fin profile, the effective width (coefficient of performance per unit size) for carrier transport in the channel between the source-drainelectrodes of the transistor doubles or even triples in FinFETs and even more in GAA FETs and CFETs, which amplifies the saturating on-current of the transistor within the same size. Meanwhile, increasing the gate control number (n) is another trend for suppressing SS and DIBL degradation at small scales. In addition, low channel doping concentration is suggested, whereby the surface conduction mode changes to the bulk conduction mode to obtain a lower threshold voltage.

FinFETs remain the industry's mainstay and have been widely used in generations of technology from the 22-nm node onward. It will push the transistor roadmap to N5 nm through self-calibrating quadruple patterning and 193-nm immersion lithography or more advanced extreme ultraviolet (EUV) patterning [237]. Beyond the N5 nm node, even with advanced processing techniques, significantly scaled FinFETs suffer from SCE and performance degradation due to their ultra-thinness and high channel size. As a natural extension of FinFET technology, gate-all-around transistors (GAA NWTs) continue implementing enhanced gate electrostatic control design and are widely recognized as one of the primary candidates for future technology nodes below N5 nm.

As one of the GAAs, the gate-all-around silicon nanowire transistor (GAA SNWT) has a 1D channel defined by Si NWs.

A fully CMOS-compatible fabrication method for Si NW-based GAA n-FETs and p-FETs was demonstrated [238], reporting near-ideal values of SS (63 mV/dec, n-type and 66 mV/dec, p-type) and DIBL (~10 mv/V) with estimated mobilities of 750 $cm^2$/V-s (n-type) and 325 $cm^2$/V-s (p-type).

For horizontally stacked NWs, a NW-last scheme, or nano-channel-last (NCL) scheme, was first proposed in 2015, achieving good SCE resistance (SS = 64 mV/V and DIBL = 24 mV/V). The NCL process is characterized by releasing the NWs while integrating a conventional bulk-silicon FinFET in a high-$\kappa$/metal gate-last process channel.

Unlike planar devices, GAA SNWTs require more careful consideration of the risk of damage by electrostatic discharge events due to the 1D channels. Thermal conductivity in the direction perpendicular to the NW channel is key to this issue. For this reason, a scheme with localized openings in buried oxide underneath [112] has been proposed. However, it has been argued that this scheme is an NW-first scheme, where the NW channels between the electrodes are defined by selective etching with too low CMOS process compatibility.

#### 2.4.4.3.3 Approach to the <N3 nm Node: 3D-stacked Si NWs

Recent reviews [233, 241] suggest that stacked NW transistors are one of the candidates for the CMOS roadmap at the <N3 nm node. By exploring space in the vertical dimension, stacked NWs can integrate more functionality in a small footprint. In 3D space, there are now two possible ways to place semiconductor Si NW channels: vertical standing or horizontal stacking configurations.

**i)** Vertical GAA SNWT

Goldberger et al. [242] demonstrated the first V-FET device in which VLS-grown ultrathin Si NWs are electrically connected through a bottom substrate (Source) and a top overhead metal (Drain) electrode pad to form a wrapped gate structure with short-channel lengths. A prototype FET device with Si-germanium core–shell NWs grown by the VLS method was also successfully demonstrated [243]. As shown in Figure 2.20, similarly to planar devices, S/D contacts of PtSi/Si were also used for V-FETs [239]. Carrier injection performance, such as hot-ion injection, was improved, greatly enhancing the tunneling contribution. For the 3 nm and higher technology nodes, Li et al. [240] have developed a novel vertical sandwich gate full-surround FET process flow (Figure 2.21), and their NW devices are fabricated by an integrated flow of

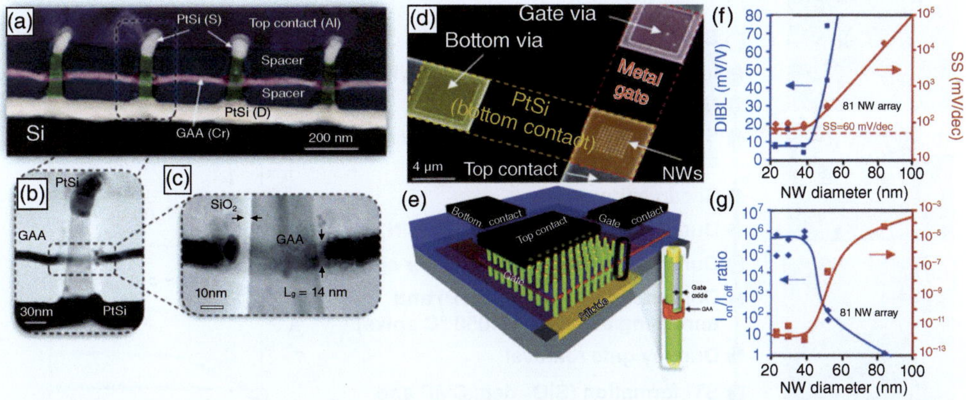


**Figure 2.20** Vertical GAA SNWT with PtSi/Si contacts that enable dense arrays. (a) TEM cross section in tilted view, pseudo-color of the device with a gate around each NW, symmetric siliconized S/D (PtSi) contacts, and 60-nm low-κ (2.7) dielectric spacer separating the S/D contacts from the gate. (b) Scaling of the TEM cross section with good planarity of the stacked layers. (c) Scaling of GAA region with 5-nm $SiO_2$ gate oxide and 14-nm gate length. (d) Aerial view of the SEM of the vertical channel FET (V-FET). (e) Conceptual diagram of the V-FET. The device is realized on a dense array of NWs with external channels connecting different layers (top, bottom, and gate contacts). (f) Evolution of the subthreshold swing and drain-induced barrier reduction as a function of NW diameter. (g) On/off current ratio and turn-off current are functions of NW diameter, where good behavior is obtained for diameters below 40 nm. Larger diameters lead to a loss of gate control over the channel, which rapidly degrades the electrical performance. *Source:* Reprinted with permission from Ref. [239]. Royal Society of Chemistry.

Si/SiGe epitaxials, selective SiGe quasi-atomic layer etching of Si, formation of a SiGe/Si core/shell/NW structure, construction of a nitride virtual gate, and replacement of the virtual gate to fabricate, as shown in Figure 2.21e, the conformal deposition of multiple ALDs that defines the wrap-around gate. Despite the increased process complexity due to the introduction of the virtual gate, the SCE resistance is excellent (DIBL = 14 mV/V, SS = 68 mV/dec), as shown in Figure 2.21f. Han et al. [244] demonstrated a vertical 1T neuron based on the vertical GAA SNWT with a footprint of less than 4F. In addition to improving the layout efficiency, the vertical structure has the advantage of scalability, which can be used in 1T neurons. Advantage of being able to monolithically integrate synaptic devices, such as resistive random access memory or memristors, on a 1T neuron. A comparative study of vertically stacked and horizontally stacked GAA NW FETs [196] demonstrated the advantages of V-FETs in reducing parasitic capacitance and shrinking contact gate pitch, providing a new approach to scaling. However, the processing of vertical transistors is nonstandardized and inherently incompatible with current standard in-plane processing platforms. Furthermore, complete 3D logic layouts or interconnect schemes for S/D and gated electrodes are still in the developmental or early stages of research.

**ii)** Layer-by-Layer Sequential Stacked NW Transistor

Due to higher CMOS compatibility, vertically stacked horizontal NWs have been studied more than V-FETs. Multiple layers of lateral NW channels are stacked along the z-axis direction to enable 3D integration. There are three stacking strategies:

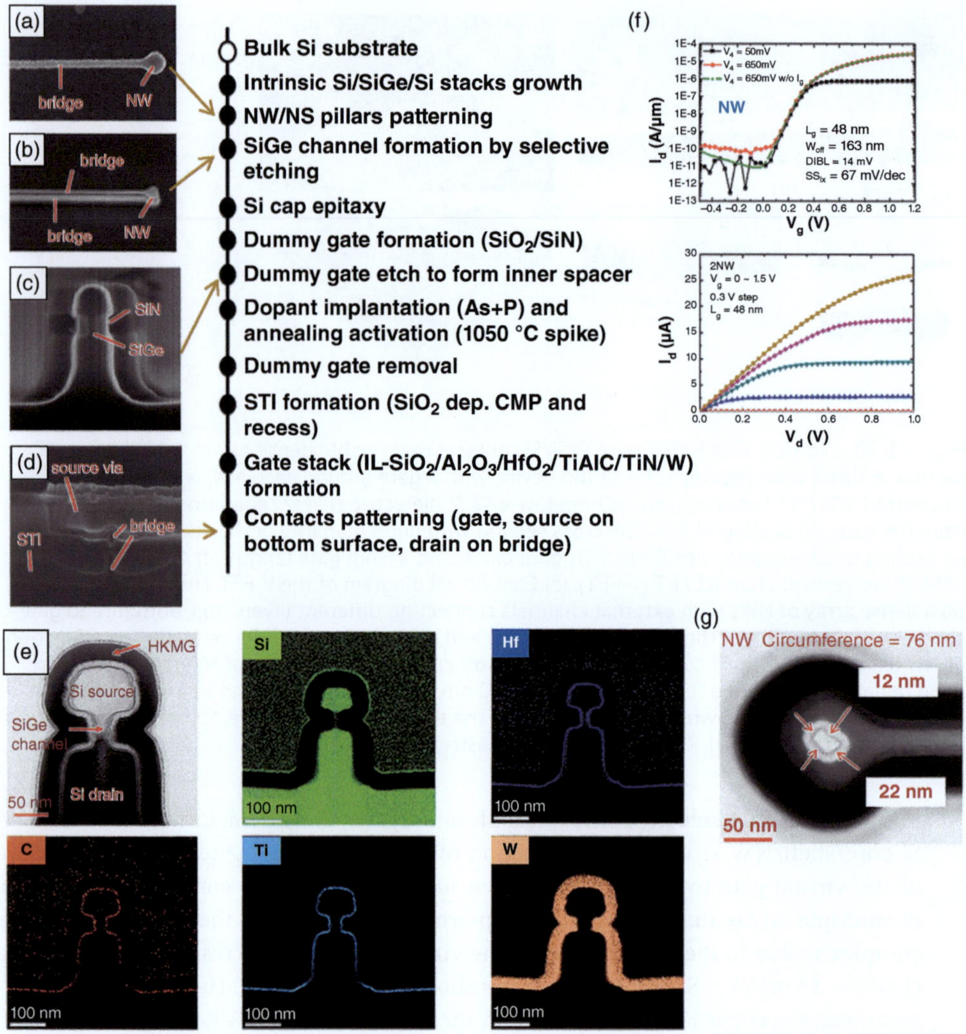


**Figure 2.21** (a–d) SEM side views of the different process steps after completion, and the arrows on the right show their positions in the process flow. (e) EDS of the NW finished device. The gate stack was the interface layer (IL) 0.7 nm/$Al_2O_3$, 0.3 nm/$HfO_2$, 2.3 nm/TiAlC, 4 nm/TiN, 3.5 nm/W, and 32 nm. The $Al_2O_3$ was only 0.3 nm to improve the interface and cannot be recognized from HRTEM. Note that because of the oxidation during the dummy gate module and $SiO_2$ IL layer growth, the Si cap layer was eventually around 1.2 nm. (f) $I_d$-$V_d$ plot, top: NW. Bottom: dual NWs in parallel. (g) HRTEM top view of NW finished device. *Source:* Reprinted with permission from Ref. [240]. Copyright 2021, American Chemical Society.

layer-by-layer sequential stacking based on the process of first-growth post-assembly, etching, and constrained in situ horizontal growth to realize this concept.

3D integration realized by layer-by-layer sequential stacking is an approach that started earlier. Using Ge/Si core/shell NWs as a representative example, Charles M. Lieber et al. [245] used TP to fabricate 10-layer vertically stacked multi-NW FETs showing good interlayer performance uniformity over the 10-layer range. The NWs were assembled at relatively high densities (~4 NW/μm). In addition, the sequential 3D integration scheme [122] based on a low-temperature transistor top process and a high-quality molecular bonding process also exhibits performance enhancement without the need for aggressive scaling.

**iii)** Vertically Stacked Multi-NWs by Etching

One etch-based method is the restricted oxidation after the Bosch process. In the Bosch process, the combined step of etching and passivation is repeated several times to obtain a sector structure periodically aligned in the z-axis direction. Both DRIE [246] and inductively coupled plasma [247] can be used for etching in the Bosch process. The obtained sectorial structures can be oxidized into vertically stacked Si NWs, provided that the process parameters are carefully optimized. Another etching-based scheme [239, 240] is the pre-deposition of a Si-SiGe superlattice on a silicon substrate, in which SiGe acts as a sacrificial layer of the periodic structure, followed by an open and anisotropic etching, where SiGe is also obtained in the position of vertically stacked Si NWs. The superlattice relaxation limits the number of layers in this approach as it is related to the dislocation density.

The ORADEP [244] was proposed by Lee et al. for the fabrication of VM-FETs with gate dielectric thermal oxides and Si/oxide/nitride/oxide/Si-based structures with vertically integrated multiple NWs for NVM to address the problem of static friction failure caused by successive iterations of sacrificial oxidation and wet etching in the Bosch process as shown in Figure 2.22, ORADEP consists of polymer passivation and successive isotropic dry etching steps. First, a C4F8-based polymer is adsorbed onto a region patterned by an oxide hard mask, where the hard mask is used as a protective layer during ORADEP to form vertically integrated multi-NWs. Next, the NWs are formed using plasma etching gas using SF6. Because the sidewalls of the patterned Si are protected from the plasma-based etching process, the NW profiles can be formed by isotropic dry etching. This process iteration, i.e. polymer passivation and plasma etching, allows for the reproducible formation of uniform NWs. JL-FETs and nonvolatile flash devices following a similar process and structure have reported better performance [196].

In summary, many etching methods can be used to define vertical stacks of Si NW structures, after which the gate is realized by an opening and casting process. Figure 2.22d depicts the process flow from stacked NW structures to vertical multichannel transistor devices, and as a representation, the scheme of Lee et al. is referenced.

Higher integration densities and precise control are thought to be achieved by in situ growth of NWs. As for how transistors grow in plain, the specific growth processes and mechanisms are discussed in 2.1.2.2, and the focus of this chapter is to present several architectures for 3D-stacked NWs and their process flows. For all architectures that rely on growth to achieve stacking, one of the key issues lies in achieving patterning in 3D space and precise and robust constraints on the growth process. We list several constraints as follows: sloping terraces and sidewalls subject to etching and superlattice sidewalls.

In 2019, Shi and Yu et al. [245] reported for the first time a new concept of 3D vectorial sidewall growth of Si NW confined to vertical or inclined sidewall surfaces (Figure 2.23). As mentioned, the In-catalyzed IPSLS process can produce in-plane, in situ-grown, high-resolution Si NWs, provided the proper constraints are met. As a proof of principle, the predefined planar

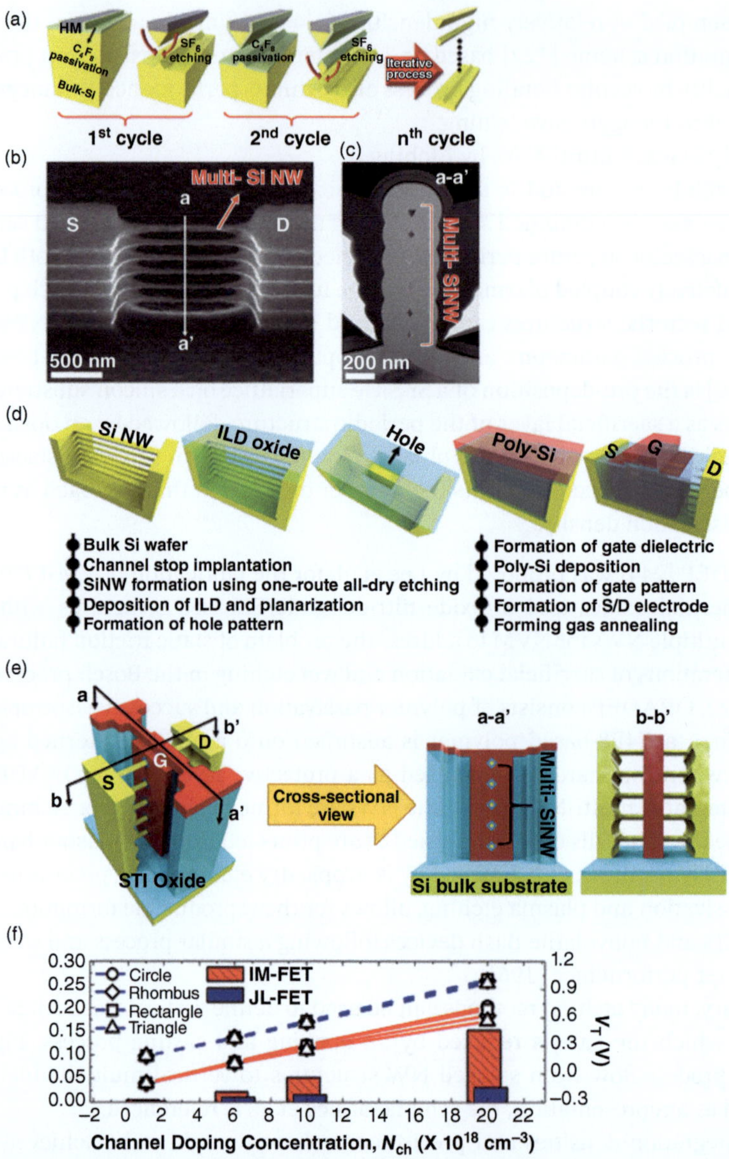


**Figure 2.22** Fabrication of 3D-stacked Si NWs: etching approach. (a) Schematic and (b) SEM image of vertically stacked multi-NWs fabricated by the one-route all-dry etching process. The absence of static friction and uniform multi-NW structure supports high reproducibility and low process variability. (c) Cross-sectional TEM image along the a-a′ direction in (b). (d) Complete process flow that is fully compatible with the Si-based CMOS process, and the dashed box summarizes the results of the etching method for obtaining Si NW structures. (e) Schematic of a vertically integrated five NW-based FET on a bulk silicon substrate, including cross-sectional images along the a-a′ (parallel to the gate length) direction and the b-b′ (parallel to the NW length) direction of the fabricated device. *Source:* Reprinted with permission from Ref. [244]. Copyright 2015, American Chemical Society. (f) JL-FETs realized by this methodology take full advantage of the performance of the structure while retaining the layer-by-layer homogeneity of the IM-FET. *Source:* Reprinted with permission from Ref. [196]. Copyright 2016, American Chemical Society.

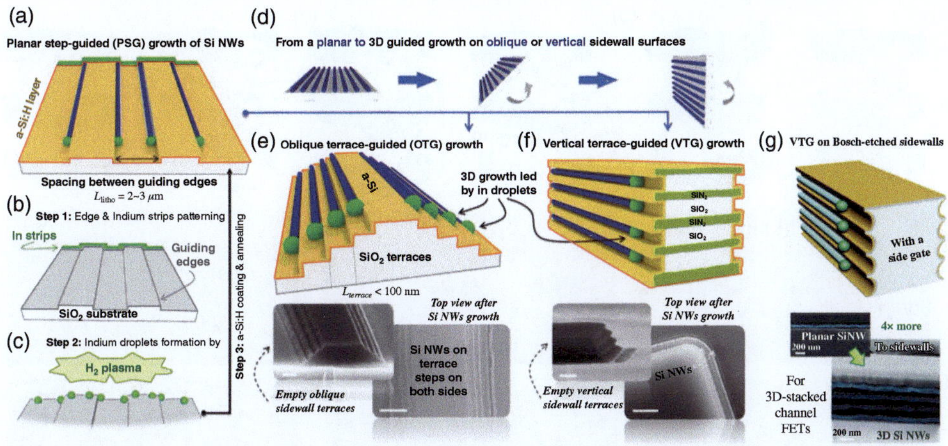


**Figure 2.23** Fabrication of 3D-stacked Si NWs: growth methods. Guided Si NW growth along planar steps and sidewall terraces. (a–c) The main fabrication procedure of planar step-guided (PSG) Si NW, headed by the formed droplets (green). (d) Transition from PSG growth to inclined or vertical 3D sidewall growth to reduce footprint. (e, f) More detailed structural configurations for oblique terrace-guided (OTG) and vertical terrace-guided (VTG) growth, as well as corresponding SEM images of the Si NW. *Source:* Reprinted with permission from Ref. [245]. Copyright 2019, WILEY-VCH Verlag GmbH & Co. KGaA, Weinheim. (g) Structural configuration of VTG growth on Bosch-etched sidewall and SEM image. *Source:* Reprinted with permission from Ref. [117]; Royal Society of Chemistry.

step edges contain enough additional a-Si to trap and guide droplets rolling in the plane, and a single step edge traps only one NW, which is called PSG (Figure 2.23a–c). The conventional lithography and etching steps realize the planes' predefinition in PSG growth. Therefore, the integration density of PSG is limited by the resolution of lithography. To further enhance the integration density, two new schemes are discussed: inclined OTG growth (Figure 2.23e) and VTG (Figure 2.23f–g) growth. In OTG growth, inclined sidewall terraces with dense arrays of small steps are prepared by a simple alternating etch process, which consists of a two-step dry etching that can be recycled multiple times: in each cycle, firstly, the $C_4F_8$ is etched downwards on the $C_4F_8$ silicon oxide substrate, and then, oxygen plasma lateral etching of the photoresist makes the photoresist so that it recedes by a constant width $W_S$.

The NW-to-NW space in the OTG growth mode is still limited by the finite platform width ($W_S$) and catalyst diameter ($D_c$), and the VTG mode growth based on vertical sidewall structures can skip this limitation. In the VTG growth, periodic multilayered structures are deposited on the silicon substrate, e.g. PECVD deposits $SiO_2/SiN_x$ structures. Then, an etching process exposes the sidewalls, and selective wet etching produces periodic grooves in the sidewalls.

The Bosch process to form sectorized structures can be a limitation for VTG growth, demonstrating the high-density integration of stacked Si NWs, where Si NWs with an average diameter of 40 nm and a pitch of 100 nm were fabricated in prototype stacked-channel transistors (switching ratios >$10^7$, hole mobility ~57 $cm^2V^{-1}s^{-1}$) [117]. These results highlight the unique potential and advantages of IPSLS for 3D-stacked NWs, especially when high integration density is required. Due to the droplets' low

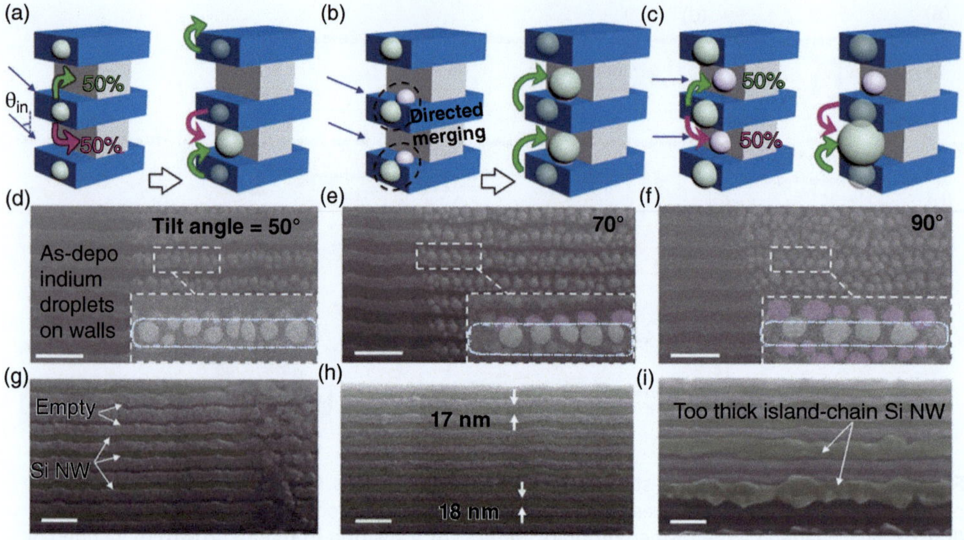


**Figure 2.24** Directional merging dynamics related to droplet tilt angle. (a–c) Schematic of the evolution and merging of In catalyst droplets formed on corrugated $SiN_x$ (wall)/$SiO_2$ (groove) sidewall recesses with increasing incidence angles of 50°, 70°, and 90° with the SEM images of the formed catalyst droplets provided in (d–f) and the corresponding SEM images of Si NW grown on the sidewall recesses in (g–i), respectively. The scale bars in (d–i) are all 100 nm. *Source:* Reprinted with permission from Ref. [247]. Copyright 2022, Wiley-VCH GmbH.

surface tension and easy evaporation, the diameter control was around 20%. However, with the revelation of the directional confluence dynamics during the growth process, the diameter control reached an error of around 1–2 nm [246, 247], comparable to state-of-the-art lithography. Figure 2.24 shows the principle: for droplet tilt angles of 60~80°, the Influx converges stably to the same side of the $SiN_x$ wall, in contrast to too small a tilt angle that leads to the behavior of In atoms riding on the top of the $SiN_x$ wall and converging randomly to one side, and tilt angles close to 90° that lead to the In atoms stalling and covering the top of the $SiN_x$ wall, providing an opportunity for droplet merging and movement, which both cases lead to fluctuating diameters. The tilt angle is determined by a combination of the geometry of the notch and the thickness of the initially deposited layer, although the principles have not been systematically presented. In addition, notches are unlocked as a new degree of freedom for growth control, and it may be possible to control the NWs' cross-sectional shape utilizing notches.

## References

1 Wagner, R.S., Ellis, W.C. Vapor-liquid-solid mechanism of single crystal growth. *Appl. Phys. Lett.* 1964, 4 (5): 89–90.

2 Givargizov, E.I. Fundamental aspects of VLS growth. In: *Vapour Growth and Epitaxy* (eds. G.W. Cullen, E. Kaldis, R.L. Parker, C.J.M. Rooymans), 20–30. Elsevier, 1975.

**3** Westwater, J., Gosain, D.P., Tomiya, S., Usui, S., Ruda, H. Growth of silicon nanowires via gold/silane vapor–liquid–solid reaction. *J. Vac. Sci. Technol. B.* 1997, 15: 554–557.

**4** Morales, A.M., Lieber, C.M.A. Laser ablation method for the synthesis of crystalline semiconductor nanowires. *Science* 1998, 279 (5348): 208–211.

**5** Zhang, Y.F., Tang, Y.H., Wang, N., Yu, D.P., Lee, C.S., Bello, I., Lee, S.T. Silicon nanowires prepared by laser ablation at high temperature. *Appl. Phys. Lett.*1998, 72: 1835–1837.

**6** Cui, Y., Lauhon, L.J., Gudiksen, M.S., Wang, J., Lieber, C.M. Diameter-controlled synthesis of single-crystal silicon nanowires. *Appl. Phys. Lett.* 2001, 78 (15): 2214–2216.

**7** Lensch-Falk, J., Hemesath, E.R., Perea, D.E., Lauhon, L.J. Alternative catalysts for VSS growth of silicon and germanium nanowires. *J. Mater. Chem.* 2009, 19 (7): 849–857.

**8** Kodambaka, S., Tersoff, J., Reuter, M.C., Ross, F.M. Diameter-independent kinetics in the vapor-liquid-solid growth of Si nanowires. *Phys. Rev. Lett.* 2006, 96 (9): 096105.

**9** Schmidt, V., Wittemann, J.V., Gösele, U. Growth, thermodynamics, and electrical properties of silicon nanowires. *Chem. Rev.* 2010, 110 (1): 361–388.

**10** Lombardi, I., Hochbaum, A.I., Yang, P., Carraro, C., Maboudian, R. Synthesis of high density, size-controlled Si nanowire arrays via porous anodic alumina mask. *Chem. Mater.* 2006, 18 (4): 988–991.

**11** Kayes, B.M., Filler, M.A., Putnam, M.C., Kelzenberg, M.D., Lewis, N.S., Atwater, H.A. Growth of vertically aligned Si wire arrays over large areas (>1cm$^2$) with Au and Cu catalysts. *Appl. Phys. Lett.* 2007, 91 (10): 103110.

**12** Kramer, A., Boeck, T., Schramm, P., Fornari, R. Investigation of Au and In as solvents for the growth of silicon nanowires on Si(1 1 1). *Phys. E Low-Dimens. Syst. Nanostruct.* 2008, 40 (7): 2462–2467.

**13** Fuhrmann, B., Leipner, H.S., Höche, H.-R., Schubert, L., Werner, P., Gösele, U. Ordered arrays of silicon nanowires produced by nanosphere lithography and molecular beam epitaxy. *Nano Lett.* 2005, 5 (12): 2524–2527.

**14** Zhang, Z., Shimizu, T., Senz, S., Gösele, U. Ordered high-density Si [100] nanowire arrays epitaxially grown by bottom imprint method. *Adv. Mater.* 2009, 21 (27): 2824–2828.

**15** Gadea, G., Morata, A., Santos, J.D., Dávila, D., Calaza, C., Salleras, M., Fonseca, L., Tarancón, A. Towards a full integration of vertically aligned silicon nanowires in MEMS using silane as a precursor. *Nanotechnology* 2015, 26 (19): 195302.

**16** Ho, T.-W., Hong, F.C.-N. A reliable method to grow vertically-aligned silicon nanowires by a novel ramp-cooling process. *Appl. Surf. Sci.* 2012, 258 (20): 7989–7996.

**17** Ahmed, N., Ramasamy, P., Bhargav, P.B., Rayerfrancis, A., Chandra, B. Development of silicon nanowires with optimized characteristics and fabrication of radial junction solar cells with < 100 nm amorphous silicon absorber layer. *Mater. Sci. Semicond. Process.* 2020, 106: 104778.

**18** Liu, J., Huang, S.-H., Chen, L.-P., He, L. Tin catalyzed silicon nanowires prepared by magnetron sputtering. *Mater. Lett.* 2015, 151: 122–125.

**19** Chou, Y.-C., Wen, C.-Y., Reuter, M.C., Su, D., Stach, E.A., Ross, F.M. Controlling the growth of Si/Ge nanowires and heterojunctions using silver–gold alloy catalysts. *ACS Nano* 2012, 6 (7): 6407–6415.

**20** Uesugi, A., Nishiyori, S., Nakagami, T., Sugano, K., Isono, Y. Integration of silicon nanowire bridges in microtrenches with perpendicular bottom-up growth promoted by surface nanoholes. *Jpn. J. Appl. Phys.* 2022, 61 (7): 075502.

**21** Fortuna, S.A., Li, X. Metal-catalyzed semiconductor nanowires: a review on the control of growth directions. *Semicond. Sci. Technol.* 2010, 25: 024005.

**22** Uesugi, A., Horita, T., Sugano, K., Isono, Y. Vapor–liquid–solid growth of silicon nanowires from surface nanoholes formed with metal-assisted chemical etching. *Jpn. J. Appl. Phys.* 2021, 60 (5): 055502.

**23** Uesugi, A., Nishiyori, S., Sugano, K., Isono, Y. Evaluation of thermoelectric properties of monolithically-integrated core-shell Si nanowire bridges. In: *2023 IEEE 36th International Conference on Micro Electro Mechanical Systems (MEMS)*, 701–704. 2023.

**24** Nebol'sin, V.A., Levchenko, E.V., Yuryev, V., Swaikat, N. Thermodynamics of silicon nanowire growth under unintended oxidation of catalytic particles. *J. Nanotechnol.* 2023, 2023 (1): 3485793.

**25** Garnett, E.C., Liang, W., Yang, P. Growth and electrical characteristics of platinum-nanoparticle-catalyzed silicon nanowires. *Adv. Mater.* 2007, 19 (19): 2946–2950.

**26** Ogata, T., Sorita, T., Kobayashi, K., Matsui, Y., Horie, K., Hirayama, M.H.M. Kinetic study of silicon nitride growth from dichlorosilane and ammonia. *Jpn. J. Appl. Phys.* 1996, 35 (3R): 1690.

**27** Bartelt, N.C., Theis, W., Tromp, R.M. Ostwald ripening of two-dimensional islands on Si(001). *Phys. Rev. B* 1996, 54 (16): 11741–11751.

**28** Dailey, E.J. Growth of epitaxial silicon and germanium nanowires using the gold catalyzed vapor-liquid-solid mechanism. *ProQuest*. 2010, 3425817.

**29** Wang, N., Cai, Y., Zhang, R.Q. Growth of nanowires. *Mater. Sci. Eng. R Rep.* 2008, 60 (1): 1–51.

**30** Schubert, L., Werner, P., Zakharov, N.D., Gerth, G., Kolb, F.M., Long, L., Gösele, U., Tan, T.Y. Silicon nanowhiskers grown on <111> Si substrates by molecular-beam epitaxy. *Appl. Phys. Lett.* 2004, 84 (24): 4968–4970.

**31** Li, F., Huang, Y., Wang, S., Zhang, S. Critical review: growth mechanisms of the self-assembling of silicon wires. *J. Vac. Sci. Technol. A* 2019, 38 (1): 010802.

**32** Wang, Y., Schmidt, V., Senz, S., Gösele, U. Epitaxial growth of silicon nanowires using an aluminium catalyst. *Nat. Nanotechnol.* 2006, 1: 186–189.

**33** Dailey, E., Madras, P., Drucker, J. Au on vapor-liquid-solid grown Si nanowires: spreading of liquid AuSi from the catalytic seed. *J. Appl. Phys.* 2010, 108 (6): 064320.

**34** Putnam, M.C., Filler, M.A., Kayes, B.M., Kelzenberg, M.D., Guan, Y., Lewis, N.S., Eiler, J.M., Atwater, H.A. Secondary ion mass spectrometry of vapor–liquid–solid grown, Au-catalyzed, Si wires. *Nano Lett.* 2008, 8 (10): 3109–3113.

**35** Boukhicha, R., Vincent, L., Renard, C., Gardès, C., Yam, V., Fossard, F., Patriarche, G., Jabeen, F., Bouchier, D. Gold nanocluster distribution on faceted and kinked Si nanowires. *Thin Solid Films* 2012, 520 (8): 3304–3308.

**36** Bootsma, G.A., Gassen, H.J. A quantitative study on the growth of silicon whiskers from silane and germanium whiskers from germane. *J. Cryst. Growth* 1971, 10 (3): 223–234.

**37** Murray, J.L., McAlister, A.J. The Al-Si (aluminum-silicon) system. *Bull. Alloy Phase Diagr.* 1984, 5 (1): 74–84.

**38** Rai-Choudhury, P., Takei, W.J. Formation of silicon whiskers by aluminum-quartz interaction. *J. Electrochem. Soc.* 1974, 121: 1228.

**39** Wacaser, B.A., Reuter, M.C., Khayyat, M.M., Wen, C.-Y., Haight, R., Guha, S., Ross, F.M. Growth system, structure, and doping of aluminum-seeded epitaxial silicon nanowires. *Nano Lett.* 2009, 9 (9): 3296–3301.

**40** Whang, S.J., Lee, S.J., Yang, W.F., Cho, B.J., Liew, Y.F., Kwong, D.L. Complementary metal-oxide-semiconductor compatible Al-catalyzed silicon nanowires: growth and the effects of surface oxidation of Al seeding layer. *Electrochem. Solid-State Lett.* 2007, 10: E11.

**41** Gebavi, H., Ristić, D., Baran, N., Marciuš, M., Gašparić, V., Syed, K., Ivanda, M. Development of silicon nanowires based on Ag-Au metal alloy seed system for sensing technologies. *Sens. Actuators Phys.* 2021, 331: 112931.

**42** Weber, L. Equilibrium solid solubility of silicon in silver. *Metall. Mater. Trans. A* 2002, 33 (4): 1145–1150.

**43** Nebol'sin, V.A., Shchetinin, A.A. Role of surface energy in the vapor–liquid–solid growth of silicon. *Inorg. Mater.* 2003, 39: 899–903.

**44** Nitta, N., Yushin, G. High-capacity anode materials for lithium-ion batteries: Choice of elements and structures for active particles. *Part. Part. Syst. Charact.* 2014, 31 (3): 317–336.

**45** Eustathopoulos, N., Drevet, B., Ricci, E. Temperature coefficient of surface tension for pure liquid metals. *J. Cryst. Growth* 1998, 191 (1): 268–274.

**46** Chung, S.-W., Yu, J.-Y., Heath, J.R. Silicon nanowire devices. *Appl. Phys. Lett.* 2000, 76 (15): 2068–2070.

**47** Al-Taay, H.F., Mahdi, M.A., Parlevliet, D., Hassan, Z., Jennings, P. Growth and characterization of silicon nanowires catalyzed by Zn metal via pulsed plasma-enhanced chemical vapor deposition. *Superlattices Microstruct.* 2014, 68: 90–100.

**48** Al-Taay, H.F., Mahdi, M.A., Parlevliet, D., Jennings, P. Fabrication and characterization of solar cells based on silicon nanowire homojunctions. *Silicon* 2017, 9 (1): 17–23.

**49** Direct Growth of Si, *Ge, and Si–Ge Heterostructure Nanowires Using Electroplated Zn: An Inexpensive Seeding Technique for Li-Ion Alloying Anodes*. Wiley Online Library.

**50** Jeon, M., Kamisako, K. Synthesis and characterization of silicon nanowires using tin catalyst for solar cells application. *Mater. Lett.* 2009, 63 (9): 777–779.

**51** Djoumi, S., Kail, F., i Cabarrocas, P.R., Chahed, L. Tin versus indium catalyst in the growth of silicon nanowires by plasma enhanced chemical vapor deposition on different substrates. *Thin Solid Films* 2022, 758: 139447.

**52** Kumar, R.R., Rao, K.N., Phani, A.R. Bismuth catalyzed growth of silicon nanowires by electron beam evaporation. *Mater. Lett.* 2012, 82: 163–166.

**53** Benabderrahmane Zaghouani, R., Yaacoubi Tabassi, M., Bennaceur, J., Srasra, M., Derouiche, H., Dimassi, W. A comparative study of the annealing atmosphere effect on bismuth particles dedicated for silicon nanowires growth catalyst. *J. Cryst. Growth* 2021, 570: 126205.

**54** Yu, L., Fortuna, F., O'Donnell, B., Jeon, T., Foldyna, M., Picardi, G., Roca i Cabarrocas, P. Bismuth-catalyzed and doped silicon nanowires for one-pump-down fabrication of radial junction solar cells. *Nano Lett.* 2012, 12 (8): 4153–4158.

**55** Nama Manjunatha, K., Paul, S. In-situ catalyst mediated growth and self-doped silicon nanowires for use in nanowire solar cells. *Vacuum* 2017, 139: 178–184.

**56** Schmidt, V., Senz, S., Gösele, U. Diameter-dependent growth direction of epitaxial silicon nanowires. *Nano Lett.* 2005, 5 (5): 931–935.

**57** Harris, B. Whisker technology. *Nat. Phys. Sci.* 1971, 232 (28): 47–47.

**58** Puglisi, R.A., Bongiorno, C., Caccamo, S., Fazio, E., Mannino, G., Neri, F., Scalese, S., Spucches, D., La Magna, A. Chemical vapor deposition growth of silicon nanowires with diameter smaller than 5 nm. *ACS Omega* 2019, 4 (19): 17967–17971.

**59** Hannon, J.B., Kodambaka, S., Ross, F.M., Tromp, R.M. The influence of the surface migration of gold on the growth of silicon nanowires. *Nature.* 2006, 440: 69–71.

**60** Mohammad, S.N. Analysis of the vapor–liquid–solid mechanism for nanowire growth and a model for this mechanism. *Nano Lett.* 2008, 8 (5): 1532–1538.

**61** Stone, H.A. Dynamics of drop deformation and breakup in viscous fluids. *Annu. Rev. Fluid Mech.* 1994, 26: 65–102.

**62** Schmid, H., Björk, M.T., Knoch, J., Riel, H., Riess, W., Rice, P., Topuria, T. Patterned epitaxial vapor-liquid-solid growth of silicon nanowires on Si(111) using silane. *J. Appl. Phys.* 2008, 103 (2): 024304.

**63** Kaplan, W.D., Chatain, D., Wynblatt, P., Carter, W.C. A review of wetting versus adsorption, complexions, and related phenomena: the Rosetta stone of wetting. *J. Mater. Sci.* 2013, 48 (17): 5681–5717.

**64** Lee, S.T., Wang, N., Lee, C.S. Semiconductor nanowires: synthesis, structure and properties. *Mater. Sci. Eng. A* 2000, 286 (1): 16–23.

**65** Wang, N., Tang, Y.H., Zhang, Y.F., Lee, C.S., Lee, S.T. Nucleation and growth of Si nanowires from silicon oxide. *Phys. Rev. B* 1998, 58 (24): R16024–R16026.

**66** Zhang, R.-Q., Lifshitz, Y., Lee, S.-T. Oxide-assisted growth of semiconducting nanowires. *Adv. Mater.* 2003, 15 (7–8): 635–640.

**67** Mohammad, S.N. Oxide-assisted growth mechanism. In: *Synthesis of Nanomaterials: Mechanisms, Kinetics and Materials Properties* (ed. S.N. Mohammad), 173–186. Springer International Publishing: Cham, 2020.

**68** Li, F.J., Zhang, S., Lee, J.-W. Rethinking of the silicon nanowire growth mechanism during thermal evaporation of Si-containing powders. *Thin Solid Films* 2014, 558: 75–85.

**69** O'Connor, K.M., Rubletz, A., Trach, J., Butler, C., Veinot, J.G.C. Understanding silicon monoxide gas evolution from mixed silicon and silica powders. *Nanoscale Horiz.* 2023, 8 (7): 892–899.

**70** Small-diameter silicon nanowire surfaces. *Science.*

**71** Chu, A.T., Thi, T.T.N., Tran, T.T., Vu, B.N., Pham, T.T., Pham, V.T., Pham, T.H., Pham, H.D. Raman photoluminescence spectra of silicon nanowires synthesized by a vapor phase transport method. *Adv. Nat. Sci. Nanosci. Nanotechnol.* 2011, 2 (3): 035004.

**72** Li, F.J., Zhang, S., Lee, J.-W., Zhao, D. Entrapment of gold catalyst in silicon/silicon-oxide nanowires. *Nanosci. Nanotechnol. Lett.* 2014, 6 (10): 922–926.

**73** Ishiyama, T., Nakagawa, S., Wakamatsu, T. Growth of epitaxial silicon nanowires on a Si substrate by a metal-catalyst-free process. *Sci. Rep.* 2016, 6 (1): 30608.

**74** Kim, B.-S., Koo, T.-W., Lee, J.-H., Kim, D.S., Jung, Y.C., Hwang, S.W., Choi, B.L., Lee, E.K., Kim, J.M., Whang, D. Catalyst-free growth of single-crystal silicon and germanium nanowires. *Nano Lett.* 2009, 9 (2): 864–869.

**75** Lee, C.H., Kim, D.R., Zheng, X. Fabricating nanowire devices on diverse substrates by simple transfer-printing methods. *Proc. Natl. Acad. Sci.* 2010, 107 (22): 9950–9955.

**76** Bogart, T.D., Lu, X., Korgel, B.A. Precision synthesis of silicon nanowires with crystalline core and amorphous shell. *Dalton Trans.* 2013, 42 (35): 12675–12680.

**77** Morita, M., Ohmi, T., Hasegawa, E., Kawakami, M., Suma, K. Control factor of native oxide growth on silicon in air or in ultrapure water. *Appl. Phys. Lett.* 1989, 55 (6): 562–564.

**78** Dashtestani, A.D., Moeinian, A., Biskupek, J., Strehle, S. Contamination-assisted rather than metal catalyst-free bottom-up growth of silicon nanowires. *Adv. Mater. Interfaces* 2021, 8 (22): 2101121.

**79** Holmes, J.D., Johnston, K., Doty, R.C., Kogelbauer, A. Control of thickness and orientation of solution-grown silicon nanowires. *Science* 2000, 287 (5457): 1471–1473.

**80** Lee, D.C., Hanrath, T., Korgel, B.A. The role of precursor-decomposition kinetics in silicon-nanowire synthesis in organic solvents. *Angew. Chem. Int. Ed.* 2005, 44 (23): 3573–3577.

**81** Heitsch, A.T., Akhavan, V.A., Korgel, B.A. Rapid SFLS synthesis of Si nanowires using trisilane with in situ alkyl-amine passivation. *Chem. Mater.* 2011, 23 (11): 2697–2699.

**82** Hanrath, T., Korgel, B.A. Supercritical fluid–liquid–solid (SFLS) synthesis of Si and Ge nanowires seeded by colloidal metal nanocrystals. *Adv. Mater.* 2003, 15 (5): 437–440.

**83** Heitsch, A.T., Fanfair, D.D., Tuan, H.-Y., Korgel, B.A. Solution–liquid–solid (SLS) growth of silicon nanowires. *J. Am. Chem. Soc.* 2008, 130 (16): 5436–5437.

**84** Vanderwielen, A.J., Ring, M.A., O'Neal, H.E. Kinetics of the thermal decomposition of methyldisilane and trisilane. *J. Am. Chem. Soc.* 1975, 97 (5): 993–998.

**85** Hessel, C.M., Heitsch, A.T., Korgel, B.A. Gold seed removal from the tips of silicon nanorods. *Nano Lett.* 2010, 10 (1): 176–180.

**86** Leonardi, A.A., Faro, M.J.L., Irrera, A. Silicon nanowires synthesis by metal-assisted chemical etching: a review. *Nanomaterials* 2021, 11 (2): 383.

**87** Vieu, C., Carcenac, F., Pépin, A., Chen, Y., Mejias, M., Lebib, A., Manin-Ferlazzo, L., Couraud, L., Launois, H. Electron beam lithography: resolution limits and applications. *Appl. Surf. Sci.* 2000, 164 (1): 111–117.

**88** Saifullah, M.S.M., Ondarçuhu, T., Koltsov, D.K., Joachim, C., Welland, M.E. A reliable scheme for fabricating sub-5 nm co-planar junctions for single-molecule electronics. *Nanotechnol.* 2002, 13 (5): 659.

**89** Yang, J.K.W., Cord, B., Duan, H., Berggren, K.K., Klingfus, J., Nam, S.-W., Kim, K.-B., Rooks, M.J. Understanding of hydrogen silsesquioxane electron resist for sub-5-nm-half-pitch lithography. *J. Vac. Sci. Technol. B Microelectron. Nanometer Struct. Process. Meas. Phenom.* 2009, 27 (6): 2622–2627.

**90** Peng, K.Q., Hu, J.J., Yan, Y.J., Wu, Y., Fang, H., Xu, Y., Lee, S.T., Zhu, J. Fabrication of single-crystalline silicon nanowires by scratching a silicon surface with catalytic metal particles. *Adv. Funct. Mater.* 2006, 16 (3): 387–394.

**91** Nahidi, M., Kolasinski, K.W. Effects of stain etchant composition on the photoluminescence and morphology of porous silicon. *J. Electrochem. Soc.* 2005, 153 (1): C19.

**92** Nahm, K.S., Hun Seo, Y., Lee, H.J. Formation mechanism of stains during Si etching reaction in HF–oxidizing agent–H2O solutions. *J. Appl. Phys.* 1997, 81 (5): 2418–2424.

**93** Seo, Y.H., Nahm, K.S., Lee, K.B. Mechanistic study of silicon etching in HF - KBrO3 - H2O solution. *J. Electrochem. Soc.* 1993: 140 (5): 1453.

**94** Miao, B., Zhang, J., Ding, X., Wu, D., Wu, Y., Lu, W., Li, J. Improved metal assisted chemical etching method for uniform, vertical and deep silicon structure. *J. Micromech. Microeng*. 2017, 27: 055019.

**95** Wendisch, F.J., Rey, M., Vogel, N., Bourret, G.R. Large-scale synthesis of highly uniform silicon nanowire arrays using metal-assisted chemical etching. *Chem. Mater*. 2020, 32 (21): 9425–9434.

**96** Kim, J., Han, H., Kim, Y.H., Choi, S.-H., Kim, J.-C., Lee, W. Au/Ag bilayered metal mesh as a Si etching catalyst for controlled fabrication of Si nanowires. *ACS Nano* 2011, 5 (4): 3222–3229.

**97** Han, H., Huang, Z., Lee, W. Metal-assisted chemical etching of silicon and nanotechnology applications. *Nano Today* 2014, 9 (3): 271–304.

**98** Leonardi, A.A., Nastasi, F., Morganti, D., Faro, M.J., Picca, R.A., Cioffi, N., Franzò, G., Serroni, S., Priolo, F., Puntoriero, F., Campagna, S., Irrera, A. New hybrid light harvesting antenna based on silicon nanowires and metal dendrimers. *Adv. Optical Mater*. 2020, 8 (24): 2001070.

**99** Nam, T.W., Grossman, J.C. Ultrafine filteration of metal catalysts enables fabrication of silicon nanowires with diameter approaching a unit cell size. *Nano Lett*. 2023, 23 (6): 2347–2353.

**100** Huang, Y., Duan, X., Wei, Q., Lieber, C.M. Directed assembly of one-dimensional nanostructures into functional networks. *Science*. 2001, 291 (5504): 630–633.

**101** Huang, Y., Duan, X., Cui, Y., Lauhon, L.J., Kim, K., Lieber, C.M. Logic gates and computation from assembled nanowire building blocks. *Science*. 2001, 294 (5545): 1313–1317.

**102** Jia, G., Westphalen, J., Drexler, J., Plentz, J., Dellith, J., Dellith, A., Andrä, G., Falk, F. Ordered silicon nanowire arrays prepared by an improved nanospheres self-assembly in combination with Ag-assisted wet chemical etching. *Photonics Nanostruct. Fundam. Appl*. 2016, 19: 64–70.

**103** Ghotbi, S., Mohammadi, S. Effect of substrate conductivity on Si self-assembled field emission arrays. In: *2021 34th International Vacuum Nanoelectronics Conference (IVNC)*, 2021.

**104** Large-scale assembly of single nanowires through capillary-assisted dielectrophoresis. *Adv. Mater*. 2015.

**105** Constantinou, M., Rigas, G.P., Castro, F.A., Stolojan, V., Hoettges, K.F., Hughes, M.P., Adkins, E., Korgel, B.A., Shkunov, M. Simultaneous tunable selection and self-assembly of Si nanowires from heterogeneous feedstock. *ACS Nano* 2016, 10 (4): 4384–4394.

**106** Fan, Z., Ho, J.C., Jacobson, Z.A., Yerushalmi, R., Alley, R.L., Razavi, H., Javey, A. Wafer-scale assembly of highly ordered semiconductor nanowire arrays by contact printing. *Nano Lett*. 2008, 8 (1): 20–25.

**107** Yerushalmi, R., Jacobson, Z.A., Ho, J.C., Fan, Z., Javey, A. Large scale, highly ordered assembly of nanowire parallel arrays by differential roll printing. *Appl. Phys. Lett*. 2007, 91 (20): 203104.

**108** Javey, A., Nam, S., Friedman, R.S., Yan, H., Lieber, C.M. Layer-by-layer assembly of nanowires for three-dimensional, multifunctional electronics. *Nano Lett*. 2007, 7 (3): 773–777.

**109** Yu, G., Li, X., Lieber, C.M., Cao, A. Nanomaterial-incorporated blown bubble films for large-area, aligned nanostructures. *J. Mater. Chem.* 2008, 18 (7): 728–734.

**110** Large-area blown bubble films of aligned nanowires and carbon nanotubes. *Nat. Nanotechnol.*

**111** Liu, X., Long, Y.-Z., Liao, L., Duan, X., Fan, Z. Large-scale integration of semiconductor nanowires for high-performance flexible electronics. *ACS Nano* 2012, 6 (3): 1888–1900.

**112** Yu, L., Alet, P.-J., Picardi, G., Roca i Cabarrocas, P. An in-plane solid-liquid-solid growth mode for self-avoiding lateral silicon nanowires. *Phys. Rev. Lett.* 2009, 102 (12): 125501.

**113** Yu, L., Roca i Cabarrocas, P. Initial nucleation and growth of in-plane solid-liquid-solid silicon nanowires catalyzed by indium. *Phys. Rev. B* 2009, 80 (8): 085313.

**114** Bismuth-catalyzed n-type doping and growth evolution of planar silicon nanowires. *Appl. Phys. Lett.*

**115** Sun, Y., Qian, W., Liu, S., Dong, T., Wang, J., Xu, J., Chen, K., Yu, L. Unexpected phosphorus doping routine of planar silicon nanowires for integrating CMOS logics. *Nanoscale* 2021, 13 (35): 15031–15037.

**116** Engineering in-plane silicon nanowire springs for highly stretchable electronics. IOPscience.

**117** Hu, R., Ma, H., Yin, H., Xu, J., Chen, K., Yu, L. Facile 3D integration of Si nanowires on Bosch-etched sidewalls for stacked channel transistors. *Nanoscale* 2020, 12 (4): 2787–2792.

**118** Yu, L., Roca i Cabarrocas, P. Morphology control and growth dynamics of in-plane solid–liquid–solid silicon nanowires. *Phys. E Low-Dimens. Syst. Nanostruct.* 2012, 44 (6): 1045–1049.

**119** Xue, Z., Chen, W., Meng, X., Xu, J., Shi, Y., Chen, K., Yu, L., Roca i Cabarrocas, P. Precise morphology control of in-plane silicon nanowires via a simple plasma pre-treatment. *Appl. Surf. Sci.* 2022, 593: 153435.

**120** Cheng, Y., Gan, X., Liu, Z., Wang, J., Xu, J., Chen, K., Yu, L. Nanostripe-confined catalyst formation for uniform growth of ultrathin silicon nanowires. *Nanomater.* 2023, 13 (1): 121.

**121** Song, X., Wu, L., Liang, Y., Liu, Z., Wang, J., Xu, J., Chen, K., Yu, L. High-performance transparent silicon nanowire thin film transistors integrated on glass substrates via a room temperature solution passivation. *Adv. Electron. Mater.* 2023, 9 (4): 2201236.

**122** Allouhi, A., Rehman, S., Buker, M.S., Said, Z. Up-to-date literature review on solar PV systems: technology progress, market status and R&D. *J. Cleaner Prod.* 2022, 362: 132339.

**123** Garnett, E.C., Brongersma, M.L., Cui, Y., McGehee, M.D. Nanowire solar cells. *Annu. Rev. Mater. Res.* 2011, 41.

**124** Design and fabrication of silicon nanowires towards efficient solar cells. *Nano Today* 2016, 11 (6), 704–737.

**125** Li, J., Yu, H., Li, Y. Solar energy harnessing in hexagonally arranged Si nanowire arrays and effects of array symmetry on optical characteristics. *Nanotechnol.* 2012, 23 (19): 194010.

**126** Sahoo, M.K., Kale, P. Integration of silicon nanowires in solar cell structure for efficiency enhancement: a review. *J. Materiomics* 2019, 5 (1), 34–48.

**127** Sarkar, K., Das, D. Investigating the Ag-assisted chemical etching kinetics in growing Si-NWs and fabricating the p-i-n core–shell heterojunction Si-NW array solar cells. *Energy Fuels* 2023, 37 (21): 16880–16892.

**128** Peng, K., Zhang, M., Lu, A., Wong, N.-B., Zhang, R., Lee, S.-T. Ordered silicon nanowire arrays via nanosphere lithography and metal-induced etching. *Appl. Phys. Lett.* 2007, 90 (16): 163123.

**129** Farangi, M., Zahedifar, M., Mozdianfard, M.R., Pakzamir, M.H. Effects of silicon nanowires length on solar cells photovoltaic properties. *Appl. Phys. A.* 2012, 109: 299–306.

**130** Bao, H., Ruan, X. Optical absorption enhancement in disordered vertical silicon nanowire arrays for photovoltaic applications. *Opt. Lett.* 2010, 35 (20): 3378–3380.

**131** Haverkort, J.E.M., Garnett, E.C., Bakkers, E.P.A.M. Fundamentals of the nanowire solar cell: optimization of the open circuit voltage. *Appl. Phys. Rev.* 2018, 5 (3): 031106.

**132** Liu, H., Du, Y., Yin, X., Bai, M., Liu, W. Micro/nanostructures for light trapping in monocrystalline silicon solar cells. *J. Nanomater.* 2022, 2022 (1): 8139174.

**133** Li, Y., Chen, Q., He, D., Li, J. Radial junction Si micro/nano-wire array photovoltaics: recent progress from theoretical investigation to experimental realization. *Nano Energy* 2014, 7: 10–24.

**134** Yuan, G., Zhao, H., Liu, X., Hasanali, Z.S., Zou, Y., Levine, A., Wang, D. Synthesis and photoelectrochemical study of vertically aligned silicon nanowire arrays. *Angew. Chem. Int. Ed.* 2009, 48 (51): 9680–9684.

**135** Petrosyan, S., Khachatryan, V., Yesayan, A. Nanowire array solar cells. In: *2022 IEEE Latin American Electron Devices Conference (LAEDC)*, 1–4, 2022.

**136** Hong, K.-H., Kim, J., Lee, J.H., Shin, J., Chung, U.-I. Asymmetric doping in silicon nanostructures: the impact of surface dangling bonds. *Nano Lett.* 2010, 10 (5): 1671–1676.

**137** Kato, S., Yamazaki, T., Kurokawa, Y., Miyajima, S., Konagai, M. Influence of fabrication processes and annealing treatment on the minority carrier lifetime of silicon nanowire films. *Nanoscale Res. Lett.* 2017, 12 (1): 242.

**138** Kim, S., Bowden, S., Honsberg, C.B. Fabrication of nanopillar structure by silica nanosphere lithography and passivation with wet chemical oxidation cleaning. In: *2016 IEEE 43rd Photovoltaic Specialists Conference (PVSC)*, 2922–2924, 2016.

**139** Dan, Y., Seo, K., Takei, K., Meza, J.H., Javey, A., Crozier, K.B. Dramatic reduction of surface recombination by in situ surface passivation of silicon nanowires. *Nano Lett.* 2011, 11 (6): 2527–2532.

**140** Jia, G., Eisenhawer, B., Dellith, J., Falk, F., Thøgersen, A., Ulyashin, A. Multiple core–shell silicon nanowire-based heterojunction solar cells. *J. Phys. Chem. C* 2013, 117 (2): 1091–1096.

**141** Chen, W., Foldyna, M., Poulain, G., i Cabarrocas, P.R. Excellent surface passivation and light absorption in crystalline Si via low-temperature Si nanowire growth. *IEEE J. Photovolt.* 2016, 6 (4): 823–829.

**142** Jura, M.P., Miller, J.B., Yim, J.W.L., Forziati, J., Murphy, B., Chleboski, R., Cooper, I.B., Rohatgi, A., Black, M.R. Conventionally-processed silicon nanowire solar cells demonstrating efficiency improvement over standard cells. In: *2014 IEEE 40th Photovoltaic Specialist Conference (PVSC)*, 0598–0600, 2014.

**143** Lin, X.X., Hua, X., Huang, Z.G., Shen, W.Z. Realization of high performance silicon nanowire based solar cells with large size. *Nanotechnol.* 2013, 24 (23): 235402.

**144** Otto, M., Kroll, M., Käsebier, T., Li, X., Gesemann, B., Füchsel, K., Ziegler, J., Sprafke, A.N., Wehrspohn, R.B. Passivation of different black silicon surfaces by ALD Deposited Al2O3. In: *2013 IEEE 39th Photovoltaic Specialists Conference (PVSC)*, 1277–1279, 2013.

**145** Ben Rabha, M., Khezami, L., Jemai, A.B., Alhathlool, R., Ajbar, A. Surface passivation of silicon nanowires based metal nano-particle assisted chemical etching for photovoltaic applications. *J. Cryst. Growth* 2017, 462: 35–40.

**146** Fang, H., Li, X., Song, S., Xu, Y., Zhu, J. Fabrication of slantingly-aligned silicon nanowire arrays for solar cell applications. *Nanotechnol.* 2008, 19 (25): 255703.

**147** Hamidinezhad, H., Hayati, A. VLS synthesis of silicon nanowires array for photovoltaic devices. *Silicon* 2022, 14 (16): 10257–10261.

**148** Raman, S., Sindhuja, M. Advances in silicon nanowire applications in energy generation, storage, sensing, and electronics: a review. *Nanotechnol.* 2023, 34 (18): 182001.

**149** Cheng, H., Shapter, J.G., Li, Y., Gao, G. Recent progress of advanced anode materials of lithium-ion batteries. *J. Energy Chem.* 2021, 57: 451–468.

**150** Feng, K., Li, M., Liu, W., Kashkooli, A.G., Xiao, X., Cai, M., Chen, Z. Silicon-based anodes for lithium-ion batteries: from fundamentals to practical applications. *Small.* 2018, 14 (8): 1702737.

**151** Liang, B., Liu, Y., Xu, Y. Silicon-based materials as high capacity anodes for next generation lithium ion batteries. *J. Power Sources* 2014, 267: 469–490.

**152** Liu, X.H., Zhong, L., Huang, S., Mao, S.X., Zhu, T., Huang, J.Y. Size-dependent fracture of silicon nanoparticles during lithiation. *ACS Nano* 2012, 6 (2): 1522–1531.

**153** Ryu, J., Bok, T., Kim, S., Park, S. Fundamental understanding of nanostructured Si electrodes: preparation and characterization. *ChemNanoMat* 2018, 4 (4): 319–337.

**154** Lim, K.W., Lee, J.-I., Yang, J., Kim, Y.-K., Jeong, H.Y., Park, S., Shin, H.S. Catalyst-free synthesis of Si-SiOx core-shell nanowire anodes for high-rate and high-capacity lithium-ion batteries. *ACS Appl. Mater. Interfaces* 2014, 6 (9): 6340–6345.

**155** Cui, L.-F., Ruffo, R., Chan, C.K., Peng, H., Cui, Y. Crystalline-amorphous core–shell silicon nanowires for high capacity and high current battery electrodes. *Nano Lett.* 2009, 9 (1): 491–495.

**156** Wang, X., Huang, L., Zhang, Y., Yin, F., Bakenov, Z., Umirov, N., Jin, M., Zhou, G. Novel silicon nanowire film on copper foil as high performance anode for lithium-ion batteries. *Ionics* 2018, 24 (2): 373–378.

**157** Lithiation behavior of silicon nanowire anodes for lithium-ion batteries: impact of functionalization and porosity. *ChemPhysChem* 2018.

**158** Kohandehghan, A., Cui, K., Kupsta, M., Memarzadeh, E., Kalisvaart, P., Mitlin, D. Nanometer-scale Sn coatings improve the performance of silicon nanowire LIB anodes. *J. Mater. Chem. A* 2014, 2 (29): 11261–11279.

**159** Chen, Y., Liu, L., Xiong, J., Yang, T., Qin, Y., Yan, C. Porous Si nanowires from cheap metallurgical silicon stabilized by a surface oxide layer for lithium ion batteries. *Adv. Funct. Mater.* 2015, 25 (43): 6701–6709.

**160** Dominguez-Adame, F., Martin-Gonzalez, M., Sanchez, D., Cantarero, A. Nanowires: a route to efficient thermoelectric devices. *Phys. E Low-Dimens. Syst. Nanostruct.* 2019, 113: 213–225.

**161** Krali, E., Durrani, Z.A.K. Seebeck coefficient in silicon nanowire arrays. *Appl. Phys. Lett.* 2013, 102 (14): 143102.

**162** Zhang, T., Wu, S., Xu, J., Zheng, R., Cheng, G. High thermoelectric figure-of-merits from large-area porous silicon nanowire arrays. *Nano Energy* 2015, 13: 433–441.

**163** Vertical silicon nanowire thermoelectric modules with enhanced thermoelectric properties. *Nano Lett.*

**164** Thermal conductivity engineering in width-modulated silicon nanowires and thermoelectric efficiency enhancement. IOPscience.

**165** Spin-on-doping for output power improvement of silicon nanowire array based thermoelectric power generators. *J. Appl. Phys.*

**166** Fonseca, L., Santos, J.-D., Roncaglia, A., Narducci, D., Calaza, C., Salleras, M., Donmez, I., Tarancon, A., Morata, A., Gadea, G., Belsito, L., Zulian, L. Smart integration of silicon nanowire arrays in all-silicon thermoelectric micro-nanogenerators. *Semicond. Sci. Technol.* 2016, 31 (8): 084001.

**167** The possibility of mW/cm$^2$-class on-chip power generation using ultrasmall Si nanowire-based thermoelectric generators. *IEEE Journals & Magazine.*

**168** Lee, K.Y., Brown, D., Kumar, S. Silicon nanowire arrays based on-chip thermoelectric generators. *IEEE Trans. Compon. Packag. Manuf. Technol.* 2015, 5 (8): 1100–1107.

**169** Tomita, M., Oba, S., Himeda, Y., Yamato, R., Shima, K., Kumada, T., Xu, M., Takezawa, H., Mesaki, K., Tsuda, K., Hashimoto, S., Zhan, T., Zhang, H., Kamakura, Y., Suzuki, Y., Inokawa, H., Ikeda, H., Matsukawa, T., Matsuki, T., Watanabe, T. Modeling, simulation, fabrication, and characterization of a 10-$\mu$W/cm$^2$ class Si-nanowire thermoelectric generator for IoT applications. *IEEE Trans. Electron Devices* 2018, 65 (11): 5180–5188.

**170** Wang, Z., Qi, Y.Y., Zhang, M.L., Ji, A., Yang, F.H., Wang, X.D. Thermoelectric device based on vertical silicon nanowires for on-chip integration. *Key Eng. Mater.* 2014, 609–610: 789–795.

**171** Chaudhry, A., Kumar, M.J. Controlling short-channel effects in deep-submicron SOI MOSFETs for improved reliability: a review. *IEEE Trans. Device Mater. Reliab.* 2004, 4 (1): 99–109.

**172** Khanal, D.R., Yim, J.W.L., Walukiewicz, W., Wu, J. Effects of quantum confinement on the doping limit of semiconductor nanowires. *Nano Lett.* 2007, 7 (5): 1186–1190.

**173** Fernández-Serra, M.V., Adessi, Ch., Blase, X. Surface segregation and backscattering in doped silicon nanowires. *Phys. Rev. Lett.* 2006, 96 (16): 166805.

**174** Diarra, M., Niquet, Y.-M., Delerue, C., Allan, G. Ionization energy of donor and acceptor impurities in semiconductor nanowires: importance of dielectric confinement. *Phys. Rev. B* 2007, 75 (4): 045301.

**175** Björk, M., Schmid, H., Knoch, J., Riel, H., Riess, W. Donor deactivation in silicon nanostructures. *Nat. Nanotechnol.* 2009, 4: 103–107.

**176** Mikolajick, T., Häublein, V., Ryssel, H. The effect of random dopant fluctuations on the minimum channel length of short-channel MOS transistors. *Appl. Phys. Mater. Sci. Process.* 1997, 64 (6): 555–560.

**177** Wu, Y., Cui, Y., Huynh, L., Barrelet, C.J., Bell, D.C., Lieber, C.M. Controlled growth and structures of molecular-scale silicon nanowires. *Nano Lett.* 2004, 4 (3): 433–436.

**178** Jin, S., Fischetti, M.V., Tang, T. Modeling of electron mobility in gated silicon nanowires at room temperature: surface roughness scattering, dielectric screening, and band nonparabolicity. *J. Appl. Phys.* 2007, 102 (8): 083715.

179 Das, K.K., Mizel, A. Radial dependence of the carrier mobility in semiconductor nanowires. *J. Phys. Condens. Matter* 2005, 17 (42): 6675.

180 Weber, W.M., Mikolajick, T. Silicon and germanium nanowire electronics: Physics of conventional and unconventional transistors. *Rep. Prog. Phys.* 2017, 80 (6): 066502.

181 Zhang, Y.-B., Sun, L., Xu, H., Han, J.-W., Wang, Y., Zhang, S.-D. Comparative study of silicon nanowire transistors with triangular-shaped cross sections. *Jpn. J. Appl. Phys.* 2015, 54 (4S): 04DN01.

182 Xuan, H., Fan, J., Yang, Y., Zhang, H., Li, M., Huang, R. Performance studies in nanowire field-effect transistors with different cross sections. In: *2014 12th IEEE International Conference on Solid-State and Integrated Circuit Technology (ICSICT)*, 1–3. 2014.

183 Gunawan, O., Sekaric, L., Majumdar, A., Rooks, M., Appenzeller, J., Sleight, J.W., Guha, S., Haensch, W. Measurement of carrier mobility in silicon nanowires. *Nano Lett.* 2008, 8 (6): 1566–1571.

184 Jia, C., Lin, Z., Huang, Y., Duan, X. Nanowire electronics: from nanoscale to macroscale. *Chem. Rev.* 2019, 119 (15): 9074–9135.

185 Cui, Y., Duan, X., Hu, J., Lieber, C.M. Doping and electrical transport in silicon nanowires. *J. Phys. Chem. B* 2000, 104 (22): 5213–5216.

186 Cui, Y., Zhong, Z., Wang, D., Wang, W.U., Lieber, C.M. High performance silicon nanowire field effect transistors. *Nano Lett.* 2003, 3 (2): 149–152.

187 Zheng, G., Lu, W., Jin, S., Lieber, C.M. Synthesis and fabrication of high-performance n-type silicon nanowire transistors. *Adv. Mater.* 2004, 16 (21): 1890–1893.

188 Rahman, A., Guo, J., Datta, S., Lundstrom, M.S. Theory of ballistic nanotransistors. *IEEE Trans. Electron Devices* 2003, 50 (9): 1853–1864.

189 Yi, K.S., Trivedi, K., Floresca, H.C., Yuk, H., Hu, W., Kim, M.J. Room-temperature quantum confinement effects in transport properties of ultrathin Si nanowire field-effect transistors. *Nano Lett.* 2011, 11 (12): 5465–5470.

190 Lee, C.-W., Afzalian, A., Akhavan, N.D., Yan, R., Ferain, I., Colinge, J.-P. Junctionless multigate field-effect transistor. *Appl. Phys. Lett.* 2009, 94 (5): 053511.

191 Park, C.-H., Ko, M.-D., Kim, K.-H., Baek, R.-H., Sohn, C.-W., Baek, C. K., Park, S., Deen, M.J., Jeong, Y.-H., Lee, J.-S. Electrical characteristics of 20-nm junctionless Si nanowire transistors. *Solid-State Electron.* 2012, 73: 7–10.

192 Narula, V., Agarwal, M., Verma, S. A pathway to improve short channel effects of junctionless based FET's after incorporating technology boosters: a review. *Eng. Res. Express* 2024, 6 (1): 012301.

193 Charles Pravin, J., Nirmal, D., Prajoon, P., Mohan Kumar, N., Ajayan, J. Investigation of 6T SRAM memory circuit using high-k dielectrics based nano scale junctionless transistor. *Superlattices Microstruct.* 2017, 104: 470–476.

194 Banerjee, K., Biswas, A. Enhanced analog/RF performance of hybrid charge plasma based junctionless C-FinFET amplifiers at 10 nm technology node. *Microelectron. J.* 2023, 131: 105662.

195 Sharma, A., Jain, A., Pratap, Y., Gupta, R.S. Effect of high-k and vacuum dielectrics as gate stack on a junctionless cylindrical surrounding gate (JL-CSG) MOSFET. *Solid-State Electron.* 2016, 123: 26–32.

196 Lee, B.-H., Hur, J., Kang, M.-H., Bang, T., Ahn, D.-C., Lee, D., Kim, K.-H., Choi, Y.-K. A vertically integrated junctionless nanowire transistor. *Nano Lett.* 2016, 16 (3): 1840–1847.

**197** Tung, R.T. (董梓則). The physics and chemistry of the Schottky barrier height. *Appl. Phys. Rev.* 2014, 1 (1): 011304.

**198** Maex, K., Rossum, M. Inspec. *Properties of Metal Silicides*; 1995.

**199** Lin, Y.-C., Lu, K.-C., Wu, W.-W., Bai, J., Chen, L.J., Tu, K.N., Huang, Y. Single crystalline PtSi nanowires, PtSi/Si/PtSi nanowire heterostructures, and nanodevices. *Nano Lett.* 2008, 8 (3): 913–918.

**200** Fritze, M., Chen, C.L., Calawa, S., Yost, D., Wheeler, B., Wyatt, P., Keast, C.L. High-speed Schottky-barrier pMOSFET with f/Sub T/=280 GHz. *IEEE Electron Device Lett.* 2004, 25 (4): 220–222.

**201** Mikolajick, T., Heinzig, A., Trommer, J., Pregl, S., Grube, M., Cuniberti, G., Weber, W.M. Silicon nanowires – a versatile technology platform. *Phys. Status Solidi RRL – Rapid Res. Lett.* 2013, 7 (10): 793–799.

**202** Weber, W.M., Heinzig, A., Trommer, J., Martin, D., Grube, M., Mikolajick, T. Reconfigurable nanowire electronics – a review. *Solid-State Electron.* 2014, 102: 12–24.

**203** Mikolajick, T., Galderisi, G., Simon, M., Rai, S., Kumar, A., Heinzig, A., Weber, W.M., Trommer, J. 20 years of reconfigurable field-effect transistors: from concepts to future applications. *Solid-State Electron.* 2021, 186: 108036.

**204** Lin, H.C., Yeh, K.L., Huang, R.G., Lin, C.Y., Huang, T.Y. Schottky barrier thin-film transistor (SBTFT) with silicided source/drain and field-induced drain extension. *IEEE Electron Device Lett.* 2001, 22 (4), 179–181.

**205** Heinzig, A., Slesazeck, S., Kreupl, F., Mikolajick, T., Weber, W.M. Reconfigurable silicon nanowire transistors. *Nano Lett.* 2012, 12 (1): 119–124.

**206** Heinzig, A., Mikolajick, T., Trommer, J., Grimm, D., Weber, W.M. Dually active silicon nanowire transistors and circuits with equal electron and hole transport. *Nano Lett.* 2013, 13 (9): 4176–4181.

**207** De Marchi, M., Sacchetto, D., Frache, S., Zhang, J., Gaillardon, P.-E., Leblebici, Y. Polarity control in double-gate, gate-all-around vertically stacked silicon nanowire FETs. In: *2012 International Electron Devices Meeting*, 8.4.1–8.4.4. 2012.

**208** Trommer, J., Heinzig, A., Slesazeck, S., Mikolajick, T., Weber, W.M. Elementary aspects for circuit implementation of reconfigurable nanowire transistors. *IEEE Electron Device Lett.* 2014, 35 (1): 141–143.

**209** Darbandy, G., Claus, M., Schröter, M. High-performance reconfigurable Si nanowire field-effect transistor based on simplified device design. *IEEE Trans. Nanotechnol.* 2016, 15 (2): 289–294.

**210** Zhang, J., Gaillardon, P.-E., De Micheli, G. Dual-threshold-voltage configurable circuits with three-independent-gate silicon nanowire FETs. In: *2013 IEEE International Symposium on Circuits and Systems (ISCAS)*, 2111–2114. 2013.

**211** Zhang, J., De Marchi, M., Gaillardon, P.-E., De Micheli, G. A Schottky-barrier silicon FinFET with 6.0 mV/dec subthreshold slope over 5 decades of current. In: *2014 IEEE International Electron Devices Meeting*, 13.4.1–13.4.4. 2014.

**212** Simon, M., Trommer, J., Liang, B., Fischer, D., Baldauf, T., Khan, M.B., Heinzig, A., Knaut, M., Georgiev, Y.M., Erbe, A., Bartha, J.W., Mikolajick, T., Weber, W.M. A wired-AND transistor: polarity controllable FET with multiple inputs. In: *2018 76th Device Research Conference (DRC)*, 1–2. 2018.

**213** Wind, L., Behrle, R., den Hertog, M.I., Murphey, C.G.E., Cahoon, J.F., Sistani, M., Weber, W.M. Nanoscale reconfigurable Si transistors: from wires to sheets and unto multi-wire channels. *Adv. Electron. Mater.* 2024, 10 (2): 2300483.

**214** Rai, S., Rupani, A., Walter, D., Raitza, M., Heinzig, A., Baldauf, T., Trommer, J., Mayr, C., Weber, W.M., Kumar, A. A physical synthesis flow for early technology evaluation of silicon nanowire based reconfigurable FETs. In: *2018 Design, Automation & Test in Europe Conference & Exhibition (DATE)*, 605–608. 2018.

**215** Voisin, B., Nguyen, V.-H., Renard, J., Jehl, X., Barraud, S., Triozon, F., Vinet, M., Duchemin, I., Niquet, Y.-M., de Franceschi, S., Sanquer, M. Few-electron edge-state quantum dots in a silicon nanowire field-effect transistor. *Nano Lett.* 2014, 14 (4): 2094–2098.

**216** Lavieville, R., Triozon, F., Barraud, S., Corna, A., Jehl, X., Sanquer, M., Li, J., Abisset, A., Duchemin, I., Niquet, Y.-M. Quantum dot made in metal oxide silicon-nanowire field effect transistor working at room temperature. *Nano Lett.* 2015, 15 (5): 2958–2964.

**217** Angus, S.J., Ferguson, A.J., Dzurak, A.S., Clark, R.G. Gate-defined quantum dots in intrinsic silicon. *Nano Lett.* 2007, 7 (7): 2051–2055.

**218** Zwanenburg, F.A., van Rijmenam, C.E.W.M., Fang, Y., Lieber, C.M., Kouwenhoven, L.P. Spin states of the first four holes in a silicon nanowire quantum Dot. *Nano Lett.* 2009, 9 (3): 1071–1079.

**219** Tsoukalas, K., Schupp, F., Sommer, L., Bouquet, I., Mergenthaler, M., Paredes, S., Vico Triviño, N., Luisier, M., Salis, G., Harvey-Collard, P., Zumbühl, D., Fuhrer, A. Prospects of silicide contacts for silicon quantum electronic devices. *Appl. Phys. Lett.* 2024, 125 (1): 013501.

**220** Salahuddin, S., Ni, K., Datta, S. The era of hyper-scaling in electronics. *Nat. Electron.* 2018, 1: 442–450.

**221** Kang, H., Cho, J., Kim, Y., Lim, D., Woo, S., Cho, K., Kim, S. Nonvolatile and volatile memory characteristics of a silicon nanowire feedback field-effect transistor with a nitride charge-storage layer. *IEEE Trans. Electron Devices* 2019, 66 (8): 3342–3348.

**222** Lim, D., Kim, M., Kim, Y., Kim, S. Memory characteristics of silicon nanowire transistors generated by weak impact ionization. *Sci. Rep.* 2017, 7 (1): 12436.

**223** Lim, D., Kim, M., Kim, Y., Cho, J., Kim, S. Nondestructive readout memory characteristics of silicon nanowire biristors. *IEEE Trans. Electron Devices* 2018, 65 (4): 1578–1582.

**224** Park, S.J., Jeon, D.-Y., Piontek, S., Grube, M., Ocker, J., Sessi, V., Heinzig, A., Trommer, J., Kim, G.-T., Mikolajick, T., Weber, W.M. Reconfigurable Si nanowire nonvolatile transistors. *Adv. Electron. Mater.* 2018, 4 (1): 1700399.

**225** Zhang, Q., Zhang, Y., Luo, Y., Yin, H. New structure transistors for advanced technology node CMOS ICs. *Natl. Sci. Rev.* 2024, 11 (3): nwae008.

**226** Zhu, H., Li, Q., Yuan, H., Baumgart, H., Ioannou, D.E., Richter, C.A. Self-aligned multi-channel silicon nanowire field-effect transistors. *Solid-State Electron.* 2012, 78: 92–96.

**227** Constantinou, M., Hoettges, K.F., Krylyuk, S., Katz, M.B., Davydov, A., Rigas, G.-P., Stolojan, V., Hughes, M.P., Shkunov, M. Rapid determination of nanowires electrical properties using a dielectrophoresis-well based system. *Appl. Phys. Lett.* 2017, 110 (13): 133103.

**228** Yu, X., Wang, Y., Zhou, H., Liu, Y., Wang, Y., Li, T., Wang, Y. Top-down fabricated silicon-nanowire-based field-effect transistor device on a (111) silicon wafer. *Small* 2013, 9 (4): 525–530.

**229** Liu, J. C., Mukhopadhyay, S., Kundu, A., Chen, S. H., Wang, H. C., Huang, D. S. A reliability enhanced 5nm CMOS technology featuring 5th generation FinFET with fully-developed EUV and high mobility channel for mobile SoC and high performance computing application. In: *IEEE Conference Publication*. 2020 IEEE International Electron Devices Meeting (IEDM), San Francisco, CA, USA, 2020, pp. 9.2.1–9.2.4.

**230** Singh, N., Agarwal, A., Bera, L.K., Liow, T.Y., Yang, R., Rustagi, S.C., Tung, C.H., Kumar, R., Lo, G.Q., Balasubramanian, N., Kwong, D.-L. High-performance fully depleted silicon nanowire (diameter/Spl Les/5 nm) gate-all-around CMOS devices. *IEEE Electron Device Lett*. 2006, 27 (5): 383–386.

**231** Li, M., Fan, J., Xu, X., Huang, R. Investigation on electrostatic discharge robustness of gate-all-around silicon nanowire transistors combined with thermal analysis. *IEEE Electron Device Lett*. 2017, 38 (12): 1653–1656.

**232** Liang, L., Hu, R., Yu, L. Toward monolithic growth integration of nanowire electronics in 3D architecture: a review. *Sci. China Inf. Sci*. 2023, 66 (10): 200406.

**233** Goldberger, J., Hochbaum, A.I., Fan, R., Yang, P. Silicon vertically integrated nanowire field effect transistors. *Nano Lett*. 2006, 6 (5): 973–977.

**234** Chen, L., Cai, F., Otuonye, U., Lu, W.D. Vertical Ge/Si core/shell nanowire junctionless transistor. *Nano Lett*. 2016, 16 (1): 420–426.

**235** Larrieu, G., Han, X. Vertical nanowire array-based field effect transistors for ultimate scaling. *Nanoscale* 2013, 5 (6): 2437–2441.

**236** Li, C., Zhu, H., Zhang, Y., Wang, Q., Yin, X., Li, J., Wang, G., Kong, Z., Ai, X., Xie, L., Liu, Y., Li, Y., Huang, W., Yan, Z., Xiao, Z., Radamson, H.H. First demonstration of novel vertical gate-all-around field-effect-transistors featured by self-aligned and replaced high-κ metal gates. *Nano Lett*. 2021, 21 (11): 4730–4737.

**237** Han, J.-K., Oh, J., Yu, J.-M., Choi, S.-Y., Choi, Y.-K. A vertical silicon nanowire based single transistor neuron with excitatory, inhibitory, and myelination functions for highly scalable neuromorphic hardware. *Small* 2021, 17 (49): 2103775.

**238** Thean, A.V.-Y., Yakimets, D., Huynh Bao, T., Schuddinck, P., Sakhare, S., Bardon, M.G., Sibaja-Hernandez, A., Ciofi, I., Eneman, G., Veloso, A., Ryckaert, J., Raghavan, P., Mercha, A., Mocuta, A. Vertical device architecture for 5nm and beyond: device & circuit implications. In: *2015 Symposium on VLSI Technology (VLSI Technology*), T26–T27. 2015.

**239** Bernard, E., Ernst, T., Guillaumot, B., Vulliet, N., Coronel, P., Skotnicki, T. Multi-Channel Field-Effect Transistor (MCFET)—Part I: Electrical Performance and Current Gain Analysis. *IEEE Xplore*. In: *IEEE Transactions on Electron Devices*. 2009, 56 (6): 1243–1251.

**240** Dupre, C., Hubert, A., Becu, S., Jublot, M., Maffini-Alvaro, V., Vizioz, C. 15nm-diameter 3D stacked nanowires with independent gates operation: ΦFET. In: *IEEE Conference Publication*. 2008 IEEE International Electron Devices Meeting, San Francisco, CA, USA, 2008, pp. 1–4.

**241** Ryckaert, J., Samavedam, S.B. The CMOS 2.0 revolution. *Nat. Rev. Electr. Eng*. 2024, 1: 139–140.

**242** Sacchetto, D., Ben-Jamaa, M.H., De Micheli, G., Leblebici, Y. Fabrication and characterization of vertically stacked gate-all-around Si nanowire FET arrays. In: *2009 Proceedings of the European Solid State Device Research Conference*, 245–248. 2009.

**243** Ng, R.M.Y., Wang, T., Liu, F., Zuo, X., He, J., Chan, M. Vertically stacked silicon nanowire transistors fabricated by inductive plasma etching and stress-limited oxidation. *IEEE Electron Device Lett.* 2009, 30 (5): 520–522.

**244** Lee, B.-H., Kang, M.-H., Ahn, D.-C., Park, J.-Y., Bang, T., Jeon, S.-B., Hur, J., Lee, D., Choi, Y.-K. Vertically integrated multiple nanowire field effect transistor. *Nano Lett.* 2015, 15 (12): 8056–8061.

**245** Wu, X., Ma, H., Yin, H., Pan, D., Wang, J., Yu, L., Xu, J., Shi, Y., Chen, K. 3D sidewall integration of ultrahigh-density silicon nanowires for stacked channel electronics. *Adv. Electron. Mater.* 2019, 5 (7): 1800627.

**246** Hu, R., Xu, S., Wang, J., Shi, Y., Xu, J., Chen, K., Yu, L. Unprecedented uniform 3D growth integration of 10-layer stacked Si nanowires on tightly confined sidewall grooves. *Nano Lett.* 2020, 20 (10): 7489–7497.

**247** Hu, R., Liang, Y., Qian, W., Gan, X., Liang, L., Wang, J., Liu, Z., Shi, Y., Xu, J., Chen, K., Yu, L. Ultra-confined catalytic growth integration of sub-10 nm 3D stacked silicon nanowires via a self-delimited droplet formation strategy. *Small* 2022, 18 (42): 2204390.

III

# Germanium

*Yi Shi*

## 3.1 Introduction

Germanium [1–4], as a typical semiconductor, has attracted much attention due to its environmentally friendly characteristics. In nature, germanium has five stable nuclides: $^{70}Ge$, $^{72}Ge$, $^{73}Ge$, $^{74}Ge$, and $^{76}Ge$. As early as 1885, Winkler prepared germanium by sulfurizing germanium and hydrogen co-heating. It is worth mentioning that germanium belongs to a semimetal [5–9], and its electrical conductivity is better than the general nonmetallic materials, the electronic arrangement of [Ar] $3d^{10}4s^24p^2$. Compared with silicon, germanium has a higher electron and hole mobility, a larger permittivity, a small bandgap of 0.67 eV at room temperature, and a large Bohr radius of about 24 nm. These properties allow germanium to be observed with quantum confinement effects [10–13] at larger particle sizes, as well as tuning light emission in the 3.3-eV range. Although there have been many semiconductor materials with similar properties, such as CdS [14, 15], InAs [16, 17], GaAs [18, 19], and PbS [20], which have shown good photochemical stability, easy spectral tuning [21–23], high quantum yields [24–26], and long fluorescence lifetimes [27–29], their toxicity is a concern. Most of these materials are highly toxic to cells and remain in the body for at least 90 days. Germanium, on the other hand, not only has the advantage of their similar properties but is even more valuable because it is nontoxic, making it a natural alternative to toxic semiconductor materials. Meanwhile, its size-dependent optical properties, good biocompatibility [30], electrochemical stability [31–33], and potential for applications in optoelectronics, bio-imaging and therapeutics, flash memory, and lithium–ion batteries are being studied extensively.

Meanwhile, nanomaterials have attracted more and more attention due to their various amazing properties, especially one-dimensional (1D) nanomaterials with excellent optical and electrical properties. Germanium nanowires (Ge NWs), as the most common 1D nanomaterials, have great potential for applications in optoelectronics and energy storage. So far, several studies have been conducted to synthesize Ge NWs in single and polycrystalline structures by chemical vapor deposition [34–36], laser ablation [37, 38], thermal evaporation [39, 40], and transmission electron microscopy (TEM) methods, respectively, which are expected to be used in large-scale applications in the fields of field effect transistors (FET), energy storage, and photodetectors.

In this chapter, we discuss the synthesis methods of germanium nanomaterials and summarize the various synthesis methods. The properties and related applications of germanium

*Single Element Semiconductors: Properties and Devices*, First Edition. Yi Shi and Shancheng Yan.

nanomaterials are discussed in depth. The outline of this chapter is as follows: Section 3.2 discusses the synthesis methods of germanium nanomaterials, including laser ablation (Section 3.2.1), supercritical fluid–liquid–solid (SFLS) (Section 3.2.2), thermal evaporation (Section 3.2.3), chemical vapor deposition (Section 3.2.4), templating (Section 3.2.5), and hydrothermal methods (Section 3.2.6). In Section 3.3, we discuss the properties and applications of germanium nanomaterials, including optical properties (Section 3.3.1), Raman spectrum (Section 3.3.2), energy storage (Section 3.3.3), and FET (Section 3.3.4).

## 3.2 Synthesis of Germanium Nanomaterials

When the material fulfills the conditions of small size and low dimensionality, it tends to make the material possess some properties that are not available in the bulk. Ge NWs, as the most promising material for the preparation of optoelectronic devices and field effect transistors, have been reported to be synthesized by a variety of methods, such as laser ablation, supercritical [41–43], thermal evaporation, chemical vapor deposition, template, hydrothermal, etc., and these researches have already yielded impressive results.

### 3.2.1 Laser Ablation

Laser ablation, which is usually carried out by both photothermal and photoablation, is a classical material preparation method. Morales and Lieber [44] proposed for the first time a method combining laser ablation cluster formation and vapor–liquid–solid (VLS) growth for the preparation of Ge NWs. They prepared Ge NWs with diameters of about 3–9 nm in a tube furnace at 820 °m by ablation of catalyst-containing clusters using a 532-nm laser with Fe as the catalyst. These NWs have a crystal structure of uniform diameter, no amorphous coating, and a distinct catalyst cluster at the end, which is determined by X-ray fluorescence analysis to have two elements, Fe and Ge, while the NWs contain only Ge. At the same time, high-resolution transmission electron microscopy (HRTEM) images of the NWs clearly demonstrate the (111) crystal plane of Ge, proving the crystalline nature of the NW.

Subsequently, Zhang et al. [45] reported a method to prepare Ge NWs without any catalyst, using only laser ablation. The method is similar to their previously reported [46] preparation of silicon NWs using laser ablation, where Ge NWs were successfully synthesized by laser ablation of a mixture prepared using particulate Ge and $GeO_2$. It is noteworthy that the highest yield of Ge NWs was achieved when the ratio of Ge and $GeO_2$ was 1 : 1. The laser they used had a wavelength of 248 nm, an energy of 450 mJ, a frequency of 10 Hz, and a laser pulse duration of 34 ns. Ge NWs with diameters of 16–500 nm were successfully grown at substrate temperatures of 690–705 °C, and the TEM images are shown in Figure 3.1a. It can be seen that the length of such NWs can be up to tens of micrometers, and most of them have smooth, curved morphology with the smallest diameter of about 12 nm, while there are some NWs with diameters as high as 190 nm in some regions of lower temperatures. Figure 3.1b shows the mass–thickness contrast TEM image of the coarse Ge NW, which consists of an amorphous $GeO_2$ sheath surrounding a crystalline Ge core with the axial orientation of (211), independent of the crystallographic orientation

(a) 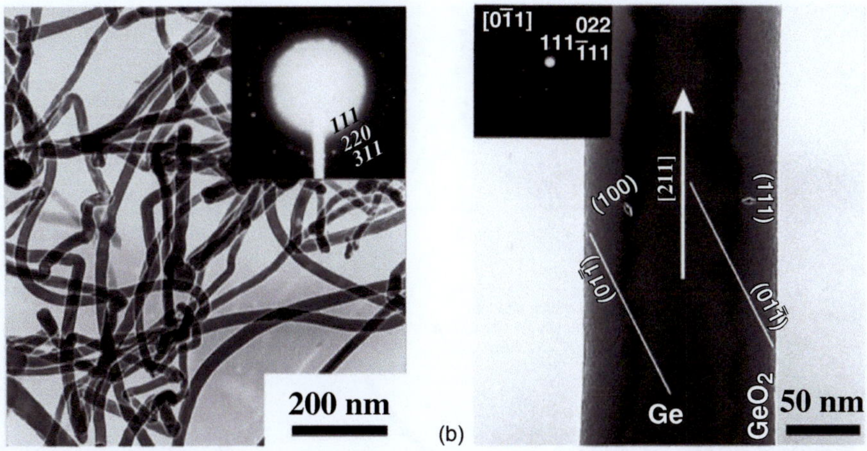 (b)


**Figure 3.1** (a) TEM image of Ge NWs and a selected-area electron-diffraction pattern (inset). (b) Mass–thickness contrast TEM image of a thick Ge NW taken along the (01$\bar{1}$) zone axis and a selected-area electron diffraction pattern (inset). *Source:* Reprinted with permission from Ref. [45]. Copyright 2000, American Physical Society.

of the Si substrate. Meanwhile, the outer surface of $GeO_2$ is very smooth. However, the interface between the amorphous sheath and the core has a wavy profile, and the profile between the core and the amorphous germanium oxide sheath has a jagged profile. The jagged edges appear to be (011), (111), or slightly tilted (100) to the face of the crystal core. It is noteworthy that the amorphous $GeO_2$ sheaths appear to have the ability to adapt to the toothed structure of the core surface and to saturate or passivate at least most of the core surface dangling bonds. In summary, $GeO_2$ sheaths play an important role in the formation of Ge NWs by saturating the surface bonds to reduce the surface energy and terminate the lateral growth with a smooth $GeO_2$ surface.

### 3.2.2 SFLS Method

As early as 1993, Heath and LeGoues [47] proposed the successful preparation of Ge NWs by reducing $GeCl_4$ and phenyl-$GeCl_3$ in an alkane solvent using a high-temperature and high-pressure environment. Among them, phenyl-$GeCl_3$ played the role of particle size control by heating the vessel to 275 °C and keeping it for 2–8 days and controlling the pressure at 100 atm, and the final NWs were formed as shown in Figure 3.2a. The diameter of this single-crystal NW is about 180 Å, and the length is around 10 μm. Size and morphology control of the sample can be achieved using this method, with the disadvantage that the yield of single-crystal Ge NWs is low. Hanrath and Korgel [48] grew Ge NWs a few micrometers long in cyclohexane that was heated and pressurized above a critical point, and the growth proceeded using a solution–liquid–solid mechanism. They used tetraethylgermanium as a precursor for germanium and employed streptavidin thiols with diameters of 2.5 or 6.5 nm to protect the nanocrystals, and the Ge NWs prepared at 400 °C are shown in Figure 3.2b. In fact, no single-crystal NWs were formed under the growth conditions at 250 °5, and some short NWs with diameters of about 40 nm appeared only when the

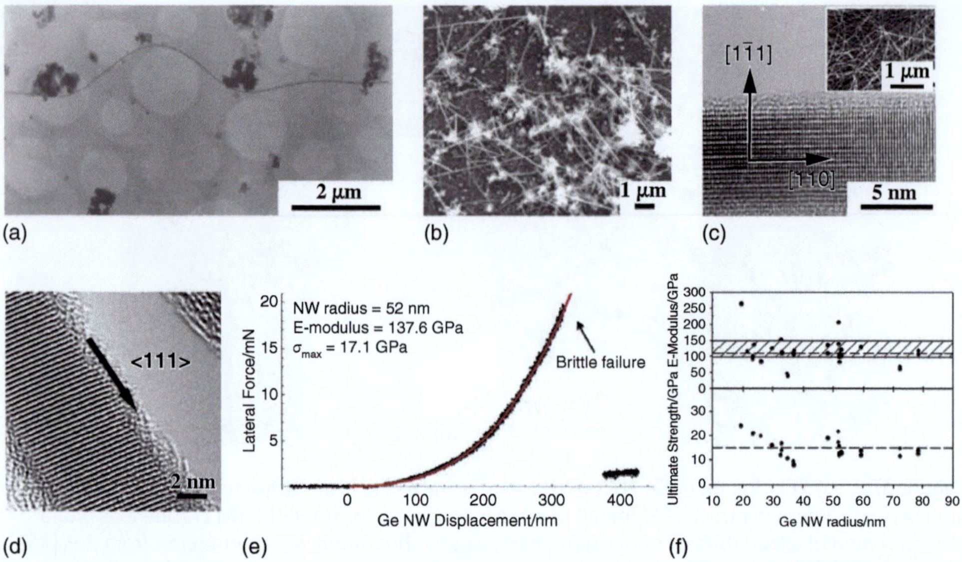


**Figure 3.2** (a) TEM micrograph of a 10-μm long, 18-nm diameter Ge single-crystal quantum wire supported on a holey, amorphous carbon substrate. The islands of material scattered throughout the micrograph consist primarily of bunches of Ge quantum dots. *Source:* Reprinted with permission from Ref. [47]. Copyright 1993, Elsevier. (b) High-resolution transmission electron microscope image of Ge NWs grown at 38MPa for 20min using tetraethylgermane at 400 °C. *Source:* Reprinted with permission from Ref. [48]. Copyright 2002, American Chemical Society. (c) HRTEM images of 8.6-nm-diameter NWs with the [110] growth direction. *Source:* Reprinted with permission from Ref. [49]. Copyright 2005, WILEY-VCH. (d) TEM image of Ni nanocrystal-seeded Ge NWs. *Source:* Reprinted with permission from Ref. [50]. Copyright 2005, American Chemical Society. (e) F-d curve from a typical NW manipulation experiment. (f) Young's modulus (upper panel) and ultimate strength (lower panel) versus Ge NW radius. *Source:* Reprinted with permission from Ref. [51]. Copyright 2006, American Chemical Society.

growth temperature was raised to 300 °m, and the yield of NWs increased dramatically only when the temperature reached 400°. When they replaced the precursor of germanium with diphenylgermane, the diameter of the NWs prepared under the same conditions decreased substantially. This phenomenon was attributed to the agglomeration of gold nanocrystals and unfavorable decomposition kinetics in the case of tetraethylgermanium as a precursor. In conclusion, this method promotes the growth of Ge NWs using spatially stabilized gold nanocrystals in a supercritical fluid environment. The crystallographic details of the NWs were further explored by Hanrath and Korgel [49]. Figure 3.2c illustrates a TEM image of Ge NWs with a diameter of 8.2 nm. Where the (220) and ($\bar{1}$11) crystal planes are parallel to the growth direction [110] of the NW, by analyzing the crystal structure of Ge NWs grown by different methods, it can be found that the growth rate is an important factor in determining the direction of growth, with high temperature favoring the [111] growth and low temperature favoring the [110] growth direction. The above types of methods can be categorized as SFLS methods since they all follow the solution–liquid–solid growth mechanism.

In 2005, Tuan et al. [50] proposed another method for the preparation of Ge NWs by thermal decomposition of diphenylgermane in supercritical toluene at a [51] presence, following the solution–solid–solid growth mechanism, which is referred to as the supercritical flow–solid–solid (SFSS) method of synthesis. Figure 3.2d demonstrates the TEM image of Ge NWs grown at 460 °C. The NWs are well crystallized with a diamond cubic structure and are almost free of defects from the surface. The diameters of the NWs are about 5–30 nm, and the average lengths are all over 10 μm. Ni nanocrystals are used as seeds in this method, but the growth temperature is much lower than the lowest eutectic temperature on the Ni/Ge phase diagram. Unlike epitaxial growth, large-diameter NWs grow at a slower rate, so most NWs have smaller diameters, which is a potential advantage of SFSS over SFLS. Ngo et al. [51] also used supercritical synthesis to prepare Ge NWs with ultrahigh strength. The F-d curve of the NWs is shown in Figure 3.2e. During the operation, the force F initially shifted with the NW displacement increases linearly, consistent with the initial bending deformation of the wire. Figure 3.2f shows the results of applying the model to Ge NWs with radii ranging from 20 to 80 nm, with a measured modulus of 112 GPa ± 43 GPa, which is in good agreement with the known bulk values of 103–150 GPa. All these results indicate that Ge NWs have excellent strength and are free from defects that impair mechanical properties.

### 3.2.3 Thermal Evaporation

Thermal evaporation method [52, 53] has been widely used by researchers as a simple and efficient synthesis method without the use of toxic gases. The synthesis of Ge NWs using thermal evaporation method requires only the use of Ge powder as a raw material and Si or $SiO_2$/Si as a substrate, which has good scalability. Gu et al. [54] evaporated germanium powder at 950 °C and transported and deposited it on Au nanoparticles (NPs) or Au thin-film substrates at 500 °C under argon gas. Figure 3.3a shows Ge NWs grown on an Au film; most of the Ge NWs are curved due to defects in the NWs, and there are visible NPs consisting of Ge and Au at the ends of the NWs. The diameter of these NWs is about 20–180 nm, whereas the Ge NWs grown on Au NPs are shown in Figure 3.3b, the Au particles are uniformly distributed on the substrate, which is more conducive for the growth of Ge NWs. The Ge NWs grown on this substrate have fewer defects, with diameters of about 20–45 nm, a relatively more uniform diameter distribution, and straighter lines. This

**Figure 3.3** (a) SEM image of Ge NWs. (b) SEM image of Ge NWs grown from 10 nm Au particles to SEM image of Ge NWs grown from these low-density 10 nm Au particles. *Source:* Reprinted with permission from Ref. [54]. Copyright 2001, American Institute of Physics. (c) Typical SEM image. *Source:* Reprinted with permission from Ref. [55]. Copyright 2008, IOP Publishing Ltd.

also indicates that the morphology of the NWs is related to the morphology of Au on the substrate. This method follows the VLS growth mechanism, where the gas transports Ge clusters onto Au NPs, allowing the growth of Au-Ge droplets and the eventual formation of Ge NWs. Sutter et al. [55] chose a Si sheet coated with Au particles of 20 nm on the surface or a thin film of Au as the substrate, and the Ge powder was gasified and then transported through Ar gas and was subjected to a temperature of 420–500 °C growth. The scanning electron microscopy (SEM) image of Ge NWs is shown in Figure 3.3c. These NWs are about 10–60 nm in diameter and tens of micrometers long. And the lattice spacing of this NW is 0.33 nm, which conforms to the (111) crystal plane of Ge. This growth method also follows the VLS mechanism, where a large amount of GeO is sublimated from $GeO_2$ at a high temperature of 700 °C, and single-crystal Ge NWs can be grown eventually due to the oxygen inhibition of Au for the GeO precursor. Ko et al. [56] also chose Au as a catalyst and prepared high-quality Ge NWs using the Sn/Ge co-evaporation method. It is worth mentioning that the inclusion of Sn powder in the source allowed the source to still be fully evaporated when the source temperature was below 600 °C and the mixed vapor was successfully transported to the substrate.

While several of the above thermal evaporation methods use Au as a catalyst, there are also catalyst-free thermal evaporation methods. Li et al. [57] prepared a large number of Ge NW arrays by catalyst-free thermal evaporation. They grew oriented Ge NW arrays by tilting the Si substrate and adjusting the angle of incidence of the Ge vapor with the substrate. Figure 3.4 shows SEM images of the NWs grown at 230 and 330 °C substrate temperatures. It can be seen that some of the NWs are in a bundled state as the deposition temperature increases, but most of the NWs are gradually separated. This is due to the fact that the surface diffusion coefficient of germanium atoms increases with increasing temperature, and the initial nucleation size becomes larger, which allows it to capture more atoms over a larger area, and the generated NWs become more dispersed. The method can be used to obtain Ge nanostructures with different crystallinity by adjusting the substrate temperature and evaporation rate. The NWs deposited at 130 and 230 are amorphous and polycrystalline, whereas the NWs deposited at 330 show a single-crystal structure. This is due to the fact that the growth rate is an important factor in determining the growth direction, with high temperature favoring the [111] growth and low temperature favoring the [110] growth direction.

### 3.2.4 Chemical Vapor Deposition

Chemical vapor deposition [58–61], as a classical synthesis method, has the advantages of low temperature, more flexible control conditions, and higher purity of the prepared NWs among all the synthesis methods. As early as 2004, Dailey et al. [62] used chemical vapor deposition to grow Ge NWs on Si with substrate temperatures in the range of 400–600 °0 and pressures in the range of $4 \times 10^{-5}$ to $1 \times 10^{-2}$ Torr. Figure 3.5a shows the Ge NWs formed on Si substrate grown at $3 \times 10^{-3}$ Torr, 400 °T for 5 min; Figure 3.5b shows Ge NWs grown at $1 \times 10^{-2}$ Torr, 400 °T for 2.5 min. It can be seen that all NWs have gold eutectic seed particles at the tips, which is consistent with the VLS growth mechanism. In the first case, the diameter of the NWs is uniform over the length of the line. However,

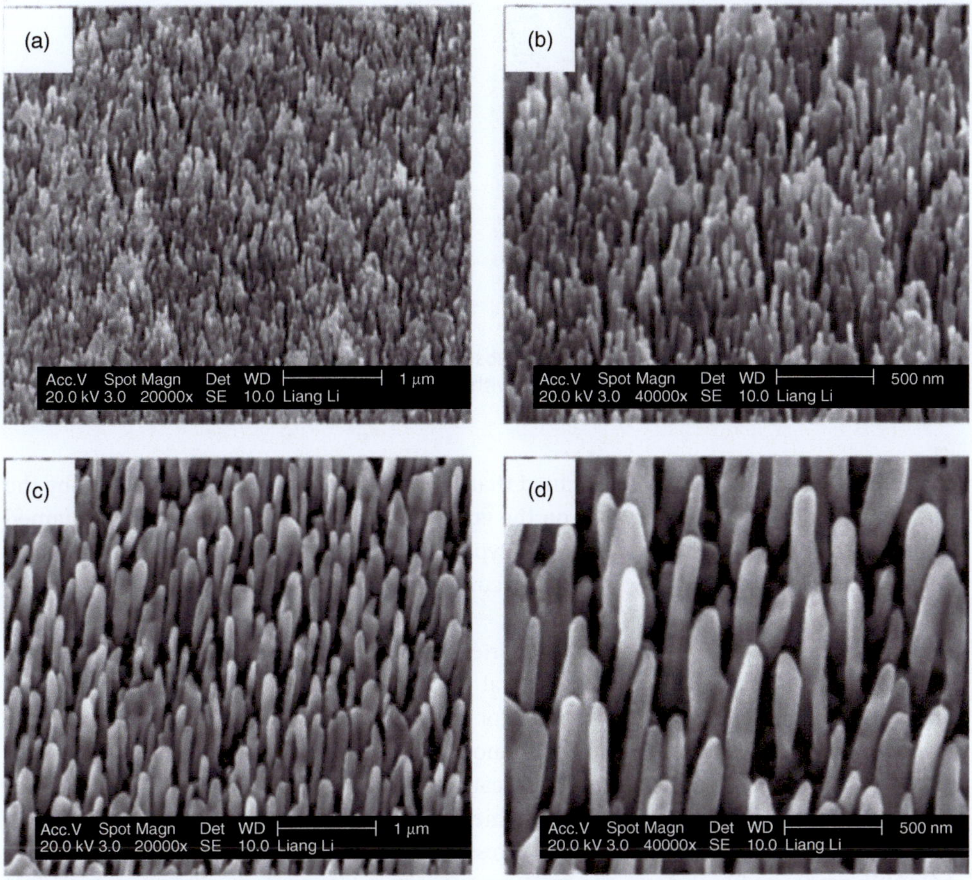


**Figure 3.4** SEM images of the Ge nanostructures deposited at a substrate temperature of (a, b) 230, (c, d) 330 °C with a flux angle of 87°. *Source:* Reprinted with permission from Ref. [57]. Copyright 2008, WILEY-VCH.

in the second case, the diameter of the NWs is slightly thinner as the distance from the Si substrate decreases. The uniformity of the diameter of the NW along the length indicates that there is little or no lateral growth of the NW, which is related to the fact that the diameter of the NW is controlled by the size of the AuGe droplets as it grows and also due to the fact that there is no loss of Au during the growth of the NW. At the same time, Mathur et al. [63] prepared dicyclopentadienylgermanium ([Ge($C_5H_5$)$_2$]) by chemical vapor deposition, which enabled the synthesis of Ge NWs and core–shell structures in high yields. The prepared substrate surface is uniformly covered with a large number of filamentary nanostructures with diameters between 15 and 20 nm and lengths up to tens of micrometers. Most of the NWs are single-crystalline and do not show defects such as layer mismatches and microtwins, which are common along the growth axis. Meanwhile, they found that the characteristic peaks of this kind of NWs are much sharper by Raman spectroscopy test, indicating that the NWs have near-perfect crystallinity. Through HRTEM images of Ge NPs

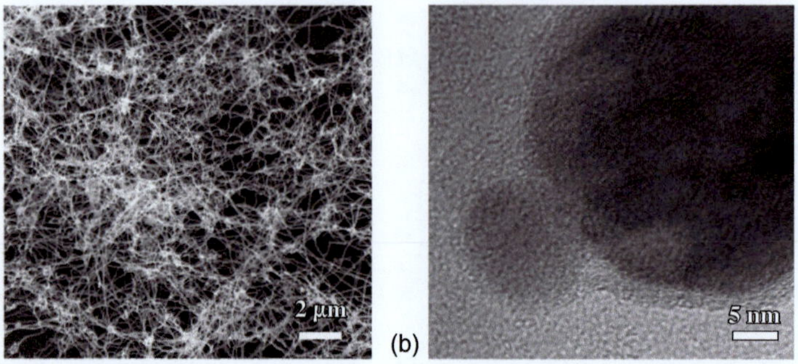


**Figure 3.5** (a) SEM image of Ge NWs on a Fe substrate. (b) HRTEM image of an epitaxially docked Ge NP on the tip of a Ge NW. *Source:* Reprinted with permission from Ref. [63]. Copyright 2004, American Chemical Society.

at the tip of the NW, it can be determined that the growth of the NW is carried out by the epitaxial docking of Ge nanocrystals on the growing NW. In contrast to Ge NWs obtained by the VLS mechanism, the growth of this type of NW does not require substrate preparation, and the size of the Fe-Ge cores in the coating can be easily adjusted by the substrate temperature.

On the other hand, Kim et al. [64] reported a chemical vapor deposition method for the preparation of high-density single-crystal Ge NWs, which is characterized by tunable conductivity and narrow size distribution for system diameter control. They synthesized single-crystal Ge NWs by nucleation on nanocrystalline seeds and subsequent 1D anisotropic growth without the use of an external catalyst. Systematic control of particle diameter and tunable particle concentration were achieved by adjusting growth conditions such as growth temperature and precursor partial pressure ratio. Figure 3.6a shows the SEM image of Ge NWs grown on Si substrate; the optimum temperature range for uniform single-crystal NW growth is narrow, and the diameter of the NWs becomes gradually smaller as the growth temperature increases. Song et al. [65] successfully grew vertical Ge NWs on GaAs(110) substrate by setting the growth temperature to 300 – 380°C, which allowed the Ge epitaxial growth to form NWs. It is worth mentioning that Ge NWs on GaAs substrate grow mainly along the [110] direction, and through this interesting phenomenon, the epitaxial growth of vertical Ge NWs on the GaAs(110) surface was realized. The yield of vertical Ge NWs epitaxially grown on the GaAs(110) surface reaches 40% at 335–365 ° 6, and the yield decreases with increasing growth temperature. Li et al. [66], on the other hand, used a low-pressure chemical vapor deposition method to grow Ge NWs on patterned Au catalysts with a 10% $GeH_4$ precursor (in a hydrogen atmosphere) under a total pressure of 5 Torr at 300 ° o. The SEM images of Ge NWs are shown in Figure 3.6b, and the diameters of the Ge NWs can be selectively grown under the Au catalysts with diameters of 5~20 nm and lengths of 1 μm of Ge NWs. The fluctuation of the NW diameter should originate from the fluctuation of the Au catalyst, and almost all of the NWs are defect-free over the entire length that can be observed. Jung et al. [67] used chemical vapor deposition to grow Ge NWs by a VLS growth mechanism using Au NPs as seeds. They used a thin Ge buffer layer grown at low temperatures below 600 °C to avoid oxidation problems. The yield of vertically oriented Ge NWs was as high as 96.3% using a thin Ge buffer layer with a root-mean-square roughness of ~2 nm.

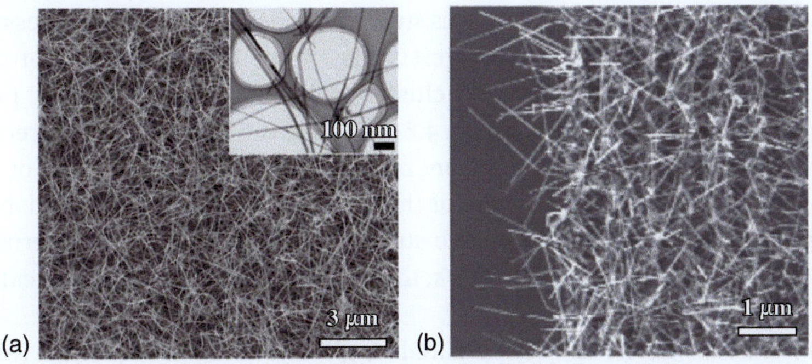


**Figure 3.6** (a) SEM image showing high-density, uniform Ge NWs grown from a planar thin $SiO_x$ film on a silicon surface. *Source:* Reprinted with permission from Ref. [64]. Copyright 2009, American Chemical Society. (b) SEM image of Ge NWs strip grown on Au catalysts strip pattern. *Source:* Reprinted with permission from Ref. [66]. Copyright 2009, The Japan Society of Applied Physics.

### 3.2.5 Hydrothermal Synthesis

Hydrothermal synthesis is also one of the common methods to synthesize 1D nanomaterials, which is favored due to its environmental friendliness and low composition. Pei et al. [68] prepared smooth Ge NWs by a simple hydrothermal deposition method using Ge and $GeO_2$ as raw materials, where Ge powder was put into 48 mL of water in the ratio of 1:1 with $GeO_2$ powder as the substrate, using Cu sheets. The temperature of 350 °C and pressure of 6.4–6.8 MPa were ensured during the growth process and kept under these conditions for 24 h. The SEM images of the NWs are shown in Figure 3.7a, where it can be seen that the NWs are grown very densely on the substrate and have a more uniform diameter of about 150 nm. It is worth mentioning that $CuGeO_3$ played a crucial role in the growth process, which induced the growth of Ge NWs. One year later, Pei et al. [69] prepared single-crystal Ge NWs by a similar method, but the synthesis temperature was reduced to 250 °C. SEM images of Ge NWs are shown in Figure 3.7b, and the products are all NW structures with uniform diameters ranging from 150 to 600 nm, and the lengths

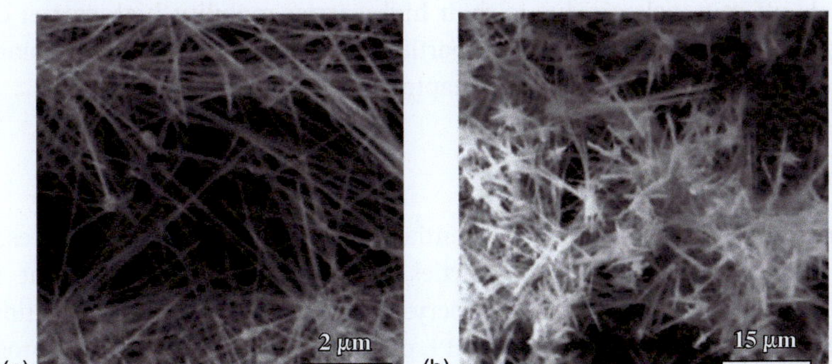


**Figure 3.7** (a) SEM image of the products. *Source:* Reprinted with permission from Ref. [68]. Copyright 2009, Elsevier. (b) SEM image of the Ge NWs. *Source:* Reprinted with permission from Ref. [69]. Copyright 2009, Elsevier.

can reach hundreds of micrometers. In this study, they also found that the pressure affects the morphology of the NWs; when the pressure is 2.1–2.5 MPa, the length of the NWs is longer, and there is a certain number of clusters, which also represents the presence of impurities. Whereas, when the pressure is 4.3~4.7 MPa, the NW length is extremely shortened and a certain number of nanorods are present. Similarly, the presence of nanoscale clusters in the product and the presence of these impurities are also unavoidable in most of the hydrothermal syntheses. The above study shows that using hydrothermal deposition method for the preparation of Ge NWs, the appropriate pressure has a great influence on the morphology of the NWs.

On the other hand, Lin et al. [70] prepared single-crystal Ge NW arrays without metal catalysts and templates. They chose Ge powder as the raw material, heated it to 450 °C in a reactor, ensured the pressure range of 12–14 MPa, and maintained the growth under stirrer agitation for 6 h. Figure 3.8 demonstrates the SEM image of the synthesized Ge NWs, from which we can see that these NWs are uniformly distributed with diameters of about 10–150 nm and lengths of up to tens of micrometers. High magnification through a microscope shows that the NWs have a uniform diameter, smooth surface, and no metal catalyst particles at the head of the line. These NWs grow in the same direction and remain well-oriented, even after ultrasonic agitation. An electromagnet drove the stirrer used in this experiment, so the environment for the growth of NWs was covered by a magnetic field. The nanoclusters on the growth tip mentioned above are electrodeposited due to the formation of polar groups by absorbing $H^+$ or $OH^-$ during the concentration process. Moreover, under the influence of the magnetic field, the polar groups will be neatly aligned, thus ultimately forming NW arrays.

## 3.3 Properties and Applications of Germanium Nanomaterials

Materials at the nanoscale, whose dimensions are close to the coherence length of an electron, will have properties that are dramatically different from those of a bulk. Ge nanomaterials are considered the most promising materials for fabricating photovoltaic nanodevices and NW field effect transistors due to their high-carrier mobility, high-carrier concentration, and faster switching speeds. The properties and applications of Ge nanomaterials will be further discussed and studied in this chapter.

### 3.3.1 Optical Properties

Germanium nanomaterials have great potential for application in optical devices, and studying their photoluminescence (PL) properties becomes particularly important. Figure 3.9 shows the PL spectra of germanium nanocrystals under 488-nm laser excitation at room temperature and the excitation spectra of PL peaks. Ge nanocrystals have a broad peak at 2.2–2.3 eV. It can be found by Kanemitsu [71] that the intensity of the PL characteristic peak of Ge nanocrystals is independent of the diameter of the crystal and is only proportional to the volume of Ge.

**Figure 3.8** The typical field-emission scanning electron microscopy images of Ge NWs arrays ranging from low to higher magnification. (a) 1000, (b) 3000, (c) 5000, (d) 10 000, (e) 50 000, (f) 50 000, (g) 100 000, (h) 100 000. *Source:* Reprinted with permission from Ref. [70]. Copyright 2010, The Royal Society of Chemistry.

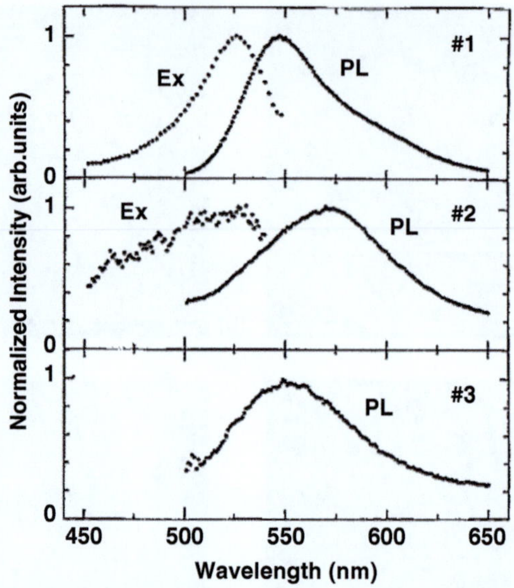


**Figure 3.9** PL spectra under 488 nm laser excitation and excitation spectra at the peak of the PL spectrum in three different samples. *Source:* Reprinted with permission from Ref. [71]. Copyright 1992, American Institute of Physics.

The lifetime τ of the PL is not linked to the photon energy and is not affected by the temperature in the temperature range of 10–300 K. The PL peaks of Ge nanocrystals have a broad peak at 2.2–2.3 eV. The fact that τ is much smaller in nanomaterials than in bulk implies that the PL decay rate is significantly higher in germanium nanomaterials. This also corroborates that bulk Ge with an indirect bandgap gradually has a direct optical bandgap as the size decreases.

Figure 3.10 demonstrates the extinction spectra of several different Ge materials. curve 1 shows the Mie theory spectrum, curve 2 shows the extinction spectrum of Ge microcrystals prepared by Hayashi et al. [72], curve 3 shows the extinction spectrum of bulk Ge, curve 4 shows the extinction spectrum of Ge nanocrystals prepared by Wilcoxon et al. [73], and curve 5 shows the extinction spectrum of Ge quantum dots prepared by Heath et al. [74]. The figure shows that the bulk Ge has a distinct peak around 300 nm and a shoulder at 550 nm, which is related to the direct band transition at Γ. As for small-sized nanocrystals, on the other hand, their extinction spectra generally have three peaks: a peak at ~360 nm, a shoulder peak at 300 nm, as well as a third peak at ~280 nm. From the shape of the curves alone, curves 4 and 5 are closer to the extinction spectra of the bulk, which is different from the data obtained from the theoretical spectral curve 1, which also implies that the effect of scattering is small.

### 3.3.2 Raman Spectrum

In order to further study the crystal structure and molecular interactions of Ge, Raman spectrum [75, 76] testing of Ge is necessary. Raman spectroscopic studies of Ge were carried

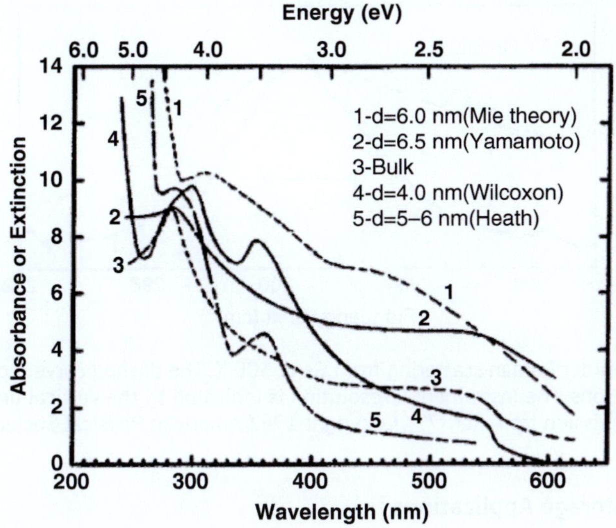


**Figure 3.10** Comparison of the extinction spectrum. *Source:* Reprinted with permission from Ref. [72–74].

out as early as 1967 by Parker et al. [77]. The first-order Raman curve of Ge is illustrated in Figure 3.11, where the dashed line represents Cowley's theoretical calculations. It can be seen that although there is a 2% difference between the actual measured peak positions and those obtained from the theoretical calculations, they are also basically in the vicinity of 300 $cm^{-1}$. The incomplete agreement between the experimental data and the theoretical calculations may also be due to Cowley's theory's overestimation of the incommensurability and the fact that impurities did not affect the test results. Kolobov et al. [78] further investigated the Raman spectra of Ge. The intensity of the laser is 50 mW, and the wavelength is 633 nm. There are three distinct characteristic peaks of 229 $cm^{-1}$, 300 $cm^{-1}$, and 435 $cm^{-1}$. Only the broad peak located at 300 $cm^{-1}$ is produced by Ge. The other peaks are from the substrate. The peak located at 490 $cm^{-1}$ is believed to be caused by Si-Si vibrations in the region of the Ge alloy, and its frequency decreases as the Ge content in the alloy increases.

On the other hand, Li et al. [57] have carried out a detailed study of Raman spectra of Ge NWs and single-crystal bulk Ge, and the results are shown in Figure 3.12a. The figure shows that as the growth temperature increases, the Raman spectral signals of Ge NWs are better and the peaks are sharper, implying that the Ge NWs are better crystallized. A sharp Ge-Ge peak at 300 $cm^{-1}$ appears in the sample above 230 °a. In contrast, the amorphous band with an extensive half-peak width appeared in the NWs grown below 130 °, indicating that the low temperature does not allow good completion of crystallization of Ge. The scattering peaks of all the NWs show different degrees of leftward shifts compared to the bulk single-crystal Ge due to the effect of phonon quantum confinement and stress. Figure 3.12b demonstrates the Raman spectra of Ge NWs after annealing at different temperatures. It can be seen that the half-peak width of the sample annealed below 550 °C does not change, which also indicates that the grown NWs are well-crystallized single crystals and that the annealing process does not have an effect on the crystallinity of Ge NWs.

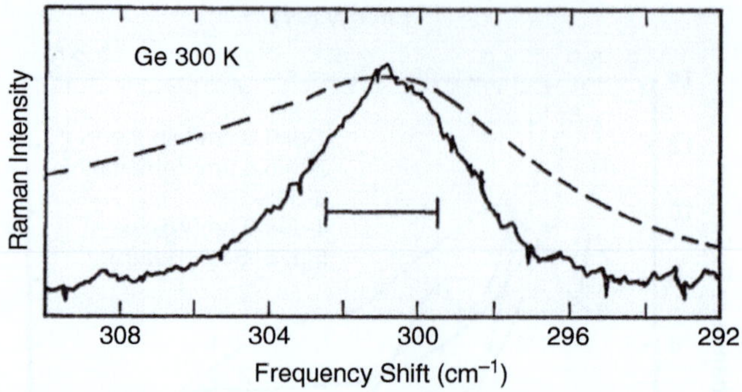


**Figure 3.11** First-order Raman scattering from Ge at 300 K. The dashed curve represents Cowley's theoretical calculations. The instrumental resolution is indicated by the vertical lines. *Source:* Reprinted with permission from Ref. [77]. Copyright 1967, American Physical Society.

### 3.3.3 Energy Storage Applications

Ge nanomaterials are widely used in energy storage [79, 80] due to their unique physical and chemical properties. Ke et al. [81] used a simple single-step synthesis method to prepare porous Ge NPs that can be filled with electrolyte liquids to facilitate lithium–ion diffusion. Figure 3.13 illustrates the electrochemical test results of Ge NPs synthesized at 450 °C and 600 °C. Figure 3.13a shows the specific capacity of porous germanium electrodes versus the number of cycles at 100 and 800 $mA \cdot g^{-1}$ with alloying capacities of 1300 and 1100 $mAh \cdot g^{-1}$, respectively, and the capacity is stable up to 40 cycles. Zhang et al. [82] prepared Ge microcubes on titanium foil using a simple hydrogen reduction method. Figure 3.13b shows a representative charge-discharge curve of Ge microcubes at 0.1 C. The discharge curve of the first cycle shows a plateau at 0.3 V due to the formation of the Ge-Li alloy, while two decaying plateaus below 0.7 V can be

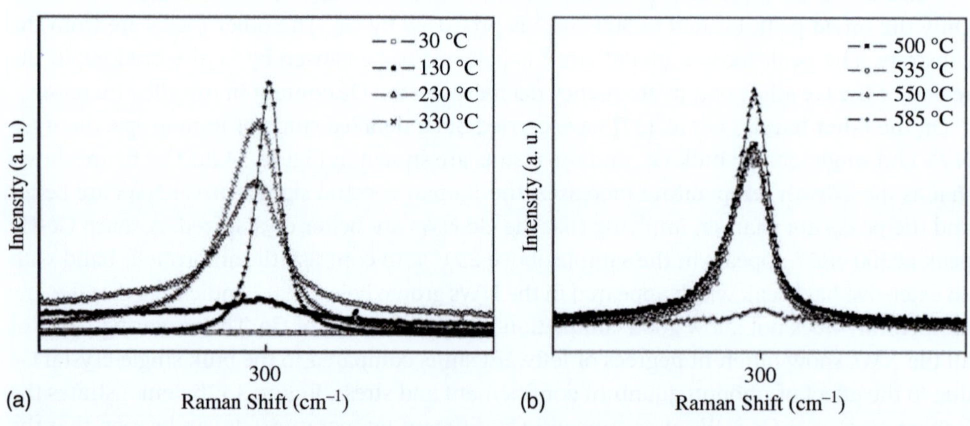


**Figure 3.12** (a) Raman spectra of the Ge nanostructures deposited at different substrate temperatures. The black pentacle curve is the spectrum of single-crystalline bulk Ge. (b) Raman spectra of Ge NWs annealed at different temperatures. Black triangles show the spectrum of single-crystalline bulk Ge. *Source:* Reprinted with permission from Ref. [57]. Copyright 2008, WILEY.

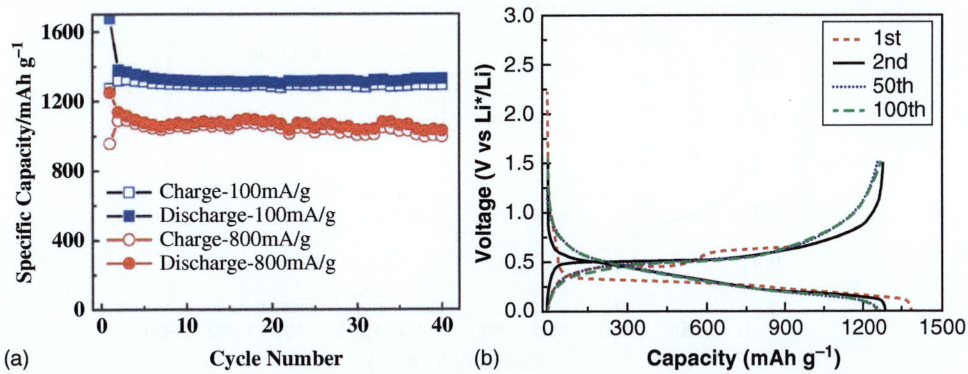


**Figure 3.13** (a) Specific capacity versus cycle number for nano/microstructure Ge electrodes under 100 mA • g$^{-1}$ and 800 mA • g$^{-1}$. *Source:* Reprinted with permission from Ref. [81]. Copyright 2014, The Royal Society of Chemistry. (b) Voltage profiles of the 1st, 2nd, 50th, and 100th cycles of the Ge microcubes in the voltage range of 0.005–1.5V versus $Li^+$/Li at 0.1 C. *Source:* Reprinted with permission from Ref. [82]. Copyright 2015, American Chemical Society.

observed in the first charge curve. In the first cycle, the Ge microcube has a discharge capacity of 1388 mAh · g$^{-1}$ and a charge capacity of 1275 mAh · g$^{-1}$. This corresponds to an initial coulombic efficiency of 91.8%, which is an excellent result. Due to the layered structure of Ge, the liquid electrolyte can fully penetrate, and all the Ge NPs synergistically participate in the electrochemical reaction, thus storing a large amount of lithium.

Moreover, due to the high crystallinity of Ge microcubes, this also ensures the reversible precipitation of lithium, resulting in an extremely high initial coulombic efficiency. The charge–discharge curve of the second cycle matches well with that of the subsequent cycles, showing highly reversible lithium insertion and extraction behavior. When fully discharged, the Ge anode provides a reversible 1250 mAh · g$^{-1}$ for up to 200 cycles, with 99% capacity retention at 0.1 C. The Ge anode has a capacity retention rate of 99% at 0.1 C. The Ge anode has a capacity retention rate of 99% at 0.1 C. It is worth mentioning that SEM images confirmed that the Ge microcubes remained morphologically intact without the addition of other substances.

The synthesis of zero-dimensional (0D) hollow and three-dimensional (3D) porous Ge NPs using the template method was first reported by Park et al. [83]. They obtain either 0D hollow NPs or 3D porous Ge NP assemblies by adjusting the weight fraction of the precursor relative to $SiO_2$. Figure 3.14a shows the voltage distribution between 0 and 1.2 V at 0.5 C rate for 0D and 3D samples with initial Coulomb efficiencies of 85 and 90% in the first test cycle. Four characteristic peaks of 0.15, 0.3, and 0.4 V are present in the inset, which match those of $Li_xGe$. The first discharge capacities of the 0D and 3D samples are 1428 and 1380 mAh · g$^{-1}$, respectively. There is only a 5% difference in the initial discharge point efficiencies of the two, despite the 0D Ge NPs having a much higher BET surface area. Figure 3.14b shows the life-cycle performance of these samples for 100 cycles at 1 C. The figure shows that the volumes of the 0D and 3D hollow samples after 100 cycles are 1162 and 1415 mAh · g$^{-1}$, which correspond to 85 and 95% initial Coulomb efficiencies,

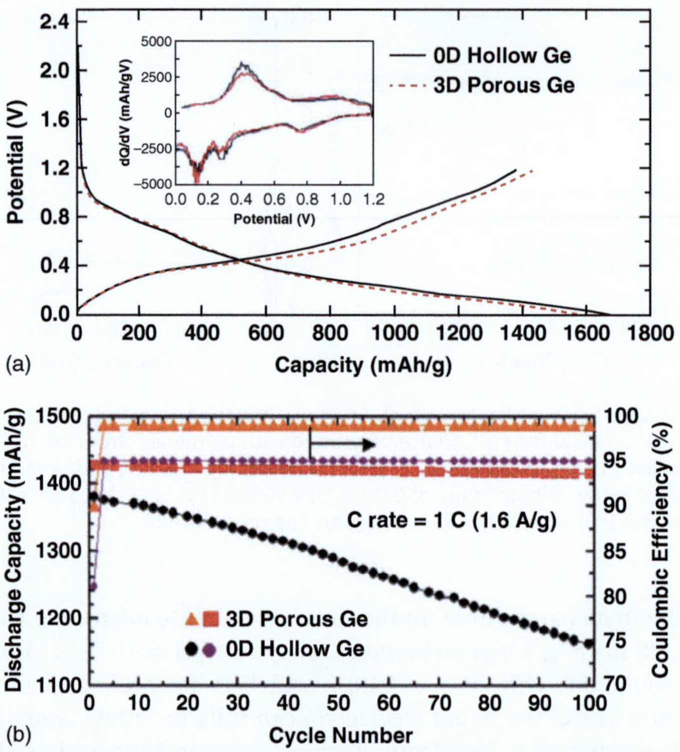


**Figure 3.14** (a) First voltage profiles of 0D hollow and 3D porous NP assemblies between 0 and 1.2 V in coin-type half cells. The rate was 0.5 C. An inset shows the differential curves of the first voltage profiles. (b) Cycle number versus discharge capacity (lithium extraction) for both samples at a rate of 1 C in coin-type half cells. *Source:* Reprinted with permission from Ref. [83]. Copyright 2010, WILEY.

respectively. The 0D hollow samples may continue to crack and fragment due to the change in volume during cycling. This cracking and fragmentation of the 0D hollow samples during cycling may continue to produce new active surfaces previously passivated by the stable surface film, which could also decrease the total capacity.

Not only can Ge NPs be used for energy storage applications, but 1D Ge nanomaterials [84–87] can also be used for energy storage. Gu et al. [88] prepared Ge NW thin films using electrodeposition and measured them to have a very high $Li^+$ charge/discharge capacity. Figure 3.15a shows the 1st, 2nd, 3rd, and 26th charge/discharge curves recorded for electrodeposited Ge NW film electrodes at 1 C rate. The significant decrease in charging capacity after the first cycle is consistent with the known formation of irreversible films at the solid-electrolyte interface of Li-ion battery electrodes. Figure 3.15b summarizes the recorded cycling data for the 26-cycle sequence. After the second charge–discharge cycle, the Coulombic efficiency was greater than 91% for each cycle. The discharge capacity decreases slowly throughout the cycling sequence, dropping less than 25% from the first discharge to the twentieth discharge, and is consistent with the Li alloy and volume expansion of the group IV $Li^+$ battery electrodes. The results of this study indicate that

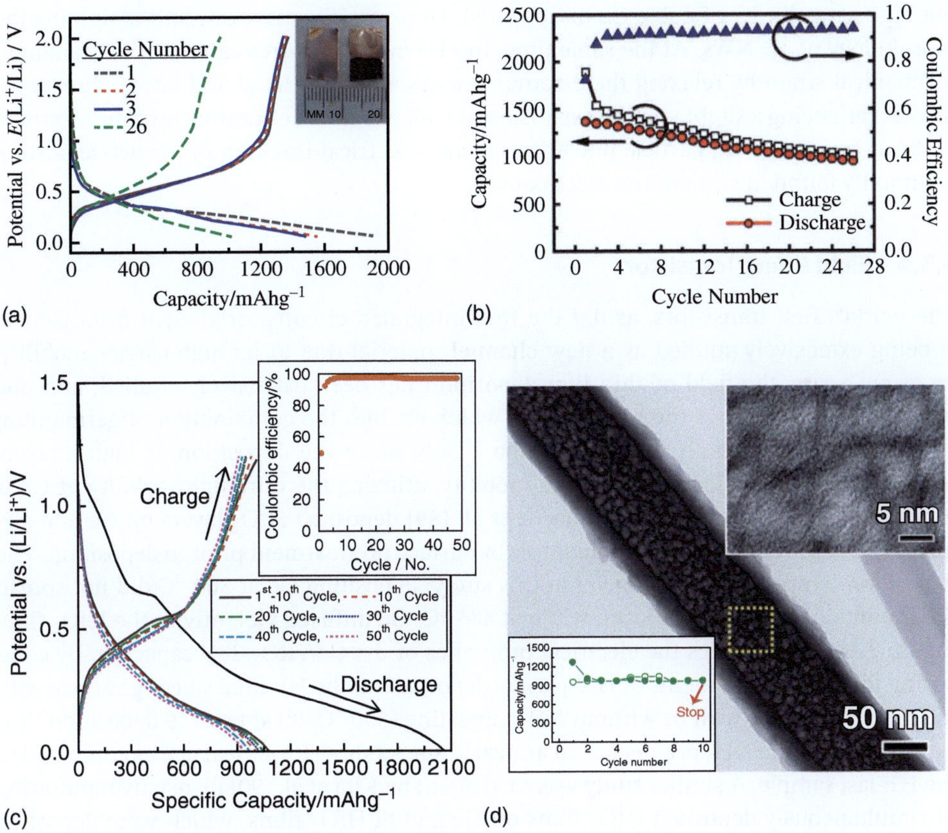


**Figure 3.15** (a) First, second, third, and twenty-sixth charge–discharge curves for a $Li^+$ anode recorded at 1 C rate. *Source:* Reprinted with permission from Ref. [88]. Copyright 2012, American Chemical Society. (b) Galvanostatic $Li^+$ (open squares) charge and (red circles) discharge cycling at 1 C rate. (c) Charging–discharging curves of the Ge NW electrode. Inset shows corresponding coulombic efficiency. (d) TEM image of a fully delithiated individual Ge NW electrode (taken after 10 cycles). *Source:* Reprinted with permission from Ref. [56]. Copyright 2011, The Royal Society of Chemistry.

the activity of the prepared Ge NW thin-film electrodes is competitive with existing alternatives for producing nanostructured Ge battery anodes.

On the other hand, Ko et al. [56] prepared core–shell structured Ge NWs using the Ge/Sn co-evaporation method, all encapsulated by an amorphous germanium oxide ($GeO_x$) layer on the surface. Figure 3.15c demonstrates the electrochemical properties of this Ge NW, with the cell cycled at a high rate of 1C at constant current. A high specific capacity of 2065 $mAh \cdot g^{-1}$ was obtained in the first discharge reaction. However, the charge/discharge capacity stabilized in the subsequent cycles. We observed a reversible capacity of more than 900 $mAh \cdot g^{-1}$ after 50 cycles, which is about 2.4 times higher than that of the carbon-based anode. It is worth mentioning that the material not only has a high reversible capacity, but the stability of the Ge NWs from the 2nd to the 50th cycle is also impressive. The TEM image of Ge NWs after ten cycles at 1 C is shown in Figure 3.15d, where the high crystallinity of cubic Ge disappeared. However, the electrode still maintains the morphology of 1D NWs. At the same time, the 1D nanostructures can effectively adapt to mechanical strain by relaxing the external stresses along the axial and lateral directions, thus experiencing a slight volume expansion, which ensures the continuity of the electrical pathway by preventing particle pulverization and electrical isolation of the active portion commonly found in germanium electrodes.

### 3.3.4 Field Effect Transistor

The world's first transistors, as did the first integrated circuits, originated from Ge. Ge is being extensively studied as a new channel material due to its high-carrier mobility. In recent years, the field of thin-film deposition has been extensively studied, and one of the brighter results is the high-k layer, which enables the combination of germanium with high-k dielectrics. If we look at high-k only, since the deposition of high-k layers on Si leads to a significant decrease in mobility, utilizing the ultrahigh mobility of Ge is expected to solve this problem. Elshocht et al. [89] deposited $HfO_2$ layers on Ge and significantly improved the surface roughness by surface pretreatment prior to deposition. The deposition temperature was crucial in this study; depositing Ge at 300 °C did not spread the mountain in the high-K layer, while at 485 °C, Ge diffused violently in the $HfO_2$ film. Figure 3.16 demonstrates the electrical properties of the Ge/$HfO_2$/TiN capacitor. By comparing the I-V measurements of $HfO_2$ layers deposited on the HF final starting surface with the same thickness, with or without $NH_3$ annealing (550 °C, 30 s) prior to deposition, it is easy to see that the $NH_3$ pre-deposited annealed stacks show lower leakage as compared to the HF-last sample. A similar study was carried out by Kita et al. [90] by reactive sputtering of simultaneously deposited $HfO_2$ films on Ge and Si $HfO_2$ films, which were deposited on Ge and Si substrates simultaneously by reactive sputtering. It was found that both the interfacial layer and the $HfO_2$ film on the Ge substrate were thinner than those on the Si substrate. The $HfO_2$/Ge system is superior to the $HfO_2$/Si system in terms of further reducing the thickness of the gate oxide in the dielectric region of the ultrathin high-k gate.

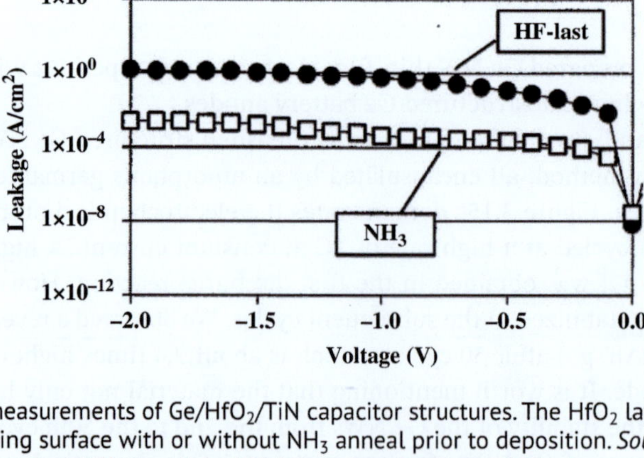


**Figure 3.16** I-V measurements of Ge/$HfO_2$/TiN capacitor structures. The $HfO_2$ layer was deposited on an HF-last starting surface with or without $NH_3$ anneal prior to deposition. *Source:* Reprinted with permission from Ref. [89]. Copyright 2004, American Institute of Physics.

In addition to being used as a substrate to realize the bonding of high-k dielectrics, Ge nanomaterials can also be directly applied to field effect transistors. Xiang et al. [91] prepared a Ge/Si core/shell NW heterostructure using a top-gate geometry. The output versus transfer curves of the device with a channel length of 1 μm are shown in Figure 3.17a,b. The drain current $I_d$ in the I-V curve first increases and then saturates with an unfavorable increase in drain voltage, similar to a conventional long-channel metal-oxide-semiconductor field effect transistor (MOSFET). The fact that $I_d$ increases when $V_g$ decreases from 1 to 2 V also shows that this is a p-type depletion field effect tube. Figure 3.17b confirms the above conclusion that, in the absence of a gate, the Fermi energy level lies below the edge of the Ge valence band. At the same time, the drain bias of the transfer curve is −1 V, indicating that this NW FET has a peak transconductance with a $g_m$ of about 26 μs. The device shows a drain current of up to 35 μA at $V_g = -2$ V, which is superior to the optimum performance of semiconductor NW FETs at that time. When the channel length was reduced to 190 nm, the $g_m$ and $I_{ON}$ of the device reached 3.3 mSμm$^{-1}$ and 2.1 mAμm$^{-1}$ due to the reduction of the channel resistance, which was already three to four times the performance

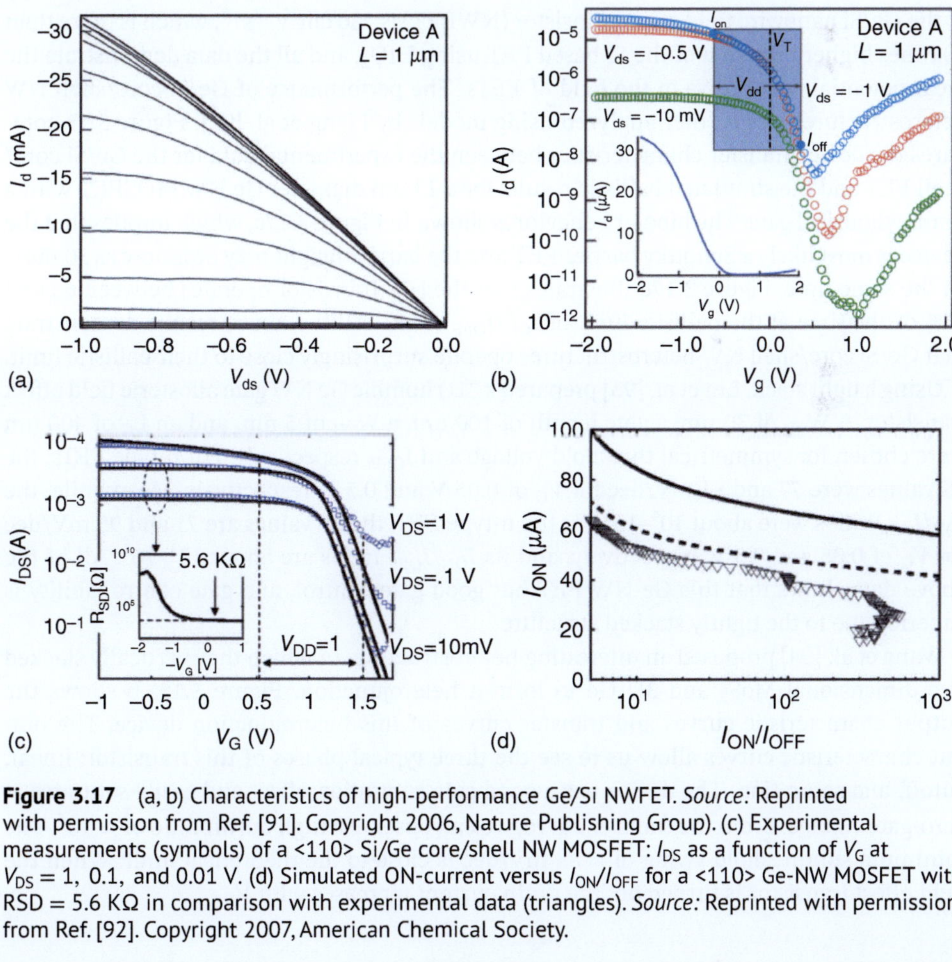


**Figure 3.17** (a, b) Characteristics of high-performance Ge/Si NWFET. *Source:* Reprinted with permission from Ref. [91]. Copyright 2006, Nature Publishing Group). (c) Experimental measurements (symbols) of a <110> Si/Ge core/shell NW MOSFET: $I_{DS}$ as a function of $V_G$ at $V_{DS}$ = 1, 0.1, and 0.01 V. (d) Simulated ON-current versus $I_{ON}/I_{OFF}$ for a <110> Ge-NW MOSFET with RSD = 5.6 KΩ in comparison with experimental data (triangles). *Source:* Reprinted with permission from Ref. [92]. Copyright 2007, American Chemical Society.

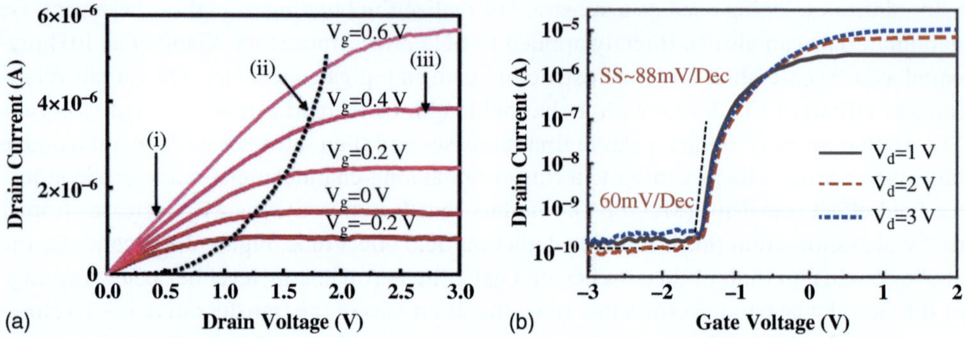


**Figure 3.18** (a) Drain current–drain voltage output characteristics of the junction field effect transistor (JFET) at different gate voltages. (b) Drain current–gate voltage transfer characteristics of the JFET at different drain voltages. *Source:* Reprinted with permission from Ref. [94]. Copyright 2021, WILEY.

of the state-of-the-art MOSFETs of the time. It is worth mentioning that the hole mobility of the Ge/Si nanowire field effect transistor (NWFET) is 730 $cm^2V^{-1}s^{-1}$, which is more than 10 times higher than that of the Si-based FET using $HfO_2$, and all the data demonstrate the great potential of Ge NWs in the field of FETs. The performance of Ge/Si core/shell NW heterostructures was further analyzed using models by Liang et al. [92]. Figure 3.17 compares the device transfer characteristics between the experimental data for the Ge/Si core/shell FET and the simulated ballistic results for a 13 nm diameter Ge NW MOSFET with a semi-cylindrical gate. The bipolar behavior is shown in Figure 3.17c, which implies that the device is most likely a Schottky barrier FET and the barrier height may be as low as 30 meV. At the same time, Figure 3.17d illustrates that the NW transistor operates between 60 and 85% of the time at the ballistic limit of $I_{ON}/I_{OFF} = 100$. All the above results demonstrate that Ge/Si core/shell NW heterostructures operate surprisingly close to their ballistic limit.

Using a tight stack, Lin et al. [93] prepared a 3D rhombic Ge NW gate allosteric field effect transistor. A $W_{CH}$ of 30 nm, a gate length of 100 nm, a $W_{CH}$ of 5 nm, and an $L_G$ of 400 nm were chosen for symmetrical threshold voltage and $I_{OFF}$, respectively. For p-type FETs, the SS values were 77 and 81 mV/dec for $V_G$ of 0.05 V and 0.5 V, respectively. Meanwhile, the $I_{ON}/I_{OFF}$ ratios were about $10^4$–$10^5$ For the n-type FET, the SS values are 71 and 95 mV/dec for $V_G$ of 0.05 and 0.5 V, respectively, and its $I_{ON}/I_{OFF}$ ratios are about $10^3$–$10^4$. All of the above data shows that this Ge-NW FET has good gate control, and gate controllability is superior due to the tightly stacked structure.

Wang et al. [94] proposed an interesting heterostructure, in which they vertically stacked two-dimensional $MoS_2$ and 3D Ge to form a heterojunction. Figure 3.18a,b shows the output characteristic curves and transfer curves of this heterojunction device. The output characteristic curves allow us to see the three typical phases of this transistor: linear, cutoff, and saturation. The device has a good gate saturation effect and easily saturates at zero-gate voltage. From the transfer curve, we can observe a high on/off ratio of $\approx 10^5$. The minimum subthreshold swing of $\approx$ 88 mV/dec is close to the theoretical limit. When the field effect transistor is turned on, the drain current increases with $V_d$.

## 3.4 Conclusion

The research on Ge still needs to be deepened, and the search for new properties and potential applications of Ge is still an essential aspect of the research. Significant breakthroughs have been made in synthesizing various Ge nanomaterials, and many NPs and NWs can be synthesized simply by conventional methods. However, highly crystalline, defect-free Ge nanomaterials are still necessary for their application in optoelectronic detection, integrated circuits, etc. This requires researchers to improve the quality of nanomaterials without the need for costly equipment and overly complicated methods. Doping is one of the important methods to improve the properties of semiconductor materials, and the physical properties of Ge NWs can be enhanced to some extent by doping with different elements. In recent years, the application of Ge NWs has mainly focused on the basic research of field effect tubes and lithium-ion batteries. Ge nanomaterials urgently need new research directions, such as photodiodes and modulators, and this research will significantly open up the scope of application of Ge nanomaterials, which is of great significance.

## References


1 Zheng, K., Benedetti, M.F., van Hullebusch, E.D. Recovery technologies for indium, gallium, and germanium from end-of-life products (electronic waste) – a review. *J. Environ. Manage.* 2023, 347: 119043.

2 Nishino, R., Tokitoh, N., Sasayama, R., Waterman, R., Mizuhata, Y. Unusual nuclear exchange within a germanium-containing aromatic ring that results in germanium atom transfer. *Nat. Commun.* 2023, 14 (1): 4519.

3 Inzani, G., Adamska, L., Eskandari-asl, A., Di Palo, N., Dolso, G.L., Moio, B., D'Onofrio, L.J., Lamperti, A., Molle, A., Borrego-Varillas, R., Nisoli, M. Field-driven attosecond charge dynamics in germanium. *Nat. Photonics* 2023, 17 (12): 1059–1065.

4 Wang, D., Zhai, C., Chen, Y., He, Y., Chen, X.D., Wang, S., Zhao, L., Frenking, G., Wang, X., Tan, G. An isolable germylyne radical with a one-coordinate germanium atom. *Nat. Chem.* 2022, 15 (2): 200–205.

5 Yang, Q., Xiao, J., Robredo, I., Vergniory, M.G., Yan, B., Felser, C. Monopole-like orbital-momentum locking and the induced orbital transport in topological chiral semimetals. *Proc. Natl. Acad. Sci. U S A* 2023, 120 (48): e2305541120.

6 Bharti, A., Dixit, G. Role of topological charges in the nonlinear optical response from Weyl semimetals. *Phys. Rev. B* 2023, 107 (22): 224308.

7 Guo, C., Asadchy, V.S., Zhao, B., Fan, S. Light control with Weyl semimetals. *eLight* 2023, 3 (1): 2.

8 Liu, J.W., Shi, F.L., Shen, K., Chen, X.D., Chen, K., Chen, W.J., Dong, J.W. Antichiral surface states in time-reversal-invariant photonic semimetals. *Nat. Commun.* 2023, 14 (1): 2027.

9 Bharti, A., Ivanov, M., Dixit, G. How massless are Weyl fermions in Weyl semimetals. *Phys. Rev. B* 2023, 108 (2): L020305.

**10** Borrelli, N.F., Hall, D.W., Holland, H.J., Smith, D.W. Quantum confinement effects of semiconducting microcrystallites in glass. *J. Appl. Phys.* 1987, 61 (12): 5399–5409.

**11** Favron, A., Gaufrès, E., Fossard, F., Phaneuf-L'Heureux, A.L., Tang, N.Y., Lévesque, P.L., Loiseau, A., Leonelli, R., Francoeur, S., Martel, R. Photooxidation and quantum confinement effects in exfoliated black phosphorus. *Nat. Mater.* 2015, 14 (8): 826–832.

**12** Bostedt, C., van Buuren, T., Willey, T.M., Franco, N., Terminello, L.J., Heske, C., Möller, T. Strong quantum-confinement effects in the conduction band of germanium nanocrystals. *Appl. Phys. Lett.* 2004, 84 (20): 4056–4058.

**13** Amara, M.-R., Said, Z., Huo, C., Pierret, A., Voisin, C., Gao, W., Xiong, Q., Diederichs, C. Spectral fingerprint of quantum confinement in single $CsPbBr_3$ nanocrystals. *Nano Lett.* 2023, 23 (8): 3607–3613.

**14** Sun, G., Tai, Z., Li, F., Ye, Q., Wang, T., Fang, Z., Jia, L., Liu, W., Wang, H. Construction of $ZnIn_2S_4$/CdS/PdS S-scheme heterostructure for efficient photocatalytic $H^2$ production. *Small.* 2023, 19 (27): 2207758.

**15** Gao, T., Li, Q.H., Wang, T.H. CdS nanobelts as photoconductors. *Appl. Phys. Lett.* 2005, 86 (17): 173105.

**16** Olsson, L.Ö., Andersson, C.B.M., Håkansson, M.C., Kanski, J., Ilver, L., Karlsson, U.O. Charge accumulation at InAs surfaces. *Phys. Rev. Lett.* 1996, 76 (19): 3626–3629.

**17** Yoon, J.I., Kim, H., Kim, M., Cho, H., Kwon, Y.A., Choi, M., Park, S., Kim, T., Lee, S., Jo, H., Kim, B. P- and N-type InAs nanocrystals with innately controlled semiconductor polarity. *Sci. Adv.* 2023, 9: eadj8276.

**18** Schulte, K.L., Johnston, S.W., Braun, A.K., Boyer, J.T., Neumann, A.N., McMahon, W.E., Young, M., Coll, P.G., Bertoni, M.I., Warren, E.L., Steiner, M.A. GaAs solar cells grown on acoustically spalled GaAs substrates with 27% efficiency. *Joule* 2023, 7 (7): 1529–1542.

**19** Granger, G., Bailly, M., Delahaye, H., Jimenez, C., Tiliouine, I., Leventoux, Y., Orlianges, J.C., Couderc, V., Gérard, B., Becheker, R., Idlahcen, S. GaAs-chip-based mid-infrared supercontinuum generation. *Light Sci. Appl.* 2023, 12 (1): 252.

**20** Mamiyev, Z., Balayeva, N.O. PbS nanostructures: a review of recent advances. *Mater. Today Sustain.* 2023, 21: 100305.

**21** Corato-Zanarella, M., Gil-Molina, A., Ji, X., Shin, M.C., Mohanty, A., Lipson, M. Widely tunable and narrow-linewidth chip-scale lasers from near-ultraviolet to near-infrared wavelengths. *Nat. Photonics* 2022, 17 (2): 157–164.

**22** Orozco-Gonzalez, Y., Kabir, M.P., Gozem, S. Electrostatic spectral tuning maps for biological chromophores. *J. Phys. Chem. B* 2019, 123 (23): 4813–4824.

**23** Gonzalez, A., Servin, M., Estrada, J.C., Quiroga, J.A. Design of phase-shifting algorithms by finetuning spectral shaping. *Opt. Express* 2011, 19 (11): 10692–10697.

**24** Zheng, J., Petty, J.T., Dickson, R.M. High quantum yield blue emission from water-soluble $Au_8$ nanodots. *J. Am. Chem. Soc.* 2003, 125: 7780–7781.

**25** Antaris, A.L., Chen, H., Diao, S., Ma, Z., Zhang, Z., Zhu, S., Wang, J., Lozano, A.X., Fan, Q., Chew, L., Zhu, M. A high quantum yield molecule-protein complex fluorophore for near-infrared II imaging. *Nat. Commun.* 2017, 8 (1): 15269.

**26** Dong, Y., Pang, H., Yang, H. B., Guo, C., Shao, J., Chi, Y., Li, C.M., Yu, T. Carbon-based dots co-doped with nitrogen and sulfur for high quantum yield and excitation-independent emission. *Angew. Chem. Int. Ed. Engl.* 2013, 52 (30): 7800–7804.

**27** Becker, W. Fluorescence lifetime imaging—techniques and applications. *J. Microsc.* 2012, 247 (2): 119–136.

**28** Sørensen, T.J., Thyrhaug, E., Szabelski, M., Luchowski, R., Gryczynski, I., Gryczynski, Z., Laursen, B.W. Azadioxatriangulenium: a long fluorescence lifetime fluorophore for large biomolecule binding assay. *Methods Appl. Fluoresc.* 2013, 1 (2): 025001.

**29** Berezin, M.Y., Achilefu, S. Fluorescence lifetime measurements and biological imaging. *Chem. Rev.* 2010, 110: 2641–2684.

**30** Marin, E., Boschetto, F., Pezzotti, G. Biomaterials and biocompatibility: an historical overview. *J. Biomed. Mater. Res. Part A* 2020, 108 (8): 1617–1633.

**31** Amanchukwu, C.V., Yu, Z., Kong, X., Qin, J., Cui, Y., Bao, Z. A new class of ionically conducting fluorinated ether electrolytes with high electrochemical stability. *J. Am. Chem. Soc.* 2020, 142 (16): 7393–7403.

**32** Schwietert, T.K., Arszelewska, V.A., Wang, C., Yu, C., Vasileiadis, A., de Klerk, N.J., Hageman, J., Hupfer, T., Kerkamm, I., Xu, Y., van der Maas, E. Clarifying the relationship between redox activity and electrochemical stability in solid electrolytes. *Nat. Mater.* 2020, 19 (4): 428–435.

**33** Hou, J.B., Zhang, K., Xiao, J.H., Xu, Z.Q., Gao, W.J., Gao, X.Y., Zhou, S.K., Jiao, Z.Z., Yi, M.R., Yin, Y.H., Wu, Z.P. Modified tungsten oxide as a binder-free anode in lithium-ion battery for improving electrochemical stability. *Tungsten* 2022, 4 (4): 356–369.

**34** Raiford, J.A., Oyakhire, S.T., Bent, S.F. Applications of atomic layer deposition and chemical vapor deposition for perovskite solar cells. *Energy Environ. Sci.* 2020, 13 (7): 1997–2023.

**35** Yi, K., Liu, D., Chen, X., Yang, J., Wei, D., Liu, Y., Wei, D. Plasma-enhanced chemical vapor deposition of two-dimensional materials for applications. *Acc. Chem. Res.* 2021, 54 (4): 1011–1022.

**36** Hou, P.X., Zhang, F., Zhang, L., Liu, C., Cheng, H.M. Synthesis of carbon nanotubes by floating catalyst chemical vapor deposition and their applications. *Adv. Funct. Mater.* 2021, 32 (11): 2108541.

**37** Yogesh, G.K., Shukla, S., Sastikumar, D., Koinkar, P. Progress in pulsed laser ablation in liquid (PLAL) technique for the synthesis of carbon nanomaterials: a review. *Appl. Phys. A* 2021, 127 (11): 810.

**38** Doble, P.A., de Vega, R.G., Bishop, D.P., Hare, D.J., Clases, D. Laser ablation–inductively coupled plasma–mass spectrometry imaging in biology. *Chem. Rev.* 2021, 121 (19): 11769–11822.

**39** Ablekim, T., Duenow, J.N., Zheng, X., Moutinho, H., Moseley, J., Perkins, C.L., Johnston, S.W., O'Keefe, P., Colegrove, E., Albin, D.S., Reese, M.O. Thin-film solar cells with 19% efficiency by thermal evaporation of CdSe and CdTe. *ACS Energy Lett.* 2020, 5 (3): 892–896.

**40** Rospars, N., Srout, M., Fu, C., Mourouga, G., Mensi, M., Ingenito, A. High performance ultra-thin lithium metal anode enabled by vacuum thermal evaporation. *Commun. Mater.* 2024, 5 (1): 179.

**41** Shah, P.S., Hanrath, T., Johnston, K.P., Korgel, B.A. Nanocrystal and nanowire synthesis and dispersibility in supercritical fluids. *J. Phys. Chem. B* 2004, 108: 9574–9587.

42 Ziegler, K.J., Lyons, D.M., Holmes, J.D., Erts, D., Polyakov, B., Olin, H., Svensson, K., Olsson, E. Bistable nanoelectromechanical devices. *Appl. Phys. Lett.* 2004, 84 (20): 4074–4076.

43 Hsieh, Y.F., Chen, L.J. Partial epitaxial growth of $Ni_2Ge$ and NiGe on Ge(111). *Thin Solid Films* 1998, 162: 287–294.

44 Morales, A.M., Lieber, C.M. A laser ablation method for the synthesis of crystalline semiconductor nanowires. *Science* 1998, 279 (9): 208–210.

45 Zhang, Y.F., Tang, Y.H., Wang, N., Lee, C.S., Bello, I., Lee, S.T. Germanium nanowires sheathed with an oxide layer. *Phys. Rev. B* 2000, 61 (7): 4518–4521.

46 Zhang, Y.F., Tang, Y.H., Lam, C., Wang, N., Lee, C.S., Bello, I., Lee, S.T. Bulk-quantity Si nanowires synthesized by SiO sublimation. *J. Crystal Growth* 2000, 212: 115–118.

47 Heath, J.R., LeGoues, F.K. A liquid solution synthesis of single crystal germanium quantum wires. *Chem. Phys. Lett.* 1993, 208 (3): 263–268.

48 Hanrath, T., Korgel, B.A. Nucleation and growth of germanium nanowires seeded by organic monolayer-coated gold nanocrystals. *J. Am. Chem. Soc.* 2002, 124 (7): 1424–1429.

49 Hanrath, T., Korgel, B.A. Crystallography and surface faceting of germanium nanowires. *Small* 2005, 1 (7): 717–721.

50 Tuan, H.-Y., Lee, D.C., Hanrath, T., Korgel, B.A. Germanium nanowire synthesis: an example of solid-phase seeded growth with nickel nanocrystals. *Chem. Mater.* 2005, 17: 5705–5711.

51 Ngo, L.T., Almécija, D., Sader, J.E., Daly, B., Petkov, N., Holmes, J.D., Erts, D., Boland, J.J. Ultimate-strength germanium nanowires. *Nano Lett.* 2006, 6 (12): 2964–2968.

52 Tsai, C.Y., Yu, S.Y., Hsin, C.L., Huang, C.W., Wang, C.W., Wu, W.W. Growth and properties of single-crystalline Ge nanowires and germanide/Ge nano-heterostructures. *CrystEngComm* 2012, 14 (1): 53–58.

53 Li, C.B., Usami, K., Muraki, T., Mizuta, H., Odal, S. The impacts of surface conditions on the vapor-liquid-solid growth of germanium nanowires on Si (100) substrate. *Appl. Phys. Lett.* 2008, 93 (4): 041917.

54 Gu, G., Burghard, M., Kim, G.T., Düsberg, G.S., Chiu, P.W., Krstic, V., Roth, S., Han, W.Q. Growth and electrical transport of germanium nanowires. *J. Appl. Phys.* 2001, 90 (11): 5747–5751.

55 Sutter, E., Ozturk, B., Sutter, P. Selective growth of Ge nanowires by low-temperature thermal evaporation. *Nanotechnology* 2008, 19 (43): 435607.

56 Ko, Y.-D., Kang, J.-G., Lee, G.-H., Park, J.G., Park, K.S., Jin, Y.H., Kim, D.W. Sn-induced low-temperature growth of Ge nanowire electrodes with a large lithium storage capacity. *Nanoscale* 2011, 3 (8): 3371–3375.

57 Li, L., Fang, X., Chew, H.G., Zheng, F., Liew, T.H., Xu, X., Zhang, Y., Pan, S., Li, G., Zhang, L. Crystallinity-controlled germanium nanowire arrays: potential field emitters. *Adv. Funct. Mater.* 2008, 18 (7): 1080–1088.

58 Kim, Y., Song, M.S., Kim, Y.D., Jung, J.H., Gao, Q., Jagadish, C., Tan, H.H. Epitaxial germanium nanowires on GaAs grown by chemical vapor deposition. *J. Korean Phys. Soc.* 2007, 51 (1): 120–124.

59 Jin, C.B., Yang, J.E., Jo, M.H. Shape-controlled growth of single-crystalline Ge nanostructures. *Appl. Phys. Lett.* 2006, 88: 193105.

**60** Kang, K., Kim, D.A., Lee, H.S., Kim, C.J., Yang, J.E., Jo, M.H. Low-temperature deterministic growth of Ge nanowires using Cu solid catalysts. *Adv. Mater.* 2008, 20 (24): 4684–4690.

**61** Tian, N., Zhou, Z.Y., Sun, S.G., Ding, Y., Wang, Z.L. Synthesis of tetrahexahedral platinum nanocrystals with high-index facets and high electro-oxidation activity. *Science* 2007, 316: 732–735.

**62** Dailey, J.W., Taraci, J., Clement, T., Smith, D.J., Drucker, J., Picraux, S.T. Vapor-liquid-solid growth of germanium nanostructures on silicon. *J. Appl. Phys.* 2004, 96 (12): 7556–7567.

**63** Mathur, S., Shen, H., Sivakov, V., Werner, U. Germanium nanowires and core-shell nanostructures by chemical vapor deposition of [$Ge(C_5H_5)_2$]. *Chem. Mater.* 2004, 16: 2449–2456.

**64** Kim, B.-S., Koo, T.-W., Lee, J.-H., Kim, D.S., Jung, Y.C., Hwang, S.W., Choi, B.L., Lee, E.K., Kim, J.M., Whang, D. Catalyst-free growth of single-crystal silicon and germanium nanowires. *Nano Lett.* 2009, 9 (2): 864–869.

**65** Song, M.S., Jung, J.H., Kim, Y., Wang, Y., Zou, J., Joyce, H.J., Gao, Q., Tan, H.H., Jagadish, C. Vertically standing Ge nanowires on GaAs (110) substrates. *Nanotechnology* 2008, 19 (12): 125602.

**66** Li, C., Usami, K., Yamahata, G., Tsuchiya, Y., Mizuta, H., Oda, S. Position-controllable Ge nanowires growth on patterned Au catalyst substrate. *Appl. Phys. Express* 2009, 2 (1): 015004.

**67** Jung, J.H., Yoon, H.S., Kim, Y.L., Song, M.S., Kim, Y., Chen, Z.G., Zou, J., Choi, D.Y., Kang, J.H., Joyce, H.J., Gao, Q., Tan, H. H., Jagadish, C. Vertically oriented epitaxial germanium nanowires on silicon substrates using thin germanium buffer layers. *Nanotechnology* 2010, 21 (29): 295602.

**68** Pei, L.Z., Zhao, H.S., Tan, W., Yu, H.Y., Chen, Y.W., Fan, C.G., Zhang, Q.F. Smooth germanium nanowires prepared by a hydrothermal deposition process. *Mater. Charact.* 2009, 60 (11): 1400–1405.

**69** Pei, L.Z., Zhao, H.S., Tan, W., Yu, H.Y., Chen, Y.W., Wang, J.F., Fan, C.G., Chen, J., Zhang, Q.F. Low temperature growth of single crystalline germanium nanowires. *Mater. Res. Bull.* 2010, 45 (2): 153–158.

**70** Lin, L.W., Tang, Y.H., Chen, C.S., Xu, H.F. Self-assembled single crystal germanium nanowires arrays under supercritical hydrothermal conditions. *CrystEngComm* 2010, 12 (10): 2975–2981.

**71** Kanemitsu, Y., Uto, H., Masumoto, Y., Maeda, Y. On the origin of visible photoluminescence in nanometer-size Ge crystallites. *Appl. Phys. Lett.* 1992, 61 (18): 2187–2189.

**72** Hayashi, S., Fujii, M., Yamamoto, K. Quantum size effects in Ge microcrystals embedded in $SiO_2$ thin films. *Jpn. J. Appl. Phys.* 1989, 28 (8): 1464–1466.

**73** Wilcoxon, J.P., Provencio, P.P., Samara, G.A. Synthesis and optical properties of colloidal germanium nanocrystals. *Phys. Rev. B* 2001, 64 (3): 035417.

**74** Heath, J.R., Shiang, J.J., Alivisatos, A.P. Germanium quantum dots: optical properties and synthesis. *J. Chem. Phys.* 1994, 101 (2): 1607–1615.

**75** Stehouwer, L.E.A., Tosato, A., Degli Esposti, D., Costa, D., Veldhorst, M., Sammak, A., Scappucci, G. Germanium wafers for strained quantum wells with low disorder. *Appl. Phys. Lett.* 2023, 123 (9): 092101.

**76** Fujii, M., Hayashi, S., Yamamoto, K. Growth of Ge microcrystals in $SiO_2$ thin film matrices: a raman and electron microscopic study. *Jpn. J. Appl. Phys.* 1991, 30 (4): 687–694.

**77** Parker, J.H., Feldman, D.W., Ashkin, M. Raman scattering by silicon and germanium. *Phys. Rev.* 1967, 155 (3): 712–714.

**78** Kolobov, A.V., Maeda, Y., Tanaka, K. Raman spectra of Ge nanocrystals embedded into $SiO_2$. *J. Appl. Phys.* 2000, 88 (6): 3285–3289.

**79** Jia, H., Kloepsch, R., He, X., Badillo, J.P., Gao, P., Fromm, O., Placke, T., Winter, M. Reversible storage of lithium in three-dimensional macroporous germanium. *Chem. Mater.* 2014, 26 (19): 5683–5688.

**80** Chockla, A.M., Klavetter, K.C., Mullins, C.B., Korgel, B.A. Solution-grown germanium nanowire anodes for lithium-ion batteries. *ACS Appl. Mater. Interfaces* 2012, 4 (9): 4658–4664.

**81** Ke, F.-S., Mishra, K., Jamison, L., Peng, X.X., Ma, S.G., Huang, L., Sun, S.G., Zhou, X.D. Tailoring nanostructures in micrometer size germanium particles to improve their performance as an anode for lithium ion batteries. *Chem. Commun.* 2014, 50 (28): 3713–3715.

**82** Zhang, C., Lin, Z., Yang, Z., Xiao, D., Hu, P., Xu, H., Duan, Y., Pang, S., Gu, L., Cui, G. Hierarchically designed germanium microcubes with high initial coulombic efficiency toward highly reversible lithium storage. *Chem. Mater.* 2015, 27 (6): 2189–2194.

**83** Park, M.H., Kim, K., Kim, J., Cho, J. Flexible dimensional control of high-capacity li-ion-battery anodes: from 0D hollow to 3D porous germanium nanoparticle assemblies. *Adv. Mater.* 2010, 22 (3): 415–418.

**84** Rudawski, N.G., Darby, B.L., Yates, B.R., Jones, K.S., Elliman, R.G., Volinsky, A.A. Nanostructured ion beam-modified Ge films for high capacity Li ion battery anodes. *Appl. Phys. Lett.* 2012, 100 (9): 083111.

**85** Kennedy, T., Mullane, E., Geaney, H., Osiak, M., O'Dwyer, C., Ryan, K.M. High-performance germanium nanowire-based lithium-ion battery anodes extending over 1000 cycles through in situ formation of a continuous porous network. *Nano Lett.* 2014, 14 (2): 716–723.

**86** Wang, D., Yang, Z., Li, F., Liu, D., Wang, X., Yan, H., He, D. Improved performance for lithium-ion batteries with nickel nanocone-arrays supported germanium anode. *Mater. Lett.* 2011, 65 (11): 1542–1544.

**87** Ling, S., Cui, Z., She, G., Guo, X., Mu, L., Shi, W. A novel type of Ge nanotube arrays for lithium storage material. *J. Nanosci. Nanotechnol.* 2012, 12 (1): 213–217.

**88** Gu, J., Collins, S.M., Carim, A.I., Hao, X., Bartlett, B.M., Maldonado, S. Template-free preparation of crystalline Ge nanowire film electrodes via an electrochemical liquid–liquid–solid process in water at ambient pressure and temperature for energy storage. *Nano Lett.* 2012, 12 (9): 4617–4623.

**89** Van Elshocht, S., Brijs, B., Caymax, M., Conard, T., Chiarella, T., De Gendt, S., De Jaeger, B., Kubicek, S., Meuris, M., Onsia, B., Richard, O. Deposition of HfO2 on germanium and the impact of surface pretreatments. *Appl. Phys. Lett.* 2004, 85 (17): 3824–3826.

**90** Kita, K., Kyuno, K., Toriumi, A. Growth mechanism difference of sputtered $HfO_2$ on Ge and on Si. *Appl. Phys. Lett.* 2004, 85 (1): 52–54.

**91** Xiang, J., Lu, W., Hu, Y., Wu, Y., Yan, H., Lieber, C.M. Ge/Si nanowire heterostructures as high-performance field-effect transistors. *Nature* 2006, 441 (7092): 489–493.

**92** Liang, G., Xiang, J., Kharche, N., Klimeck, G., Lieber, C.M., Lundstrom, M. Performance analysis of a Ge/Si core/shell nanowire field-effect transistor. *Nano Lett.* 2007, 7 (3): 642–646.

**93** Lin, Y.W., Chang, H.H., Huang, Y.H., Sun, C.J., Yan, S.C., Lin, S.W., Luo, G.L., Wu, C.T., Wu, Y.C., Hou, F.J. Tightly stacked 3D diamond-shaped Ge nanowire gate-all-around FETs with superior nFET and pFET performance. *IEEE Electron Device Lett.* 2021, 42 (12): 1727–1730.

**94** Wang, B., Wang, L., Zhang, Y., Yang, M., Lin, D., Zhang, N., Jiang, Z., Liu, M., Zhu, Z., Hu, H. Mixed-dimensional $MoS_2$/Ge heterostructure junction field-effect transistors for logic operation and photodetection. *Adv. Funct. Mater.* 2021, 32 (10): 2110181.

# IV

# Phosphorus

*Shancheng Yan*

## 4.1 Introduction

Black phosphorus (BP) is the most stable of the three common phosphorus isomers (red phosphorus [RP], white phosphorus, and BP). BP has four crystal structures – orthorhombic, rhombic, simple cubic, and amorphous – with orthorhombic crystal structure at room temperature and pressure. Similar to graphite, BP has a lamellar structure, but the difference is that the phosphorus atoms in the same layer are not in the same plane, showing a kind of honeycomb-folded structure. The layers are strongly covalently bonded, with individual pairs of electrons remaining, so that each atom is saturated, and the atoms between the layers act by van der Waals forces [1]. By disrupting the weak van der Waals interactions between the layers, BP material can be obtained from bulk crystals with single or few layers [2].

As a p-type 2D material, BP has won extensive attention from researchers for its unique band structure and optoelectronic properties [3]. BP nanosheets have the same flexibility and mechanical peeling ability as graphene and other 2D materials, and multichannel dynamic sensing tests show that BP has high sensitivity, selectivity, and fast response compared with other 2D-sensing materials [4–6].

With the continuous optimization of the preparation process of the 2D material BP, more researchers are applying it in life and scientific research, and thus it has great potential for applications in energy storage devices, semiconductor devices, biosensors, optical communications, and so on [7, 8].

## 4.2 Synthesis of BP

The future realization of BP for applications in fields such as electronics and photonics depends to a large extent on the development of methods that enable the reliable, large-scale synthesis, and preparation of 2D BP. The preparation of 2D BP is divided into two parts: first, the preparation of bulk BP, and second, the preparation of 2D BP. The preparation of lump BP mainly includes high-pressure method, ball milling method, and mineralization method.

*Single Element Semiconductors: Properties and Devices*, First Edition. Yi Shi and Shancheng Yan.

### 4.2.1 High Pressure

The high-pressure method was the first method used for the preparation of BP and was developed by Bridgman [9] in 1914. First, white phosphorus was put into a cylindrical steel container with a length of about 15 cm and an inner diameter of 1.5 cm, then the container with white phosphorus was put into a high-pressure device with kerosene and pressurized to $0.6 \times 10^9$ Pa at room temperature, then the device was heated up to 200°C and the pressure was raised to $1.2 \times 10^9$ Pa, which could obtain BP. Subsequently, in 1953, Keyes [10] conducted the same experiment and also succeeded in preparing BP. In 2012, Sun et al. [11] proposed a new viewpoint; they used the high-temperature and high-pressure method, under the pressure of 2–5.0 GPa and the temperature of 200–800°C, and utilized a cubic top hammer high-pressure device and a cubic sample chamber, reacted for 15 min, and transformed the 3-mm-thick, 10-mm diameter white phosphorus cylindrical block and RP. A cylindrical block of 3-mm thick, 10-mm diameter cylindrical block of white phosphorus and RP was successfully converted into a cylindrical block of BP, and the schematic diagram of the high-pressure device is shown in Figure 4.1.

The preparation of BP by the high-pressure method is more reproducible, and the conversion of RP to BP can be realized in a very short time. However, its preparation method

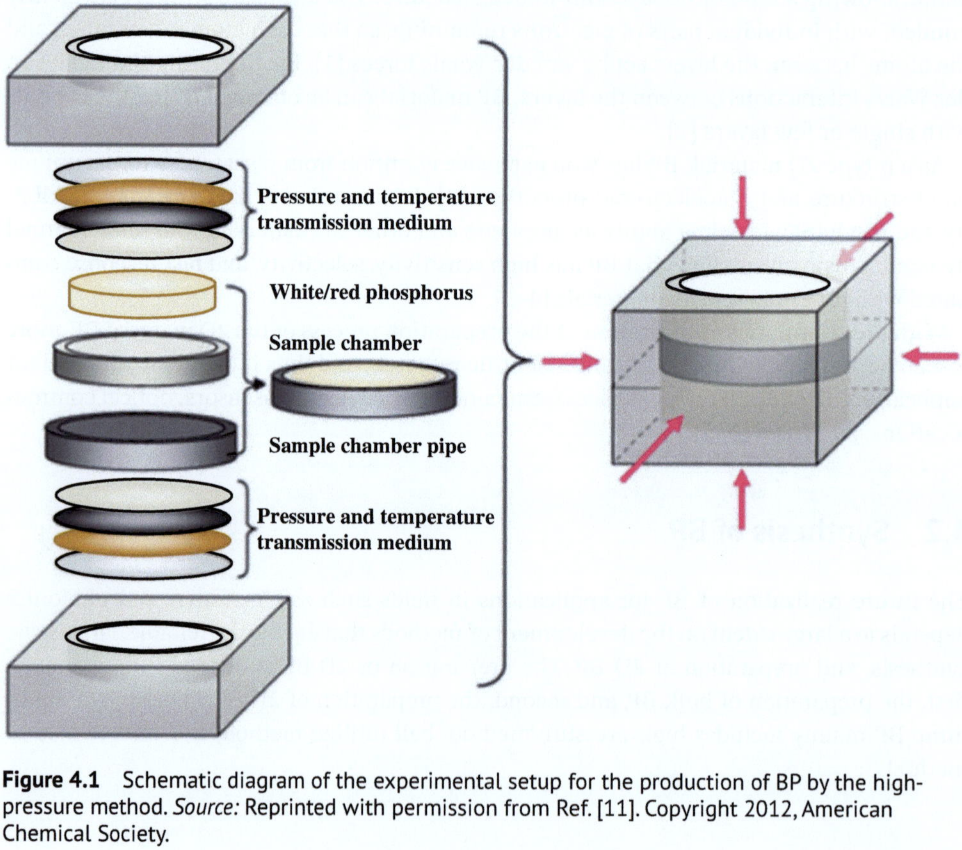


**Figure 4.1** Schematic diagram of the experimental setup for the production of BP by the high-pressure method. *Source:* Reprinted with permission from Ref. [11]. Copyright 2012, American Chemical Society.

requires the use of ultrahigh-pressure equipment, which to a certain extent leads to a high cost of BP production and is not conducive to large-scale commercial production.

### 4.2.2 Ball Milling

Ball milling is a common method of grinding micro- and nanomaterials, which utilizes mechanical collisions, where the grinding balls impact, crush, and abrade the material, as well as the collision of the balls with the walls of the vessel, which can generate transient high temperatures and pressures, which can cause a phase transition [12]. The transformation between RP and BP requires high temperatures and pressures, so ball milling [13, 14] can lead to the transformation of RP to BP, which has been used as one of the methods to synthesize BP.

### 4.2.3 Mineralization

Mineralization method is a method developed in recent years to prepare BP. Au, Sn, $SnI_4$, and RP are mixed in a certain mass ratio, vacuum encapsulated in a quartz tube, heated to a certain temperature, and held for a certain period to obtain BP, but a small amount of unconverted RP and impurities such as reaction-generated metal phosphide may exist in the final product [15].

In 2008, Nilges et al. [16] made $SnI_4$ mineralizer by adding tin powder and iodine to toluene. Highly crystalline BP crystals of sizes greater than 1 cm were obtained by mixing the mineralizer $SnI_4$ and RP in silica ampoules under low-pressure conditions, ramping up the temperature from 673 to 873 K, and holding it for 24 h. The crystals were then heated from 673 to 873 K and held for 24 h. The crystals were then heated to 873 K under low-pressure conditions. In 2014, Köpf et al. [17] also used $SnI_4$ as a mineralizing agent for the preparation of BP as shown in Figure 4.2. Due to the high price of Au, Köpf did not use Au in the preparation to reduce the cost; BP crystals could also be prepared by improving the

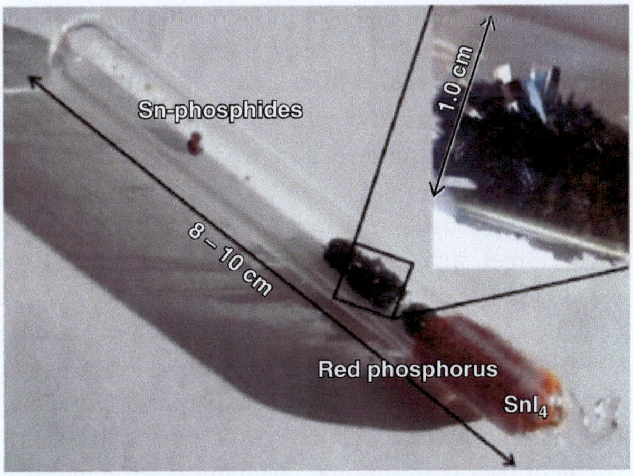


**Figure 4.2** Schematic of BP preparation by mineralization method using $SnI_4$ as a catalyst. *Source:* Reprinted with permission from Ref. [17]. Copyright 2014, Elsevier.

temperature conditions. In 2018, Zhang et al. [18] used the same $SnI_4$ mineralizer to obtain BP by heating the material three times and cooling it twice as well. The temperature was first increased from 298 K to 943 K, held for 7.5 h, then increased again to 823 K, and then held for 7.5 h, followed by cooling. The yield of BP prepared by this method was improved.

Mineralization method is currently the most promising BP preparation method for industrial application because of its simple preparation process, green, nontoxic, and low cost, but its reaction materials need to be sealed, and the reaction time is long, which needs to be improved.

## 4.3 Synthesis of BP Nanosheets

In the study of 2D materials, the preparation of high-quality samples plays a key role in the study of the properties and applications of 2D materials [19]. There are various methods for the preparation of 2D materials, mainly including the "top-down" method and the "bottom-up" method. The preparation method of BP nanosheets is described in this section.

### 4.3.1 "Top-down" Approach

The "top-down" method refers to the breaking of the layer-to-layer force of 2D materials by molecular intercalation or mechanical force to realize the peeling of large-size 2D materials into single-layer or few-layer 2D materials, e.g. mechanical peeling method, electrochemical peeling method, and liquid-phase peeling method.

#### 4.3.1.1 Mechanical Exfoliation Method

Mechanical exfoliation is a method of peeling off the layered black phosphene from the bulk material by applying mechanical force to the material, which is capable of destroying the weak van der Waals forces between the phosphene layers and the layers [20]. Mechanical exfoliation is precisely one of the simplest and most commonly used methods to separate a single layer of 2D material from a bulk by mechanical cutting. A number of research teams have successfully prepared high-quality monolayers of BP by tearing the BP from the bulk with transparent adhesive tapes.

In 2004, Novoselov and Morozov succeeded in producing monolayer graphene by using adhesive tapes to remove a small slice of graphite samples and then repeatedly peeling it off [21]. Whereas BP structures are similar to graphene in that they both interact through interlayer van der Waals forces, such forces are relatively weaker in BP structures, and thus BP is more readily available as a monolayer or few-layer structure. In 2014, Zhang [22] successfully obtained few layers of BP with a thickness of a few nanometers by mechanical stripping method for the first time, proving its unique p-type conductive properties, and investigated the field effect transistors (FETs) prepared based on BP, and the test showed that its carrier mobility could reach 103 $cm^2 \cdot V^{-1} \cdot s^{-1}$. In the same year, Andres [23] achieved an improvement over conventional mechanical stripping techniques by selecting tapes with appropriate adhesive properties, and the researchers utilized the selected tapes for conventional mechanical stripping to obtain fewer layers of BP nanosheets, which dramatically increased throughput and reduced contamination of the samples by the adhesive on the tape surface. The researchers also demonstrated the acquisition of two-layer BP nanosheets using optical microscopy, Raman spectroscopy, and transmission electron microscopy to quickly and reliably determine the number of layers. Thickness is a key parameter in the electrical, optical, and thermal properties of 2D materials, so it is important to control the thickness of 2D materials to achieve modulation of their properties. In the same year, Lu et al. [24] achieved controlled growth of different layers of BP by obtaining few layers of BP by mechanical stripping method and then further thinning them using argon plasma. The thickness of BP can be clearly determined by Raman spectroscopy combined with optical comparison of atomic force microscope. Moreover, it was found that the structure of BP was not damaged by the characterization. This indicates that the control of the number of BP layers can be realized by using the argon plasma method. Although the mechanical stripping method can obtain high-quality BP samples, it is not conducive to large-scale batch industrialized production because of its cumbersome steps, harsh production environment, long preparation time, and low efficiency.

#### 4.3.1.2 Electrochemical Stripping

Electrochemical stripping is a method of stripping materials by applying a voltage to an ionic liquid, such as cationic intercalation, which causes the structure of the sample to expand [25].

In 2017, Ambrosi et al. [26] proposed a simple, fast, and top-down electrochemical method. The researchers established a two-electrode system using a BP sheet as an anode and platinum as a cathode. With the constant application of voltage, the BP crystals slowly decomposed, the solution slowly turned orange, and fine particulate matter was found at the bottom of the electrochemical cell. Subsequently, the researchers filtered the solution, dispersed the stripped material in dimethylformamide (DMF), and sonicated it to obtain BP. The stripped material was characterized using X-ray photoelectron spectroscopy, Raman spectroscopy, and scanning transmission electron microscopy. The researchers found that the thickness of the stripped layer was significantly reduced compared with that of the original material. In 2018, Xiao et al. [27] placed BP powder into a porous plastic tube with one platinum plate as the cathode and the other platinum plate as the anode. Then, the two platinum plate electrodes were immersed in propylene carbonate (PC)-containing tetrabutylammonium hexafluorophosphate with 30-V voltage action for 12 h. After electrochemical swelling action, the swollen powder was put into pure PC solution for ultrasonic cleaning for half an hour, and the filtering and cleaning operation was repeated three times, and it was put into a vacuum oven at 60°C to dry for 8 h. The BP nanosheets were obtained after drying. Using this electrochemical cathodic stripping method, oxidation of the electrodes can be completely avoided in the process, and high-quality BP nanosheets can be obtained. In 2019, Wang et al. [28] used a two-electrode system, employing BP crystals and platinum foil as cathode and anode, respectively, to prepare an electrolyte by dissolving tetra-n-butylammonium bisulfate in anhydrous, deoxygenated propylene glycol malonate. The voltage was decreased to −10.0 V for 20 min to facilitate the wetting of the electrodes, and then the voltage was fixed at −30.0 V. The electrochemical delamination and stripping process occurred, and the BP nanosheets with larger sizes could be obtained isolated in $O_2$ and $H_2O$.

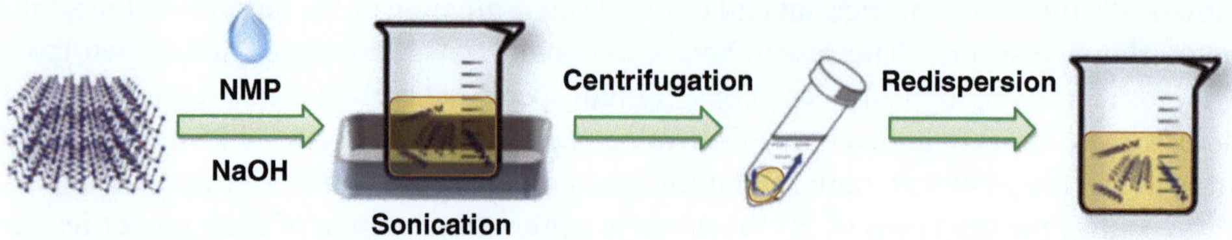


**Figure 4.3** Schematic diagram of the BP-stripping process in NMP. *Source:* Reprinted with permission from Ref. [30]. Copyright 2015, WILEY-VCH.

#### 4.3.1.3 Liquid-phase Stripping

Liquid-phase stripping is a method of obtaining 2D materials by placing an object in a solvent that interacts with it, making the intervals between the layers larger through the solvent, and then separating the crystals in the various liquids using ultrasonic energy [29].

In 2015, Guo et al. [30] used the liquid-phase stripping method to remove larger samples by ultrasonically immersing bulk BP in *N*-methylpyrrolidone (NMP) at 30°C for 24 h. After sonication, the larger samples were then removed by high-speed centrifugation, and 3–5 layers of BP nanosheets with dimensions of 200 nm × 200 nm, as well as monolayers and bilayers of BP nanosheets with dimensions of 20 nm × 20 nm, can be obtained, as shown in Figure 4.3. Subsequently, Yasaei et al. [31] compared the stripping results of BP crystals under different solvents as dispersant conditions, which showed that nonplasmonic, polar solvents such as DMF and dimethylsulfoxide had the best solubility among all solvents and were good solvents for the synthesis of BP nanosheets of atomic thickness and were able to produce a homogeneous and stable dispersion after ultrasound treatment. The solution was centrifuged, and its supernatant was collected by pipette to obtain monolayer and multilayer BP. In 2016, Chen et al. [32] took advantage of the hydrophilic nature of BP and used water as a solvent for the preparation of BP to synthesize centimeter-sized bulk BP crystals by using highly efficient mineralization-assisted gas-phase reforming method and stripping these BP bodies in water, which could lead to the preparation of few-layer BP nanoflakes. The samples obtained by this method are highly crystalline, free of impurities, and very stable in water, which can be used for further processing and applications. Although the liquid-phase stripping method can obtain a large number of 2D materials with single or few layers, the quality of the samples is uncontrollable and poor, and the solvent residue is easy to be left.

### 4.3.2 "Bottom-up" Approach

The "bottom-up" approach refers to the assembly of smaller structural units into larger structural systems by means of vapor phase, vacuum evaporation, etc., e.g. vapor phase deposition, solvothermal methods.

#### 4.3.2.1 Chemical Vapor Deposition

Chemical vapor deposition (CVD), a method of growing 2D materials on substrate surfaces after volatilizing or reacting precursors at high temperatures, is widely used in the preparation of 2D materials because of its high efficiency and low cost [33]. Researchers usually adopt the method of converting RP to BP to realize the growth of BP. They first prepared a thin film of RP by vapor deposition and then converted the RP to BP at high temperature and high pressure. In 2015, Li et al. [34] first proposed a method to synthesize BP films on flexible substrates. They first deposited an RP film on a flexible polyester substrate using a tube furnace and then converted the RP into BP by placing the RP film into a high-pressure chamber at 400°C for conversion. In this way, they produced circular BP films with a size of 4 mm and a thickness of 40 nm. In the following year, Zhang et al. [35] used a two-step heating chemical vapor transport (CVT) reaction method to simplify the subsequent isolation and purification steps of BP single crystals. In the same year, Smith et al. [33] proposed a method for the growth of large-area 2D BP nanosheets by selecting silicon as the substrate and vacuum heating RP powder or BP powder to 873 K in a tube furnace for a reaction time of 30 min, and first, RP films were prepared. Then, the substrate of the RP film and the mineralizing agent $SnI_4$ prepared in advance were transferred to a conical-bottomed glass centrifuge tube, and the centrifuge tube was put into a sealed and vacuumed pressure vessel and transferred to a tube furnace, and the vacuum was heated up to 1223 K, and the RP was converted into BP, so that the 2D-BP nanosheets with an average area of more than 3 $\mu m^2$ could be obtained. In 2018, Li et al. [36] proposed a method to prepare polycrystalline BP thin films by RP using CVD. The process flow diagram is shown in Figure 4.4a. The researchers first realized the deposition of RP films on a sapphire substrate by a deposition device, which is shown in Figure 4.4b. The researchers also realized the deposition of RP films on a sapphire substrate by a deposition device. Then they transferred the RP film onto a hexagonal boron nitride (Hbn) substrate and finally converted the RP into BP at 700°C and 1.5 GPa to obtain BP films with a thickness of 50 nm and a crystal size of 40–70 μm on the substrate.

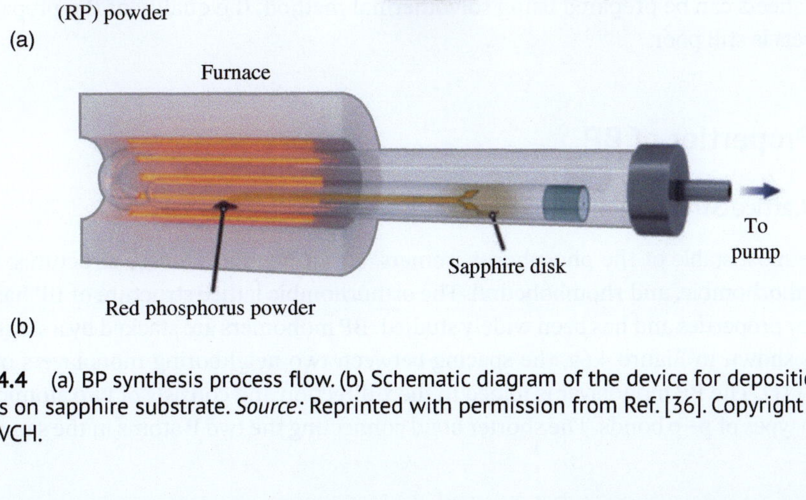


**Figure 4.4** (a) BP synthesis process flow. (b) Schematic diagram of the device for deposition of RP films on sapphire substrate. *Source:* Reprinted with permission from Ref. [36]. Copyright 2018, WILEY-VCH.

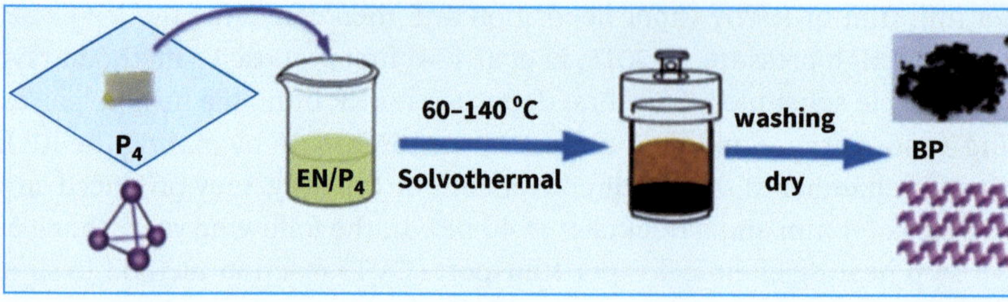


**Figure 4.5** Schematic diagram of solvothermal synthesis of BP. *Source:* Reprinted with permission from Ref. [40]. Copyright 2020, NATL ACAD SCIENCES.

#### 4.3.2.2 Solvothermal

Solvothermal method, also known as hydrothermal method, refers to the product obtained by a violent chemical reaction under high temperature and high pressure in an airtight container. Depending on the type of reaction, it can be divided into hydrothermal oxidation, hydrothermal reduction, and hydrothermal synthesis [37].

In 2016, Zhang et al. [38] reported a solvothermal method for the preparation of BP nanosheets. RP was dispersed into ethanol and then transferred to a high-pressure reactor and reacted at a temperature of 400°C for several hours. The BP nanosheet samples were separated at the top by ultrasonication. The size of the obtained nanosheets was around 1.0 μm, and the thickness varied from 0.5 to 4 nm. In 2017, Zhao et al. [39] improved the above method; they found that ammonium fluoride can reduce the surface activation energy of the material, and with the assistance of ammonium fluoride, the BP material with better quality can be exfoliated. Therefore, ammonium fluoride was added in the hydrothermal process to prepare nanosheets with few layers of BP.

In 2018, Tian et al. [40] successfully synthesized BP nanosheets with orthorhombic crystal structure by using simple raw materials, white phosphorus and ethylenediamine, in an autoclave reactor lined with polytetrafluoroethylene at a temperature of 100°C for 12 h. The reaction was carried out at a low reaction temperature, and BP nanosheets were synthesized in gram quantities, which showed excellent due to the partial oxidation of the surface of the BP stability due to the partial oxidation of the BP surface, as shown in Figure 4.5. Although BP nanosheets can be prepared using solvothermal method, the quality of the prepared BP nanosheets is still poor.

## 4.4 Properties of BP

### 4.4.1 Lattice Structure of BP

BP is the most stable of the phosphorus isomers, and it has three lattice structures: simple cubic, orthorhombic, and rhombohedral. The orthorhombic lattice structure of BP has semiconductor properties and has been widely studied. BP monomers are stacked by a single layer of BP, as shown in Figure 4.6a, the spacing between two neighboring monolayers of BP is 0.53 nm [41]. The BP monolayer is folded in the x-direction and consists of two atomic layers with two types of p–p bonds. The shorter bond connecting the two P atoms in the same plane

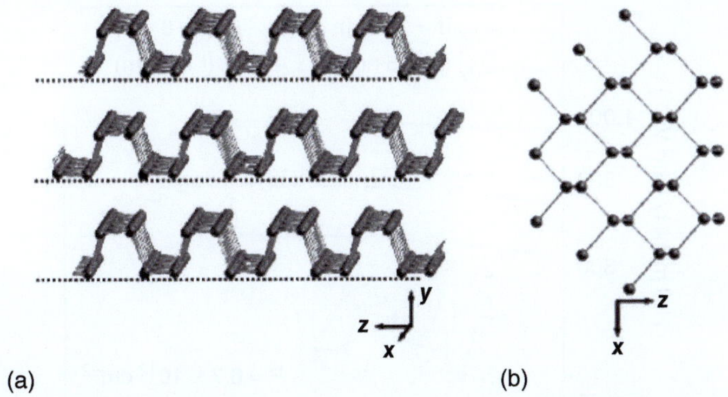


**Figure 4.6** (a) Side view of crystalline BP. (b) Top view of monolayer BP. *Source:* Reprinted with permission from Ref. [41]. Copyright 2015, Angewandte Chemie International Edition.

has a bond length of 0.2224 nm, and the longer bond connecting the top and bottom P atoms in the monolayer has a bond length of 0.2244 nm, as shown in Figure 4.6b, the BP exhibits a hexagonal structure along the z-direction, and the bonding angles formed by the chemical bonds are 96.3° and 102.1°, respectively. The unique structure of BP makes it anisotropic, and the different properties along different directions make BP have a broad application prospect.

### 4.4.2 Energy Band Structure of BP

BP is a 2D layered semiconductor material with a direct bandgap. Tran et al. simulated the energy band structure of BP with different numbers of layers by first nature principle [42], as shown in Figure 4.7. The researchers found that the bandgap value of BP in a single layer is 2.0 eV, which decreases with the increase of the number of layers in a power function law. The bandgap value of bulk BP is 0.3 eV, which has the advantage of adjustable bandgap, making it more promising for optoelectronic devices [43].

### 4.4.3 Anisotropy of BP

The anisotropy of BP is an important feature that distinguishes it from other 2D materials. The different lattice constants of BP in different directions are the fundamental reasons

**Figure 4.7** Variation of bandgap with thickness of oligolayer BP. *Source:* Reprinted with permission from Ref. [42]. Copyright 2014, Physical Review B.

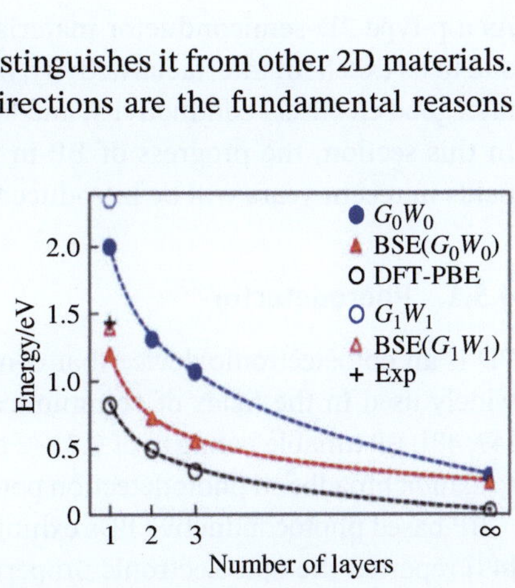

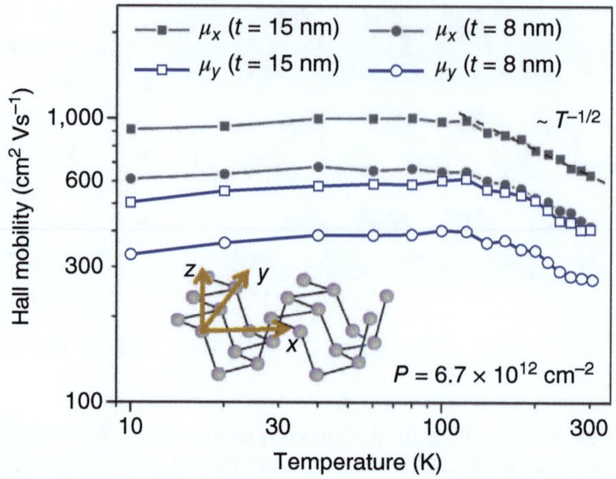


**Figure 4.8** Hall mobilities measured at a constant hole doping concentration of $6.7 \times 10^{12}$ cm$^{-2}$ along the x ($\mu_x$) and y ($\mu_y$) directions for BP thin films with a thickness of 8 and 15 nm, respectively. $\mu_x$ is about 1.8 times as large as $\mu_y$, and thicker films consistently show larger mobility than that of thinner films. Inset: schematic view of a single-layer BP showing the crystalline directions. *Source:* Reprinted with permission from Ref. [45]. Copyright 2020, Springer Nature.

for the anisotropy of BP [44]. The effect of BP anisotropy on Hall mobility was reported by Xia et al. [45], as shown in Figure 4.8. The Hall mobility of 15-nm BP along the x and y directions was measured to be 1000 cm$^2$·V$^{-1}$·s$^{-1}$ and 600 cm$^2$·V$^{-1}$·s$^{-1}$, respectively. Liu et al. investigated the anisotropy of the mobility of BP by theoretical calculations [46] and found that the anisotropy ratio of the mobility of a single layer of BP along the x and y directions was about 3–4. Due to its anisotropic properties, BP has great potential for applications in energy storage devices, semiconductor devices, biosensors, and optical communications.

## 4.5 Applications of BP

As a p-type 2D-semiconductor material, BP has attracted much attention in recent years due to its excellent interfacial activity, high photovoltaic performance, fast electron transfer rate, good electrical conductivity, and biocompatibility, which can be used in various fields. In this section, the progress of BP in photodetectors (PDs), FETs, solar cells, and other fields in recent years will be introduced.

### 4.5.1 Photodetector

PD is an optoelectronic device that converts optical signals into electrical signals, which is widely used in the fields of communication, medical treatment, transportation, and so on [47, 48]. BP tunable bandgap of 0.3 ~ 2 eV and carrier mobility of 1000 cm$^2$·V$^{-1}$·s$^{-1}$ enable it to exhibit broadband photodetection potential from UV to mid-infrared wavelength range [7].

BP-based photoconductive PDs exhibit excellent detection performance. In 2015, Wu et al. [49] reported the optoelectronic properties of high-quality, layer-less BP PD; the schematic

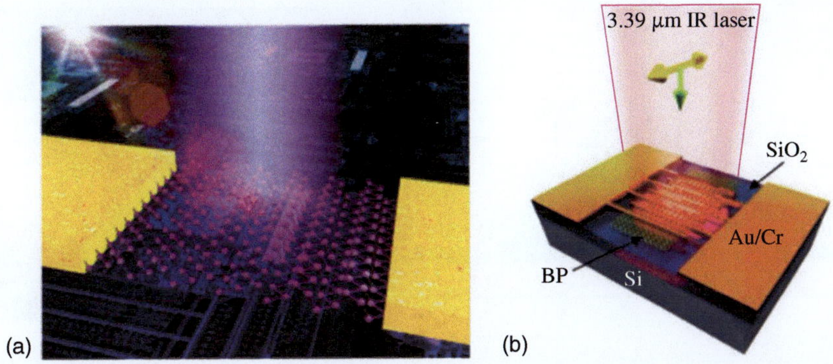


**Figure 4.9** (a) Schematic diagram of a photoelectric FET. *Source*: Reprinted with permission from Ref. [49]. Copyright 2015, American Chemical Society. (b) Schematic diagram of the BP PD operating at 3.39 μm. *Source:* Reprinted with permission from Ref. [50]. Copyright 2016, American Chemical Society.

diagram of a photoelectric FET is shown in Figure 4.9a. The researchers used a mechanical stripping method to obtain thin sheets of BP and then rotationally coated polymethylmethacrylate (PMMA) to avoid the rapid degradation of BP in the environment. Next, they used electron beam lithography for the preparation of electrodes to deposit 5/80-nm-thick Ti/Au contacts. The device achieved a responsivity of $9 \times 10^4$ W in the near-ultraviolet band at 275.6 nm, which is five orders of magnitude higher than that in the visible-near-infrared band. The high UV responsivity of the device is attributed to the resonant interband jumps between the valence and conduction bands (CBs) of two special nested bands. In 2016, Guo et al. [50] reported a mid-infrared PD with high gain based on BP. The structure of the device is shown in Figure 4.9b. The researchers first peeled off the BP film from a bulk BP crystal and then transferred the BP film to a Si substrate and measured the thickness of the BP film with an atomic force microscope. Next, they prepared a PMMA resist layer using an electron beam lithography system. Finally, the researchers evaporated 3-nm Cr with 35-nm Au fork-finger electrodes to form contacts. At a wavelength of 3.39 nm, the detector achieved a responsivity of 82 $A{\cdot}W^{-1}$. The high carrier mobility of BP and the unique polarization response due to its crystal symmetry enabled the high responsivity of the detector in the mid-infrared wavelength band. In addition, 2D PDs usually operate in photoconductive mode, with high dark current and slow response speed [51, 52]. Researchers have improved their detection performance by constructing semiconductor heterojunctions and chemical doping.

Semiconductor heterostructure generates a built-in electric field that suppresses dark current and separates photogenerated electron-hole pairs for fast photovoltaic conversion [49, 53, 54]. In 2018, Cao et al. [55] constructed photoconductive PDs based on heterojunctions of BP and iron disulfide ($ReS_2$) by using a mechanical exfoliation method to obtain BP flakes, which were then transferred onto a transparent polydimethylsiloxane (PDMS) film, as shown in Figure 4.10a. Under an optical microscope, the BP flakes were transferred onto a 30-nm-thick $H_fO_2$/Si substrate together with the PDMS film, and then the $ReS_2$ flakes were transferred onto the PDMS film in the same way. After that, the researchers utilized electron beam lithography to determine the contact pattern. Finally, a 10/50-nm-thick Ni/Au layer was deposited on an electron beam evaporation system. Under UV illumination, the devices

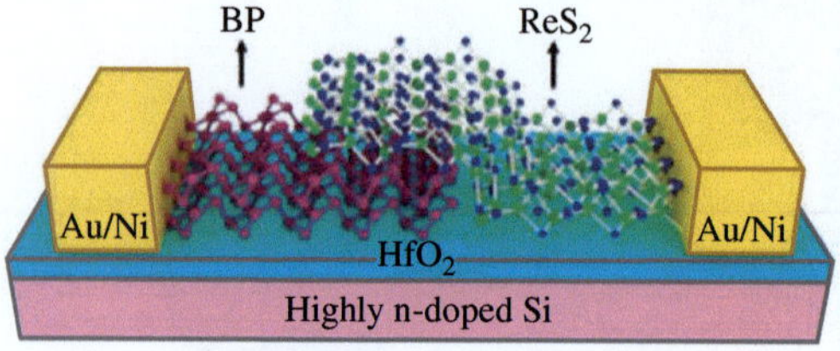


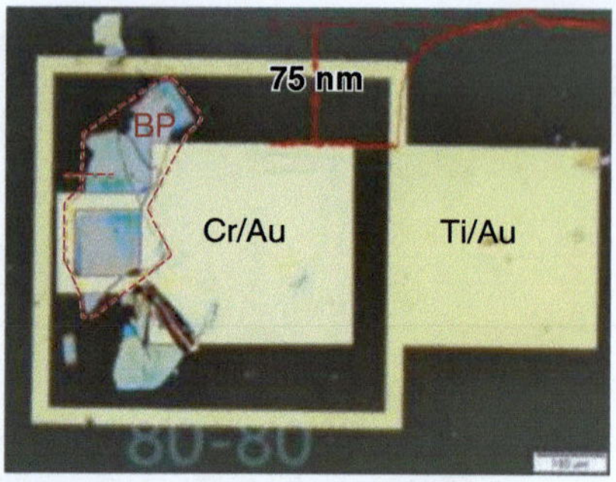


**Figure 4.10** (a) Schematic diagram of the BP/$ReS_2$ heterostructure on $H_fO_2$/Si substrate. *Source:* Reprinted with permission from Ref. [55]. Copyright 2018, ROYAL SOC CHEMISTRY. (b) Optical microscopy images of UV/IR PDs prepared based on β-$Ga_2O_3$/BP p-n heterojunctions. *Source:* Reprinted with permission from Ref. [56]. Copyright 2020, WILEY-VCH.

showed a high photoresponsivity of 4120 $A{\cdot}W^{-1}$. In addition, the photoresponsivity was 11811 $A{\cdot}W^{-1}$ at a BP length of 1 μm and a back-gate voltage of 4.4 V with different channel lengths. In 2020, He et al. [56] heteroepitaxially grew heterojunction thin films of β-$Ga_2O_3$/BP by metal-organic CVD and applied them to UV and IR dual-frequency photoconductive PDs; its optical microscope image is shown in Figure 4.10b. The researchers used organic CVD to heteroepitaxially grow 100-nm-thick β-$Ga_2O_3$ films on a sapphire substrate and plasma-enhanced CVD to deposit a 60-nm-thick $SiO_2$ dielectric layer on the β-$Ga_2O_3$ films and then selectively etch to expose the active region. After transferring the BP obtained from mechanical stripping to the active zone by the mineralizer-assisted gas-phase transformation method, the researchers deposited Cr/Au (20/100 nm) electrodes on the left side using electron beam evaporation. After selectively etching the $SiO_2$ dielectric layer and exposing β-$Ga_2O_3$, the researchers deposited Ti/Au (20/100 nm) electrodes on the right side using electron beam evaporation. The prepared devices exhibited good photoresponse under both UV and IR light, with responsivities of 88.5 $mA{\cdot}W^{-1}$ and 1.24 $mA{\cdot}W^{-1}$, respectively, which was attributed to the fact that the built-in electric field in the depletion zone of β-$Ga_2O_3$/BP heterojunction heterostructures accelerates the separation of photogenerated electron-hole pairs, thus improving the device performance.

Integrating BP into silicon photonic waveguides can also improve detector performance. In 2015,Youngblood et al. [57] reported layer-less BP gated transistor-based PD integrated silicon photonic waveguide for mid-infrared photodetection; the device schematic is shown in Figure 4.11a. The researchers used e-beam lithography to prepare the underlying silicon photonic layer substrate, and then $SiO_2$ was deposited on the substrate by e-beam evaporation. The source and drain contacts were identified using e-beam lithography, and the top few-layer graphene gates were transferred using a wet-transfer method, and the pattern was formed by e-beam lithography and oxygen plasma etching. Finally, the top gate contact was patterned using poly (methyl methacrylate) and 5-nm Ti/30-nm Au was deposited by electron beam evaporation. At 0.4-V bias voltage, the device achieves a response rate of up to 135 $mA{\cdot}W^{-1}$ at room temperature, with a response bandwidth of more than 3 GHz. 135 and 657 $mA{\cdot}W^{-1}$ are obtained at room temperature for the devices with thicknesses

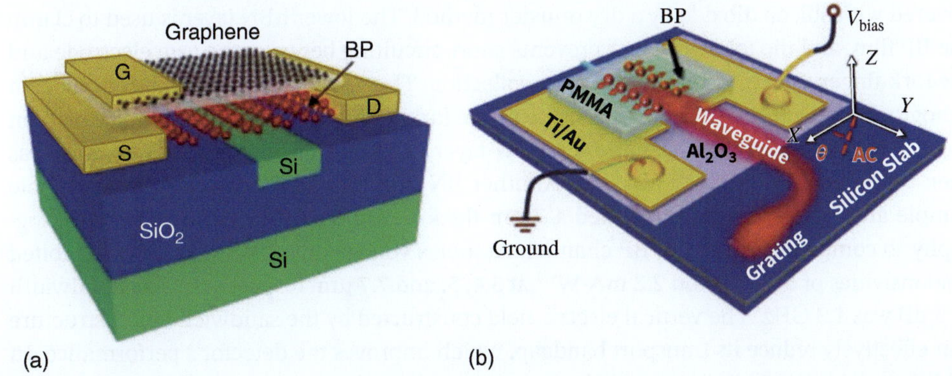


**Figure 4.11** (a) Three-dimensional illustration of the device configuration, featuring a few-layer graphene top-gate. *Source:* Reprinted with permission from Ref. [57]. Copyright 2015, NATURE RESEARCH. (b) Schematic of the structure of the silicon/BP hybrid waveguide PD. *Source:* Reprinted with permission from Ref. [58]. Copyright 2019, WILEY-VCH.

of 11.5 and 100 nm, respectively, and the photocurrent of the device is controlled by the photovoltaic effect, with a response bandwidth of more than 3 GHz. In 2019, Yin et al. [58] integrated BP in silicon photonic waveguide for mid-infrared photoconductive PD; the structure of this device is shown in Figure 4.11b. In this device, doped insulating silicon was selected with a silicon ridge height of 220 nm and a silicon core height of 340 nm. The researchers then deposited an $Al_2O_3$ layer using atomic layer deposition to cover a silicon ridge waveguide as an insulator between the BP and the silicon core and placed Ti/Au (15/80 nm) metal sheets symmetrically on both sides of the silicon ridge. The BP film was then transferred to the top of the silicon ridge waveguide by using an imprint transfer. Finally, in order to prevent the BP film from being oxidized, a PMMA layer was rotationally coated on the chip. The measured responsivity was as high as 306.7 $mA \cdot W^{-1}$ at a bias voltage of 0.4 V, and the 3-dB bandwidth was as high as 1.33 GHz. The BP can cover the surface of the silicon-based waveguide very well, and thus the optical absorption on the surface of the silicon-based waveguide is enhanced.

In addition, BP-based transistorized PDs have shown excellent detection performance. In 2017, Chen et al. [59] reported the preparation of a transistorized mid-infrared PD using a hBN/BP/hBN sandwich heterostructure. The device structure is shown in Figure 4.12, where a BP film is sandwiched by an hBN sheet and then transferred onto a silicon substrate

**Figure 4.12** Structure of the mid-infrared PD based on tunable BPs. *Source:* Reprinted with permission from Ref. [59]. Copyright 2017, Springer Nature.

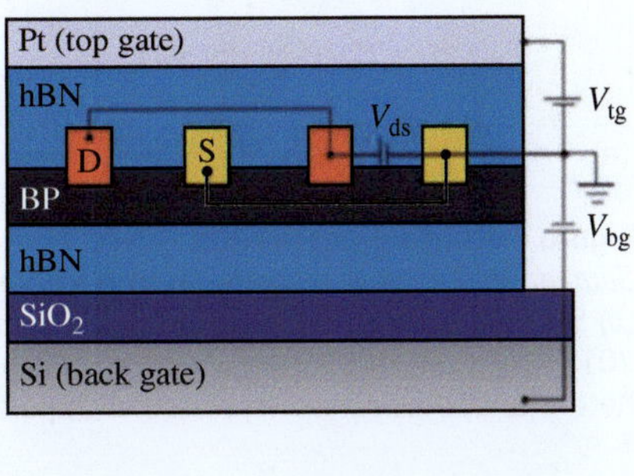

covered with silicon dioxide by a dry transfer method. The lower hBN layer is used to clamp the BP film, and the top hBN layer prevents short-circuiting between the gate electrode and the fork-finger electrode for photocurrent collection. The researchers used an electron beam lithography system to determine the shape of the fork-finger electrode and then used an ion etching system to etch the exposed top hBN layer. A chromium/gold (3/27 nm) film was then evaporated to form the contacts. Another BN thin film was transferred to the whole sample area. Finally, they evaporated a 7 nm-thick Pt top port after electron beam lithography to completely cover the BP channel. At a bias voltage of 1.2 V, the devices exhibited responsivities of 518, 30, and 2.2 $mA\cdot W^{-1}$ at 3.4, 5, and 7.7 μm, respectively. The bandwidth at 3 dB was 1.3 GHz. The vertical electric field constructed by the sandwich heterostructure can effectively reduce its transport bandgap, which improves the detector's performance. In addition, the hBN medium not only improves the collection of carriers but also prevents the black phosphor oxidation.

### 4.5.2 Field Effect Transistor

BP is an attractive channel material with lamellar structure and pleated honeycomb [60], which can realize high-performance FETs [61, 62]. Exfoliated, few-layer BP flakes exhibit room temperature mobility up to 1000 $cm^2\cdot V^{-1}\cdot s^{-1}$, on/off current ratios of about $10^4$–$10^5$, and anisotropic transport properties [22]. All these unique properties make BP a promising 2D material for FETs.

In 2014, Zhang et al. [22] fabricated BP-based FETs with different thicknesses for the first time using few layers of BP. The experiments showed a clear sample thickness dependence of the BP-based FETs. The researchers characterized the switching behavior of the less-layered BP transistors at room temperature, and the current switching ratios of the BP-based FETs were almost four orders of magnitude higher compared to graphene, which is almost close to that of $MoS_2$ devices.

The switching regulation of the transistor and the improvement of the subthreshold of the transistor can be realized by modulating the depletion region of the p-junction formed by the heterostructure. In 2016, Jeon et al. [63] reported a p-n junction diode heterostructured with BP nanosheets and ZnO nanowires (NWs), as shown in Figure 4.13a, which was applied in a FET. The device exhibits good static rectification characteristics with a

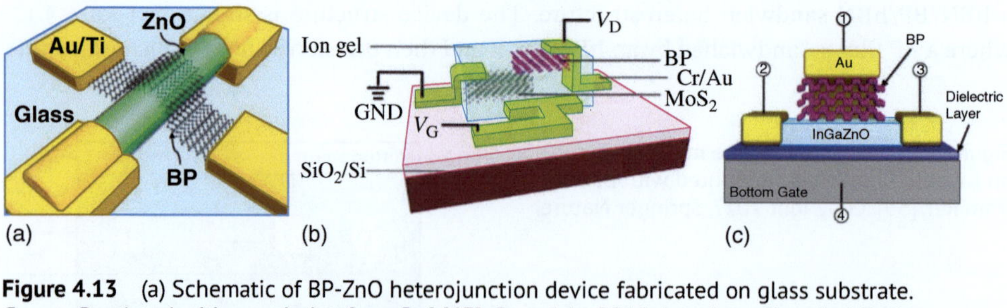


**Figure 4.13** (a) Schematic of BP-ZnO heterojunction device fabricated on glass substrate. *Source:* Reprinted with permission from Ref. [63]. Copyright 2016, American Chemical Society. (b) Schematic Structure of BP-$MoS_2$ FET. *Source:* Reprinted with permission from Ref. [64]. Copyright 2017, American Institute of Physics. (c) Schematic of InGaZnO transistor based on BP/InGaZnO vdW heterojunction. *Source:* Reprinted with permission from Ref. [65]. Copyright 2020, WILEY.

switching ratio of 104. A van der Waals heterojunction is formed between BP and ZnO by directly imprinting 17-nm BP nanosheets onto 85-mm-thick ZnO NWs. This type of 1D–2D hybrid junction is rare for heterojunction transistors. Later, in 2017, Xu et al. [64] designed a FET based on a BP-molybdenum disulfide heterojunction. The structure of the device is shown in Figure 4.13b. The obtained transistor achieves a subthreshold of 65 mV-dec$^{-1}$ at room temperature and 51 mV·dec$^{-1}$ at 160 K, and maintains a low subthreshold over more than two orders of magnitude of drain current range. The interband tunneling of the transistor is controlled by modulating the energy band alignment through the doping of an ion-gel dielectric on the BP side and electrostatic modulation on the molybdenum disulfide side. In 2020, Bi et al. [65] reported a vertically stacked BP-indium gallium-zinc oxide (InGaZnO) van der Waals heterojunction based on the BP/InGaZnO heterojunction, as shown in Figure 4.13c. The heterojunction was designed and fabricated with a thin BP/InGaZnO heterojunction based on the BP/InGaZnO heterojunction. Steel gallium-zinc oxide (InGaZnO) tunneling field effect transistors and heterojunction field effect transistors based on thin BP/InGaZnO heterostructured transistors exhibit good transistor performance with switching ratios exceeding 105 and field effect mobility as high as 23.5 cm$^2$·V$^{-1}$·s$^{-1}$. A large room-temperature negative differential resistance characteristic with a peak-to-valley ratio of 2.1 and a tunneling current density of 160 mA·mm$^{-2}$ was obtained in a BP/nGaZnO heterojunction. On this basis, a subthreshold swing of 11 mV·dec$^{-1}$ for subthermal ions was also realized in the transistor.

### 4.5.3 Solar Cell

A solar cell is a device that converts light energy into electrical energy through the photovoltaic or photochemical effect. BP has attracted much attention for its unique 2D layered structure with excellent optical and electrical properties. With a direct bandgap of 0.3–2.0 eV with respect to the layer thickness, bipolar operation, and natural p-type semiconductor properties, BP is a promising material for solar cells, and 2D BP has already shown great potential for application in the field of solar cells [66–68]. Researchers have applied them to the electron transport layer, hole transport layer and active layer of solar cells in order to improve the performance and stability of solar cells.

In 2014, Buscema et al. [69] experimentally demonstrated the photovoltaic effect of a few-layer BP 2D semiconductor p-n junction, which was experimentally shown to have a zero-bias photocurrent and significant photogenerated volt effect open-circuit voltage to generate electricity at a wavelength of 940 nm under illumination. Since then, BP has been used as a selective material to generate a variety of solar cells, with electrodes and heterostructures into p-n junctions, which generate and separate carriers in solar cells to achieve high power conversion and wide wavelength response.

The high carrier mobility and tunable energy bandgap of BP can enhance the transport performance of the electron transport layer, thus improving the performance of solar cells. In 2019, Batmunkh et al. [67] exfoliated highly crystallized and atomically thick few-layer BP nanosheets under near-infrared (NIR) pulsed laser irradiation using a vortex device and applied them to the electron transport layer of a low-temperature-prepared chalcogenide solar cell. The structure of the device is shown in Figure 4.14a. With the addition of BP, the average energy conversion efficiency of the chalcogenide solar cell was increased from

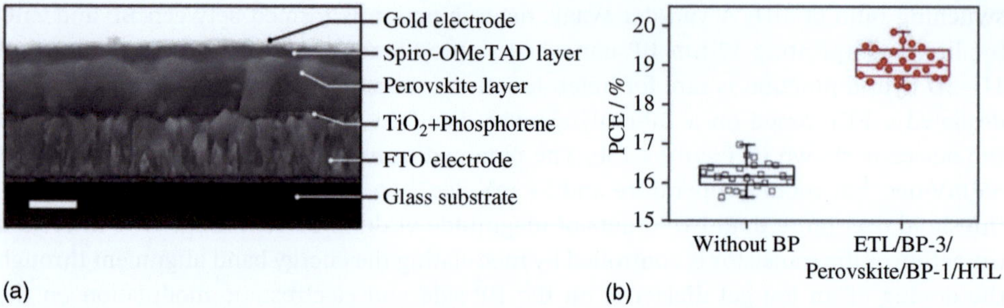


**Figure 4.14** (a) SEM cross-sectional image of planar n-i-p chalcogenide solar cell prepared on the basis of $TiO_2$-BP electrode. *Source:* Reprinted with permission from Ref. [67]. Copyright 2019, WILEY-VCH. (b) PV performance statistics with and without BP double insertion device. *Source:* Reprinted with permission from Ref. [70]. Copyright 2020, WILEY-VCH.

14.32 to 16.53%, and the maximum energy conversion efficiency reached 17.85%, which is comparable to that of the device prepared by the conventional high-temperature scheme. The study of density functional theory shows that the improvement of the device energy conversion efficiency is due to the high carrier mobility and more suitable energy band arrangement of BP. This study not only provides ideas for the novel synthesis of 2D BP but also opens up a new way for the application of BP as an electron transport layer in photovoltaic devices. In 2020, Zhang et al. [70] prepared highly efficient and stable chalcogenide solar cells by applying BP as a bifunctional nanomaterial for carrier extraction at the interfaces of electron transport layer/chalcogenide and chalcogenide/hole transport layer. As shown in Figure 4.14b, the device energy conversion efficiency of the chalcogenide solar cell was gradually increased from 16.95 (without BP) to 19.83%. Through the study of interfacial engineering, it was found that the introduction of BP could enhance the light absorption and reduce the defect density, thus improving the photovoltaic performance of chalcogenide solar cells.

Second, BP can also be doped in the hole-transporting layer for improving the energy band structure and hydrophobicity, thus increasing the crystallinity of chalcogenide and enhancing the performance of solar cells. In 2017, Muduli et al. [71] obtained few-layer BP nanosheets as a hole-transporting material for chalcogenide solar cells by liquid-phase stripping method. The results showed that the energy conversion efficiency of the solar cell with BP nanosheets introduced into Spiro-OMeTAD reached 16.4% compared with Spiro-OMeTAD ($\eta = 13.1\%$), which is 25% higher than the energy conversion efficiency of the solar cell without added BP nanosheets. In 2019, Li et al. [72] added BP between a hole transport layer (CuSCN) and an active layer ($CsPbI_3$), and the solar device achieved a photocurrent output of 19.3 $mA{\cdot}cm^{-2}$. It is found that the energy level structure and work function arrangement of the multilayer BP can be adjusted by improving the number of layers, which induces the formation of the built-in electric field and facilitates the transfer of carriers at the interface. In addition, the multilayer BP can increase the relative permittivity of the BP/calcite heterostructure in the localized dielectric environment due to its effective dielectric shielding effect, which reduces the exciton binding energy and improves the exciton dissociation efficiency of the inorganic calixarene, leading to a significant increase in the free-carrier density of the inorganic calixarene. The energy conversion efficiency of the

BP-$CsPbI_3$ heterostructured chalcogenide solar cell prepared with inorganic CuSCN hole transport layer reaches 14.17%. Subsequently, in 2021, Dong et al. [73] employed few-layer BP nanosheets doped with poly(triarylamine) (PTAA) as a hole transport layer in a chalcogenide solar cell. The BP-doped PTAA improved the charge extraction at the interface of chalcogenide and hole transport layer, decreased the surface potential of the PTAA film, and increased the conductivity of the hole transport layer. Compared with pristine PTAA, the chalcogenide crystals grown on the hole transport layer of BP-doped PTAA were larger in size and had fewer boundaries. The larger grain size and fewer boundaries imply fewer defects, and the charge compounding in the chalcogenide solar cell is suppressed. The results show that the energy conversion efficiency of the BP-doped device is dramatically increased from the original 18.26–20.45%, and the energy conversion efficiency of the optimal device is 20.49%. In addition, the doping of BP makes the PTAA more hydrophobic, which improves the device stability significantly.

In addition, BP can be doped into the active layer of solar cells to improve the performance of solar cells by suppressing $Pb^O$ defects. Wang et al. [74] incorporated BP into $CH_3NH_3PbI_3$ inclusion crystals, which exhibited significant photostability enhancement in addition to the photovoltaic performance of the devices. The BP-doped solar cells maintained an initial efficiency of about 94% after continuous illumination in a dry $N_2$ glove box for 1000 h, while the efficiency of the undoped solar cells decreased to 30%. Although the influence of BP on the structure and morphology of the clathrate crystal is not very large, the $Pb^O$ defects are effectively suppressed, and the hot-carrier complexation is delayed, thus improving the photostability of the device, which provides a new strategy for the improvement of the photostability of solar cells.

### 4.5.4 Biomedical and Sensing

BP nanosheets have a large specific surface area and high chemical activity due to their folded structure, which can be efficiently drug-carrying through van der Waals force interactions or $\pi$–$\pi$ coupling, and their high photothermal conversion efficiency makes them promising for photothermal therapy applications. Stability and biocompatibility of BP are crucial for biomedical applications [75].

In 2018, Guo et al. [76] synthesized black phosphorus quantum dots (BPQDs) with high stability and low toxicity by coupling BP nanosheets using the coupling agent polyethylene glycol, as shown in Figure 4.15. The obtained BPQDS had a small kinetic diameter and could be efficiently excreted through the kidneys, avoiding the potential physiological toxicity of BPQDS. The final product of BP is biocompatible and nonbiotoxic $P_xO_y{}^{n-}$ ion, which is a material with strong ability to target cancer cells. During application, BP can be degraded to phosphate ions [77], thus affecting the hydrolysis of intracellular adenosine triphosphate, inducing cell death, and strongly inhibiting the proliferation of cancer cells, while maintaining biocompatibility with normal cells, which is an effective anticancer strategy for tumor immunotherapy. In 2022, Sun et al. [78] fabricated a photoelectrochemical (PEC) biosensor for the ultrasensitive detection of small noncoding RNA (piRNA) based on BP/RP heterojunction-$Bi_2Te_3$ heterodimers, which combines the photosensitive heterodimer with an enzymatic signal amplification strategy for the detection of colorectal cancer-associated piRNA-823. The synergistic type II heterostructures of RP and BP,

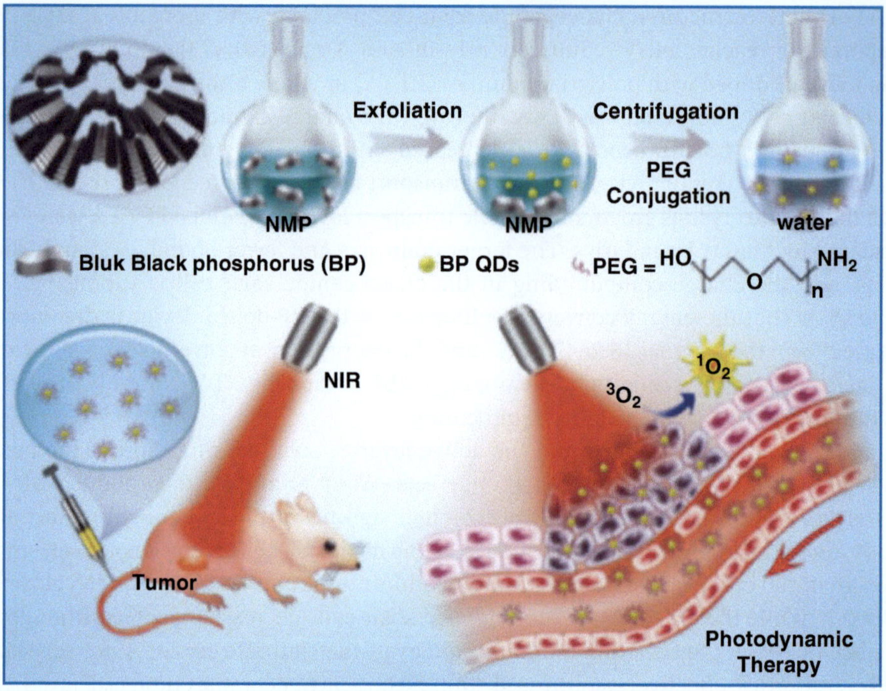


**Figure 4.15** Schematic illustration of the synthesis of BPQDs and their potential application in photo dynamic therapy. *Source:* Reprinted with permission from Ref. [76]. Copyright 2018, WILEY.

the photothermal effect of BP, and the excellent electron transfer ability and thermoelectric effect of $Bi_2Te_3$ result in the excellent photoelectric conversion performance of BP/RP heterojunction-$Bi_2Te_3$ heteroconjugates. The PEC biosensor exhibits excellent sensitivity, stability, reproducibility, and selectivity, and the sensor has great potential for ultrasensitive liquid biopsy of piRNA and cancer diagnosis.

BPQDs have excellent biocompatibility and biosafety, providing a platform for bioimaging and biosensing applications. Jiang et al. [79] explored a $Zn^{2+}$ ligand, adsorbed on the surface of BP through cation-$\pi$ bond interactions to passivate the lone pair of electrons of phosphorus, and used ultrasound-assisted solvent-thermal reaction to prepare Zn-BPQDs, which improved the strong fluorescence performance and stability of BPQDs; CoOOH nanosheets were prepared by hydrothermal reaction using propionic acid as the capping ligand, and Zn-BPQDs were combined with Zn-BPQDs to construct a strong fluorescent visual bio-detection probe for Zn-BPQDs/CoOOH nanosheets complexes, and the addition of glutathione as a reducing agent triggered the change of fluorescence intensity and visual color rule of Zn-BPQDs, which was further investigated for visual bio-detection and multifunctional bioimaging of Zn-BPQDs. According to the characterization analysis, it was found that Zn-BPQDs nanoparticles (NPs) were uniformly dispersed on the surface of CoOOH nanosheets to provide a large specific surface area for small NPs with hexagonal CoOOH nanosheets, and with the increase of CoOOH concentration, the efficiency of fluorescence bursting was significantly improved, and the change of fluorescence color could be observed visually. Therefore, Zn-BPQDs can be used as efficient fluorescent reagents for long-term tracking and bioimaging in vivo and in vitro.

In 2018, Chen et al. [80] proposed that BP nanosheets with a large number of periodic atomic grooves on the surface are ideal binding sites for proteins and other linear materials and can be used as a biodegradable delivery system for small interfering RNA of telomerase reverse transcriptase, a legal person for cancer gene therapy, and modified BP nanosheets with bifunctionalization of poly(ethylene glycol) and poly(ethylene imide) to use them as a novel carrier for small interfering RNA, which exhibited high loading capacity and strong cellular uptake. The BP nanomaterials are considered to be one of the most effective drug carriers due to their large specific surface area, unique surface chemistry, and optical properties.

### 4.5.5 Photocatalytic Degradation

Photocatalytic technologies, including photocatalytic hydrogenevolution reaction (HER), photocatalytic nitrogen reduction reaction (NRR), and $CO_2$ reduction reaction ($CO_2$RR), provide a new technological solution for clean hydrogen energy preparation, low energy consumption nitrogen fixation, and $CO_2$ reutilization, which is important for the realization of carbon-neutral carbon cycle, through solar energy. Driving water decomposition, $CO_2$ reduction, and $N_2$ reduction reactions provide a new technological solution for clean hydrogen energy preparation with low energy consumption for nitrogen fixation and $CO_2$ reuse, which is of great significance for realizing carbon neutrality and carbon cycle. The photocatalytic process based on semiconductor materials can be divided into the following steps: the material is irradiated by light with energy greater than or equal to its $E_g$, which generates photogenerated electron-hole pairs; the electrons jump to the CB, and holes appear in the valence band, and the electrons and holes occur on the surface of the material in the reduction and oxidation reactions, respectively [81, 82].

Experimental studies have shown that the photocatalytic performance of pure BP is relatively low or even inactive [83]. In 2021, Sun et al. [84] prepared quasi-monolayer BP from the precursor RP using a mild bottom-up approach and proposed a linear diamine-induced "solvation conjugation" mechanism using polyphosphate clusters to account for the phase transition of phosphorus in solution and for the first time demonstrated the advantages of quasi-monolayer BP in photocatalytic HER. The precursor RP was converted into ethylenediamine, 1,3-propylenediamine, or 1,4-butanediamine modified quasi-monolayers of BP, abbreviated as eda-BP, pda-BP and bda-BP, respectively, using a mild wet chemical reaction. In addition, pure BP was prepared by CVT method, and prepared into BP nanosheets (cvt-BP) by ultrasonic stripping method using NMP as solvent and compared its photocatalytic performance with that of quasi-monolayer BP. The pristine BP nanosheets presented a lower hydrogen production activity (7.9 $\mu mol \cdot h^{-1} \cdot g^{-1}$), whereas the photocatalytic performance of the quasi-monolayer BP was significantly improved after diamine modification, as shown in Figure 4.16a–e.

The advantages of monolayer BP (unique electronic structure, tunable bandgap and optical properties, and high carrier mobility) make it a promising and highly active $CO_2$ photoreduction catalyst with high efficiency of solar energy conversion [85]. In 2020, Zhu et al. [86] obtained, for the first time, in situ stabilized BP monolayer with hydroxyl modification

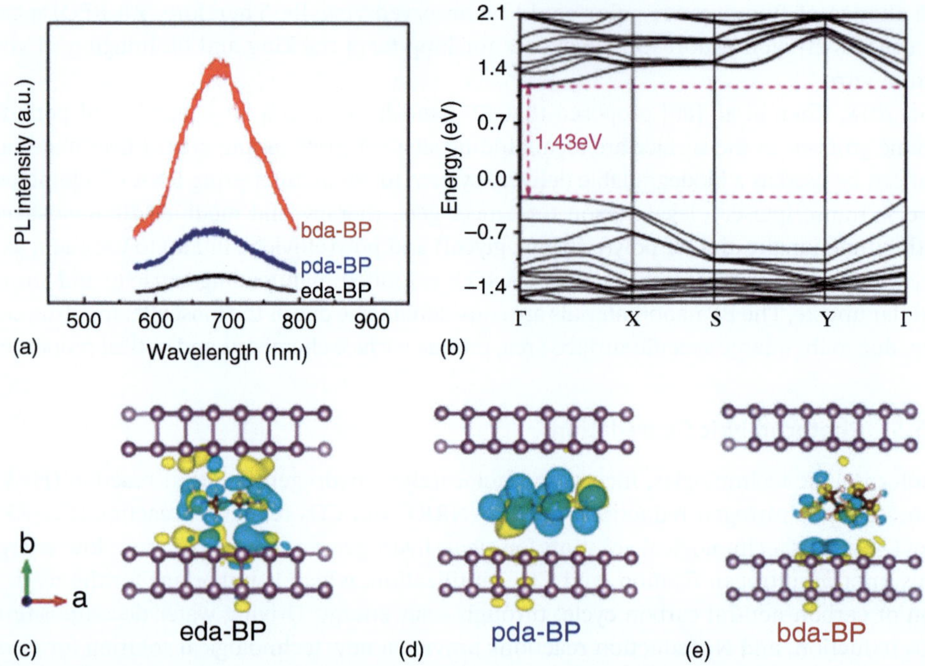


**Figure 4.16** (a) Photoluminescence spectra of eda-BP, pda-BP, and bda-BP showing photoluminescence peaks at ~692 nm. (b) Simulated band structure of eda-BP. (c) Charge density difference of the eda-BP. (d) pda-BP and (e) bda-BP systems. *Source:* Reprinted with permission from Ref. [84]. Copyright 2021, ELSEVIER SCIENCE SA.

(M-BP-OH) using liquid nitrogen-assisted green stripping method and successfully applied it as a photocatalyst for $CO_2$ reduction. The photocatalytic performances of the bulk BP and M-BP-OH materials were evaluated by photocatalytic $CO_2$RR under visible light irradiation, and the bulk BP exhibited a moderate $CO_2$RR activity with a CO yield of 25.3 $\mu mol \cdot h^{-1} \cdot g^{-1}$, while the CO yield of M-BP-OH reached 112. 6 $\mu mol \cdot h^{-1} \cdot g^{-1}$.

BP has many properties suitable for ammonia synthesis, such as an abundance of surface and edge sites [87, 88]. At the same time, the weak hydrogen absorption on bare BP can also inhibit the competitive hydrogen precipitation reaction [89]. These features indicate that BP is suitable for NRR, the core process of ammonia production. At the same time, since BP is a direct bandgap semiconductor, with a high carrier mobility and a wide range of light absorption, it has a great potential in light-assisted catalysis. In 2020, Liu et al. [90] utilized BP nanosheets as metal-free 2D catalysts for PEC NRR. The PEC synthesis of ammonia from BP nanosheets was carried out in a hermetically sealed H-type quartz reactor under a 0.1 mol/L HCl atmosphere saturated with $N_2$. Experiments showed that ammonia can be formed by electrocatalytic (electrocatalysis (EC) NRR in dark and PEC NRR in light at a bias of −0.4 V. In contrast, photocatalytic NRR in light without an external potential produced only a small amount of ammonia, suggesting that BP plays an important role in both EC NRR and PEC NRR, and the latter is more efficient than the former.

## 4.6 Blue Phosphorus

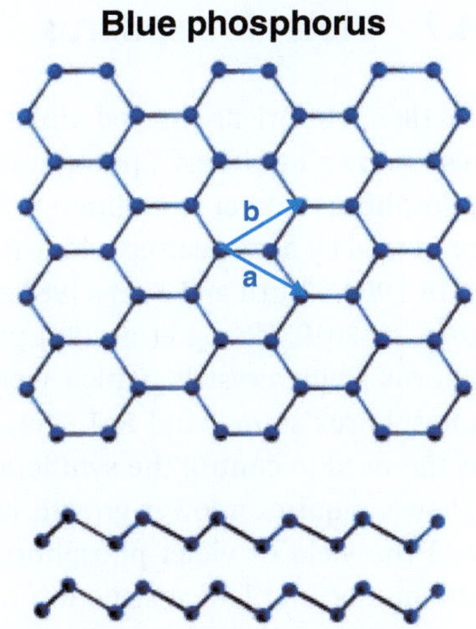


**Figure 4.17** Atomic model of blue phosphorus. a and b are the jagged edges of blue phosphorus, a = b = 3.28. *Source:* Reprinted with permission from Ref [92]. Copyright 2023, ROYAL SOC CHEMISTRY.

In 2014, Zhu and Tománek [91] investigated a new phosphorus phase based on density functional calculations that shares the same layered structure and high stability as the BP isomer, and the researchers named this in-plane hexagonal structure and body layer stacking blue phosphorus; its color is determined by its fundamental bandgap. It was found that blue phosphorus has a wider bandgap and has a different structure from BP, which has deeper "armchair" ridges, while blue phosphorus has a slightly flatter "zigzag" ridge similar to that of silylene, as shown in Figure 4.17, where the light blue arrowed portion of the figure shows the single-cell structure of blue phosphorus. The black rhombus in the figure shows the single-cell structure of blue phosphorus. a and b have Z-shaped edges, with a = b = 3.28 Å. Blue phosphorus atoms are covalently bonded within the layer with a bond length of 2.27 Å, The floor spacing is 5.63 Å. Blue phosphorus has a wider bandgap, and the bandgap of blue phosphorus single-molecule film can be adjusted from 2 to 1.2 eV in the block, and it is easy to peel, which makes it more suitable for application in electronic devices.

Compared to BP, blue phosphorus is less anisotropic in terms of mechanical, thermal, and electrical properties [93, 94]; its mechanical properties are more similar to those of isotropic materials, which is more conducive to device design and performance adjustment. However, blue phosphorus is also unstable and research on blue phosphorus started relatively late, limited by technical conditions, hindering the application of blue phosphorus in practice.

Monolayers of blue phosphorus crystals grow on several noble metals, but individual blue phosphorus bulk crystals do not exist. The synthesis of pure 2D atomically layered blue phosphorus remains a challenge.

In 2016, Zhang et al. [95] proposed a simple method for the molecular beam epitaxial synthesis of monolayers of blue phosphorus on Au(111) substrates using BP as a precursor and a combination of low-temperature scanning tunneling microscopy and density functional theory calculations. It is determined that the electronic bandgap of the monolayer of blue phosphorus on the Au(111) surface is 1.10 eV, capable of realizing the epitaxial growth of atomic-layered blue phosphorus in large quantities and with high purity. In 2020, Zhang et al. [96] synthesized atom-only ultrathin blue phosphorus single-molecule membranes by Si intercalation with fine-tuned interfaces and embedded blue phosphorus alloys on the Au(111) surface. And the evolution of the energy band structure from blue phosphorus-Au alloy to Si-intercalated blue phosphorus was confirmed by using density functional theory calculations.

## 4.7 Violet Phosphorus

In 1865, Hittorf discovered violet phosphorus from molten lead [97], violet phosphorus also known as Hittorf's phosphorus (or a-metallic phosphorus), is another stable layered phosphorus isomer in addition to BP, or a sub-stable intermediate from RP to BP. It can be produced by heat treating white/RP at 500°C or higher.

In 1969, Thurn and Krebs [98] synthesized violet phosphorus and determined its structure. In 2019, Zhang et al. [99] proposed a synthesis method that has been violet phosphorus single crystals, which were grown by CVT method using amorphous RP as the phosphorus source and $SnI_4$+$SnI_4$ as the transport agent. The key step in the synthesis is the need to control the synthetic growth temperature; and the synthesis of violet phosphorus requires a lower growth temperature gradient and a higher growth temperature; and the yield of violet phosphorus is roughly around 36%. The researchers determined the monoclinic lattice structure of violet phosphorus using X-ray diffraction with space group P2/n (13), lattice constants of, a = 9.210 Å, b = 9.128 Å, c = 21.893 Å, β = 97.776°, and a density of 2.369 g/cm$^{-3}$. Its asymmetric unit is shown in Figure 18a, and the lattice structure is shown in Figure 4.18b. The purple region in Figure 4.18b corresponds to the violet phosphorus repeating unit structure with a layer spacing of 1.1 nm in Figure 4.18a. The lattice structure is slightly different from that of the monoclinic crystal system, P2/c, reported by Thern and Krebs in 1969.

Violet phosphorus is a promising 2D semiconductor material for photonic and electronic applications [100]. In 2023, Wang et al. [101] prepared large-sized single crystals of violet phosphorus from molten lead by improving the parameters of the synthesis process. The crystal structure of the prepared violet phosphorus was found to be consistent with that of the violet phosphorus synthesized using CVT method. The large-size violet phosphorus

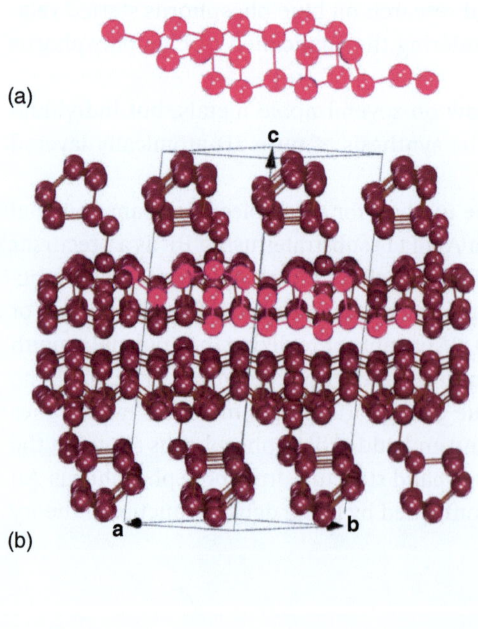


**Figure 4.18** (a) Asymmetric unit of violet phosphorus. (b) Layered structure of violet phosphorus. *Source:* Reprinted with permission from Ref. [99]. Copyright 2020, John Wiley and Sons Ltd.

has a transverse area of $1.30 \pm 0.41\ mm^2$, which is favorable for the stripping of violet phosphorus and can be used in the fabrication of electronic and optoelectronic devices.

## 4.8 Conclusion

Since its introduction, BP has attracted much attention from researchers due to its tunable energy bandgap, high carrier mobility, good switching ratio, and anisotropy, which make it a good compromise material between graphene and transition metal sulfides. In recent years, BP, as a popular material, has been widely used in the fields of photovoltaic devices, biomedicine, and optical communications. This paper focuses on the preparation of bulk BP and 2D BP and their applications in PDs, FET solar cells, biomedicine, photocatalysis, and other fields. Two phosphorus isomers, blue phosphorus and violet phosphorus, are also introduced. Various phosphorus isomers have made considerable achievements with the continuous expansion of preparation methods and structural properties, but there are still plenty of opportunities for further development.

## References


1 Jamieson, J.C. Crystal structures adopted by black phosphorus at high pressures. *Science* 1963, 139 (3561): 1291–1292.
2 Novoselov, K.S., Mishchenko, A., Carvalho, A., Castro Neto, A. 2D materials and van der Waals heterostructures. *Science* 2016, 353 (6298): aac9439.
3 Fei, R., Yang, L. Strain-engineering the anisotropic electrical conductance of few-layer black phosphorus. *Nano Lett.* 2014, 14 (5): 2884–2889.
4 Du, K., Lv, Q., Liang, Z., Liu, G., Hussain, S., Liu, J., Qiao, G. Trends in the preparation and passivation techniques of black phosphorus nanostructures for optoelectronics applications: a review. *ACS Appl. Nano Mater.* 2023, 6 (5), 3159–3183.
5 Zhang, H. Ultrathin two-dimensional nanomaterials. *ACS Nano* 2015, 9 (10): 9451–9469.
6 Wan, J., Lacey, S.D., Dai, J., Bao, W., Fuhrer, M.S., Hu, L. Tuning two-dimensional nanomaterials by intercalation: materials, properties and applications. *Chem. Soc. Rev.* 2016, 45 (24): 6742–6765.
7 Wu, S., Hui, K.S., Hui, K.N. 2D black phosphorus: from preparation to applications for electrochemical energy storage. *Adv. Sci.* 2018, 5 (5): 1700491.
8 Xing, B., Guan, L., Yu, Y., Niu, X., Yan, X., Zhang, S., Yao, J., Wang, D., Sha, J., Wang, Y. $HfO_2$-passivated black phosphorus field effect transistor with long-termed stability and enhanced current on/off ratio. *Nanotechnology* 2019, 30 (34): 345208.
9 Bridgman, P.W. Two new modifications of phosphorus. *J. Am. Chem. Soc.* 1914, 36 (7): 1344–1363.
10 Keyes, R.W. The electrical properties of black phosphorus. *Phys. Rev.* 1953, 92 (3): 580.
11 Sun, L.-Q., Li, M.-J., Sun, K., Yu, S.-H., Wang, R.-S., Xie, H.-M. Electrochemical activity of black phosphorus as an anode material for lithium-ion batteries. *J. Phys. Chem. C* 2012, 116 (28): 14772–14779.

**12** Zhou, F., Li, S., Ouyang, L., Liu, J., Liu, J., Huang, Z., Zhu, M. Facile synthesis of black phosphorene via a low melting media assisted ball milling. *Chem. Eng. J.* 2022, 444: 136593.

**13** Zhou, F., Ouyang, L., Zeng, M., Liu, J., Wang, H., Shao, H., Zhu, M. Growth mechanism of black phosphorus synthesized by different ball milling techniques. *J. Alloys Compd.* 2019, 784: 339–346.

**14** Lin, C., Yang, L., Ouyang, L., Liu, J., Wang, H., Zhu, M. A new method for few-layer graphene preparation via plasma-assisted ball milling. *J. Alloys Compd.* 2017, 728: 578–584.

**15** Bouša, D., Otyepková, E., Lazar, P., Otyepka, M., Sofer, Z. Surface energy of black phosphorus alloys with arsenic. *ChemNanoMat* 2020, 6 (5): 821–826.

**16** Nilges, T., Kersting, M., Pfeifer, T. A fast low-pressure transport route to large black phosphorus single crystals. *J. Solid State Chem.* 2008, 181 (8): 1707–1711.

**17** Köpf, M., Eckstein, N., Pfister, D., Grotz, C., Krüger, I., Greiwe, M., Hansen, T., Kohlmann, H., Nilges, T. Access and in situ growth of phosphorene-precursor black phosphorus. *J. Cryst. Growth* 2014, 405: 6–10.

**18** Zhang, D., Yuan, Z., Zhang, G., Tian, N., Liu, D., Zhang, Y. Preparation and characterization of black phosphorus. *Acta Chim. Sin.* 2018, 76 (7): 537.

**19** Zhu, J.-X., Liu, X.-D., Xue, M.-Z., Chen, C.-X. Phosphorene: synthesis, structure, properties and device applications. *Acta Phys.-Chim. Sin.* 2017, 33 (11): 2153–2172.

**20** Wu, Z., Huang, T., Sathishkumar, G., He, X., Wu, H., Zhang, K., Rao, X., Kang, E.T., Xu, L. Phytic acid-promoted exfoliation of black phosphorus nanosheets for the fabrication of photothermal antibacterial coatings. *Adv. Healthc. Mater.* 2024, 13 (4): 2302058.

**21** Novoselov, K.S., Geim, A.K., Morozov, S.V., Jiang, D.-E., Zhang, Y., Dubonos, S.V., Grigorieva, I.V., Firsov, A.A. Electric field effect in atomically thin carbon films. *Science* 2004, 306 (5696): 666–669.

**22** Li, L., Yu, Y., Ye, G.J., Ge, Q., Ou, X., Wu, H., Feng, D., Chen, X.H., Zhang, Y. Black phosphorus field-effect transistors. *Nat. Nanotechnol.* 2014, 9 (5): 372–377.

**23** Castellanos-Gomez, A., Vicarelli, L., Prada, E., Island, J.O., Narasimha-Acharya, K., Blanter, S.I., Groenendijk, D.J., Buscema, M., Steele, G.A., Alvarez, J. Isolation and characterization of few-layer black phosphorus. *2D Materials* 2014, 1 (2): 025001.

**24** Lu, W., Nan, H., Hong, J., Chen, Y., Zhu, C., Liang, Z., Ma, X., Ni, Z., Jin, C., Zhang, Z. Plasma-assisted fabrication of monolayer phosphorene and its Raman characterization. *Nano Res.* 2014, 7: 853–859.

**25** Rabiei Baboukani, A., Khakpour, I., Wang, C. In: *Phosphorene Exfoliated from Black Phosphorous Via Bipolar Electrochemistry, Electrochemical Society Meeting Abstracts 235*. The Electrochemical Society, Inc., 807–807. 2019.

**26** Ambrosi, A., Sofer, Z., Pumera, M. Electrochemical exfoliation of layered black phosphorus into phosphorene. *Angew. Chem. Int. Ed.* 2017, 56 (35): 10443–10445.

**27** Xiao, H., Zhao, M., Zhang, J., Ma, X., Zhang, J., Hu, T., Tang, T., Jia, J., Wu, H. Electrochemical cathode exfoliation of bulky black phosphorus into few-layer phosphorene nanosheets. *Electrochem. Commun.* 2018, 89: 10–13.

**28** Wang, T., Jin, X., Yang, J., Wu, J., Yu, Q., Pan, Z., Shi, X., Xu, Y., Wu, H., Wang, J. Oxidation-resistant enable highly ambient-stable ultrafast pulse generation at a 2 μm Tm/Ho-doped fiber laser. *ACS Appl. Mater. Interfaces* 2019, 11 (40): 36854–36862.

**29** Jia, C., Zhao, L., Cui, M., Yang, F., Cheng, G., Yang, G., Zeng, Z. Surface coordination modification and electrical properties of few-layer black phosphorus exfoliated by the liquid-phase method. *J. Alloys Compd.* 2019, 799: 99–107.

**30** Guo, Z., Zhang, H., Lu, S., Wang, Z., Tang, S., Shao, J., Sun, Z., Xie, H., Wang, H., Yu, X.F. From black phosphorus to phosphorene: basic solvent exfoliation, evolution of Raman scattering, and applications to ultrafast photonics. *Adv. Funct. Mater.* 2015, 25 (45): 6996–7002.

**31** Yasaei, P., Kumar, B., Foroozan, T., Wang, C., Asadi, M., Tuschel, D., Indacochea, J.E., Klie, R.F., Salehi-Khojin, A. High-quality black phosphorus atomic layers by liquid-phase exfoliation. *Adv. Mater* 2015, 27 (11): 1887–1892.

**32** Chen, L., Zhou, G., Liu, Z., Ma, X., Chen, J., Zhang, Z., Li, F., Cheng, H., Ren, W., Scalable clean exfoliation of high-quality few-layer black phosphorus for a flexible lithium ion battery. *Adv. Mater. (Deerfield Beach, Fla.)* 2015, 28 (3): 510–517.

**33** Smith, J.B., Hagaman, D., Ji, H.-F. Growth of 2D black phosphorus film from chemical vapor deposition. *Nanotechnology* 2016, 27 (21): 215602.

**34** Li, X., Deng, B., Wang, X., Chen, S., Vaisman, M., Karato, S.-I., Pan, G., Lee, M.L., Cha, J., Wang, H. Synthesis of thin-film black phosphorus on a flexible substrate. *2D Materials* 2015, 2 (3): 031002.

**35** Zhang, Z., Xin, X., Yan, Q., Li, Q., Yang, Y., Ren, T.-L. Two-step heating synthesis of sub-3 millimeter-sized orthorhombic black phosphorus single crystal by chemical vapor transport reaction method. *Sci. China Mater.* 2016, 2 (59): 122–134.

**36** Li, C., Wu, Y., Deng, B., Xie, Y., Guo, Q., Yuan, S., Chen, X., Bhuiyan, M., Wu, Z., Watanabe, K., Taniguchi, T., Wang, H., Cha, J.J., Snure, M., Fei, Y., Xia, F. Synthesis of crystalline black phosphorus thin film on sapphire. *Adv. Mater.* 2018, 30 (6): 1703748.

**37** Zhou, Y., Chen, J., Yue, L., Ren, D. Preparation of high-intensity fluorescent black phosphorus nanosheets. *Physica Status Solidi (a)* 2023, 220 (13): 2200806.

**38** Zhang, Y., Rui, X., Tang, Y., Liu, Y., Wei, J., Chen, S., Leow, W.R., Li, W., Liu, Y., Deng, J. Wet-chemical processing of phosphorus composite nanosheets for high-rate and high-capacity lithium-ion batteries. *Adv. Energy Mater.* 2016, 6 (10): 1502409.

**39** Zhao, G., Wang, T., Shao, Y., Wu, Y., Huang, B., Hao, X. A novel mild phase-transition to prepare black phosphorus nanosheets with excellent energy applications. *Small* 2017, 13 (7): 1602243.

**40** Tian, B., Tian, B., Smith, B., Scott, M., Lei, Q., Hua, R., Tian, Y., Liu, Y. Facile bottom-up synthesis of partially oxidized black phosphorus nanosheets as metal-free photocatalyst for hydrogen evolution. *Proc. Natl. Acad. Sci.* 2018, 115 (17): 4345–4350.

**41** Zhang, X., Xie, P.H., Liu, Z., Tan, C., Luo, D.Z., Li, D.H., Lin, J., Sun, D.L., Black phosphorus quantum dots. *Angew. Chem. Int. Ed.* 2015, 54 (12): 3653–3657.

**42** Tran, V., Soklaski, R., Liang, Y., Yang, L. Layer-controlled band gap and anisotropic excitons in few-layer black phosphorus. *Phys. Rev. B* 2014, 89 (23): 235319.

**43** Fox, M. *Optical Properties of Solids, Vol. 3.* Oxford university press, 2010.

**44** Tian, Y., Wang, H., Li, H., Guo, Z., Tian, B., Cui, Y., Li, Z., Li, G., Zhang, H., Wu, Y. Recent advances in black phosphorus/carbon hybrid composites: from improved stability to applications. *J. Mater. Chem. A* 2020, 8 (9): 4647–4676.

**45** Xia, F., Wang, H., Jia, Y. Rediscovering black phosphorus as an anisotropic layered material for optoelectronics and electronics. *Nat. Commun.* 2014, 5 (1): 4458.

**46** Liu, Y., Low, T., Ruden, P.P. Mobility anisotropy in monolayer black phosphorus due to scattering by charged impurities. *Phys. Rev. B* 2016, 93 (16): 165402.

**47** Hengkang, Z., Ting, J., Guohui, L. Research progress of photodetectors based on two-dimensional material. *Semicond. Mater.* 2019, 45 (01): 43–51.

**48** Li, L., Pi, L., Li, H., Zhai, T. Photodetectors based on two-dimensional semiconductors: progress, opportunity and challenge. *Chin Sci. Bull.* 2017, 62: 3134–3153.

**49** Wu, J., Koon, G.K.W., Xiang, D., Han, C., Toh, C.T., Kulkarni, E.S., Verzhbitskiy, I., Carvalho, A., Rodin, A.S., Koenig, S.P. Colossal ultraviolet photoresponsivity of few-layer black phosphorus. *ACS Nano* 2015, 9 (8): 8070–8077.

**50** Guo, Q., Pospischil, A., Bhuiyan, M., Jiang, H., Tian, H., Farmer, D., Deng, B., Li, C., Han, S.-J., Wang, H. Black phosphorus mid-infrared photodetectors with high gain. *Nano Lett.* 2016, 16 (7): 4648–4655.

**51** Wang, J., Fang, H., Wang, X., Chen, X., Lu, W., Hu, W. Recent progress on localized field enhanced two-dimensional material photodetectors from ultraviolet—Visible to infrared. *Small* 2017, 13 (35): 1700894.

**52** Wu, F., Li, Q., Wang, P., Xia, H., Wang, Z., Wang, Y., Luo, M., Chen, L., Chen, F., Miao, J. High efficiency and fast van der Waals hetero-photodiodes with a unilateral depletion region. *Nat. Commun.* 2019, 10 (1): 4663.

**53** Lee, C.-H., Lee, G.-H., Van Der Zande, A.M., Chen, W., Li, Y., Han, M., Cui, X., Arefe, G., Nuckolls, C., Heinz, T.F. Atomically thin p–n junctions with van der Waals heterointerfaces. *Nat. Nanotechnol.* 2014, 9 (9): 676–681.

**54** Fang, H., Battaglia, C., Carraro, C., Nemsak, S., Ozdol, B., Kang, J.S., Bechtel, H.A., Desai, S.B., Kronast, F., Unal, A.A. Strong interlayer coupling in van der Waals heterostructures built from single-layer chalcogenides. *Proc. Natl. Acad. Sci.* 2014, 111 (17): 6198–6202.

**55** Cao, S., Xing, Y., Han, J., Luo, X., Lv, W., Lv, W., Zhang, B., Zeng, Z. Ultrahigh-photoresponsive UV photodetector based on a BP/$ReS_2$ heterostructure p–n diode. *Nanoscale* 2018, 10 (35): 16805–16811.

**56** He, T., Li, C., Zhang, X., Ma, Y., Cao, X., Shi, X., Sun, C., Li, J., Song, L., Zeng, C. Metalorganic chemical vapor deposition heteroepitaxial β-$Ga_2O_3$ and black phosphorus Pn heterojunction for solar-blind ultraviolet and infrared dual-band photodetector. *Physica Status Solidi* (a) 2020, 217 (2): 1900861.

**57** Youngblood, N., Chen, C., Koester, S.J., Li, M. Waveguide-integrated black phosphorus photodetector with high responsivity and low dark current. *Nat. Photonics* 2015, 9 (4): 247–252.

**58** Yin, Y., Cao, R., Guo, J., Liu, C., Li, J., Feng, X., Wang, H., Du, W., Qadir, A., Zhang, H. High-speed and high-responsivity hybrid silicon/black-phosphorus waveguide photodetectors at 2 μm. *Laser Photonics Rev.* 2019, 13 (6): 1900032.

**59** Chen, X., Lu, X., Deng, B., Sinai, O., Shao, Y., Li, C., Yuan, S., Tran, V., Watanabe, K., Taniguchi, T. Widely tunable black phosphorus mid-infrared photodetector. *Nat. Commun.* 2017, 8 (1): 1672.

**60** Kou, L., Chen, C., Smith, S.C., Phosphorene: fabrication, properties, and applications. *J. Phys. Chem. Lett.* 2015, 6 (14): 2794–2805.

**61** Gao, T., Li, X., Xiong, X., Huang, M., Li, T., Wu, Y. Optimized transport properties in lithium doped black phosphorus transistors. *IEEE Electron Device Lett.* 2018, 39 (5): 769–772.

**62** Gao, A., Zhang, Z., Li, L., Zheng, B., Wang, C., Wang, Y., Cao, T., Wang, Y., Liang, S.J., Miao, F. Robust impact-ionization field-effect transistor based on nanoscale vertical graphene/black phosphorus/indium selenide heterostructures. *ACS Nano* 2019, 14 (1): 434–441.

**63** Jeon, P.J., Lee, Y.T., Lim, J.Y., Kim, J.S., Hwang, D.K., Im, S. Black phosphorus–zinc oxide nanomaterial heterojunction for p–n diode and junction field-effect transistor. *Nano Lett.* 2016, 16 (2): 1293–1298.

**64** Xu, J., Jia, J., Lai, S., Ju, J., Lee, S. Tunneling field effect transistor integrated with black phosphorus-MoS2 junction and ion gel dielectric. *Appl. Phys. Lett.* 2017, 110 (3): 033103.

**65** Bi, J., Zou, X., Lv, Y., Li, G., Liu, X., Liu, Y., Yu, T., Yang, Z., Liao, L. InGaZnO tunnel and junction transistors based on vertically stacked black phosphorus/InGaZnO heterojunctions. *Adv. Electron. Mater.* 2020, 6 (8): 2000291.

**66** Batmunkh, M., Bat-Erdene, M., Shapter, J.G. Black phosphorus: synthesis and application for solar cells. *Adv. Energy Mater.* 2018, 8 (5): 1701832.

**67** Batmunkh, M., Vimalanathan, K., Wu, C., Bati, A.S., Yu, L., Tawfik, S.A., Ford, M.J., Macdonald, T.J., Raston, C.L., Priya, S. Efficient production of phosphorene nanosheets via shear stress mediated exfoliation for low-temperature perovskite solar cells. *Small Methods* 2019, 3 (5): 1800521.

**68** You, P., Tang, G., Yan, F. Two-dimensional materials in perovskite solar cells. *Mater. Today Energy* 2019, 11: 128–158.

**69** Buscema, M., Groenendijk, D.J., Steele, G.A., Van Der Zant, H.S., Castellanos-Gomez, A. Photovoltaic effect in few-layer black phosphorus PN junctions defined by local electrostatic gating. *Nat. Commun.* 2014, 5 (1): 4651.

**70** Zhang, M., Ye, M., Wang, W., Ma, C., Wang, S., Liu, Q., Lian, T., Huang, J., Lin, Z. Synergistic cascade carrier extraction via dual interfacial positioning of ambipolar black phosphorene for high-efficiency perovskite solar cells. *Adv. Mater.* 2020, 32 (28): 2000999.

**71** Muduli, S.K., Varrla, E., Kulkarni, S.A., Han, G., Thirumal, K., Lev, O., Mhaisalkar, S., Mathews, N. 2D black phosphorous nanosheets as a hole transporting material in perovskite solar cells. *J. Power Sources* 2017, 371: 156–161.

**72** Li, B., Zhang, Y., Fu, L., Zhang, L., Liu, Z., Yin, L. Two-dimensional black phosphorous induced exciton dissociation efficiency enhancement for high-performance all-inorganic $CsPbI_3$ perovskite photovoltaics. *J. Mater. Chem. A* 2019, 7 (39): 22539–22549.

**73** Dong, W., Xiong, S., Yang, J., Qiao, W., Zeng, Q., Wang, X., Yao, Y., Bao, Q. Black phosphorus doped poly (triarylamine) as hole transport layer for highly efficient perovskite solar cells. *Org. Electron.* 2021, 89: 106052.

**74** Wang, Y., Zhang, H., Zhang, T., Shi, W., Kan, M., Chen, J., Zhao, Y. Photostability of $MAPbI_3$ perovskite solar cells by incorporating black phosphorus. *Solar RRL* 2019, 3 (9): 1900197.

**75** Li, R., Zha, Z., Miao, Z., Xu, C.-Y., Emerging 2D pnictogens for biomedical applications. *Chin. Chem. Lett.* 2022, 33 (5): 2345–2353.

**76** Guo, T., Wu, Y., Lin, Y., Xu, X., Lian, H., Huang, G., Liu, J.Z., Wu, X., Yang, H.H. Black phosphorus quantum dots with renal clearance property for efficient photodynamic therapy. *Small* 2018, 14 (4): 1702815.

**77** Ma, C., Zhang, J., Zhang, Y., Ma, C., Ma, M., Ran, F., Liu, X., Zhang, H. Phosphate imbalance conducting by BPs-based cancer-targeting phosphate anions carrier induces necrosis. *Chin. Chem. Lett.* 2021, 32 (4): 1550–1554.

**78** Sun, Z., Tong, Y., Li, J., Wang, Y., Gao, F., Li, H., Wang, C., Du, L., Jiang, Y. Ultrasensitive photoelectrochemical biosensor based on black/red phosphorus heterojunction@ $Bi_2Te_3$ hybrid and enzymatic signal amplification for the detection of colorectal cancer-related piRNA-823. *Sensors Actuat B: Chem.* 2022, 368: 132244.

**79** Jiang, X., Jin, H., Gui, R. Visual bio-detection and versatile bio-imaging of zinc-ion-coordinated black phosphorus quantum dots with improved stability and bright fluorescence. *Biosens Bioelectron.* 2020, 165: 112390.

**80** Chen, L., Chen, C., Chen, W., Li, K., Chen, X., Tang, X., Xie, G., Luo, X., Wang, X., Liang, H. Biodegradable black phosphorus nanosheets mediate specific delivery of hTERT siRNA for synergistic cancer therapy. *ACS Appl. Mater. Interfaces* 2018, 10 (25): 21137–21148.

**81** Shen, Z.K., Yuan, Y.J., Pei, L., Yu, Z.T., Zou, Z. Black phosphorus photocatalysts for photocatalytic $H^2$ generation: a review. *Chem. Eng. J.* 2020, 386: 123997.

**82** Bian, S., Wen, M., Wang, J., Yang, N., Chu, P.K., Yu, X.-F. Edge-rich black phosphorus for photocatalytic nitrogen fixation. *J. Phys. Chem. Lett.* 2020, 11 (3): 1052–1058.

**83** Ozawa, A., Yamamoto, M., Tanabe, T., Hosokawa, S., Yoshida, T. Black phosphorus synthesized by solvothermal reaction from red phosphorus and its catalytic activity for water splitting. *J. Mater. Chem. A* 2020, 8 (15): 7368–7376.

**84** Sun, Z., Miao, H., Khurram, M., Zhang, Z., Zhu, Y., Yan, Q. Bottom-up approach to quasi-monolayer black phosphorus advancing photocatalytic $H_2$ evolution. *Chem. Eng. J.* 2021, 421: 127841.

**85** Han, C., Li, J., Ma, Z., Xie, H., Waterhouse, G.I., Ye, L., Zhang, T., Black phosphorus quantum dot/g-C3N4 composites for enhanced $CO_2$ photoreduction to CO. *Sci. China Mater.* 2018, 9 (61): 1159–1166.

**86** Zhu, X., Huang, S., Yu, Q., She, Y., Yang, J., Zhou, G., Li, Q., She, X., Deng, J., Li, H. In-situ hydroxyl modification of monolayer black phosphorus for stable photocatalytic carbon dioxide conversion. *Appl. Catal. B: Environ.* 2020, 269: 118760.

**87** Zhang, L., Ding, L.X., Chen, G.F., Yang, X., Wang, H., Ammonia synthesis under ambient conditions: selective electroreduction of dinitrogen to ammonia on black phosphorus nanosheets. *Angew. Chem.* 2019, 131 (9): 2638–2642.

**88** Liu, Y.T., Li, D., Yu, J., Ding, B. Stable confinement of black phosphorus quantum dots on black tin oxide nanotubes: a robust, double-active electrocatalyst toward efficient nitrogen fixation. *Angew. Chem.* 2019, 131 (46): 16591–16596.

**89** Lu, J., Zhang, X., Liu, D., Yang, N., Huang, H., Jin, S., Wang, J., Chu, P.K., Yu, X.F. Modulation of phosphorene for optimal hydrogen evolution reaction. *ACS Appl. Mater. Interfaces* 2019, 11 (41): 37787–37795.

**90** Liu, D., Wang, J., Bian, S., Liu, Q., Gao, Y., Wang, X., Chu, P.K., Yu, X.F. Photoelectrochemical synthesis of ammonia with black phosphorus. *Adv. Funct. Mater.* 2020, 30 (24): 2002731.

**91** Zhu, Z., Tománek, D. Semiconducting layered blue phosphorus: a computational study. *Phys. Rev. Lett.* 2014, 112 (17): 176802.

**92** Tian, H., Wang, J., Lai, G., Dou, Y., Gao, J., Duan, Z., Feng, X., Wu, Q., He, X., Yao, L. Renaissance of elemental phosphorus materials: properties, synthesis, and applications in sustainable energy and environment. *Chem. Soc. Rev.* 2023, 52 (16): 5388–5484.

**93** Zhang, J.L., Zhao, S., Telychko, M., Sun, S., Lian, X., Su, J., Tadich, A., Qi, D., Zhuang, J., Zheng, Y. Reversible oxidation of blue phosphorus monolayer on Au(111). *Nano Lett.* 2019, 19 (8): 5340–5346.

**94** Sun, Y., Wang, L., Wang, C., Tang, C. Mechanical properties of 2D blue phosphorus and temperature effect. *Nanotechnology* 2020, 32 (8): 085702.

**95** Zhang, J.L., Zhao, S., Han, C., Wang, Z., Zhong, S., Sun, S., Guo, R., Zhou, X., Gu, C.D., Yuan, K.D. Epitaxial growth of single layer blue phosphorus: a new phase of two-dimensional phosphorus. *Nano Lett.* 2016, 16 (8): 4903–4908.

**96** Zhang, J.L., Zhao, S., Sun, S., Ding, H., Hu, J., Li, Y., Xu, Q., Yu, X., Telychko, M., Su, J. Synthesis of monolayer blue phosphorus enabled by silicon intercalation. *ACS Nano* 2020, 14 (3): 3687–3695.

**97** Hittorf, W. Zur kenntniss des phosphors. *Annalen der Physik* 1865, 202 (10): 193–228.

**98** Thurn, H., Krebs, H. Über struktur und eigenschaften der halbmetalle. XXII. Die kristallstruktur des hittorfschen phosphors. *Acta Crystallographica Section B: Struct. Crystallogr. Cryst Chem.* 1969, 25 (1): 125–135.

**99** Zhang, L., Huang, H., Zhang, B., Gu, M., Zhao, D., Zhao, X., Li, L., Zhou, J., Wu, K., Cheng, Y. Structure and properties of violet phosphorus and its phosphorene exfoliation. *Angew. Chem.* 2020, 132 (3): 1090–1096.

**100** Lu, Y.L., Dong, S., He, H., Li, J., Wang, X., Zhao, H., Wu, P. Enhanced photocatalytic activity of single-layered Hittorf's violet phosphorene by isoelectronic doping and mechanical strain: a first-principles research. *Comput. Mater. Sci.* 2019, 163: 209–217.

**101** Wang, Y., Jin, M., Gu, M., Zhao, X., Xie, J., Meng, G., Cheng, Y., He, J., Zhang, J. Synthesis of violet phosphorus with large lateral sizes to facilitate nano-device fabrications. *Nanoscale* 2023, 15 (29): 12406–12412.

V

# Tellurium

*Shancheng Yan*

## 5.1 Introduction

Tellurium (Te) belongs to the sixth group of the genus Sulfur, atomic number 52, atomic mass 127.6. Tellurium has two kinds of isotopes. One is crystalline tellurium, with a general metallic luster, density of 6.25 g/cm$^2$, melting point of 452°C, and boiling point of 1390°C; and the other is amorphous tellurium, a grayish-black powder, density of 6 g/cm$^2$, melting point of 449.5 ± 0.3°C, and boiling point of 989.8 ± 3.8°C. Tellurium is a well-known p-type semiconductor, a member of the sulfur family of elements, with a narrow bandgap of 0.35 eV at room temperature [1]. The properties of tellurium nanocrystals depend mainly on the crystal structure of elemental tellurium, where individual helical chains of tellurium atoms are stacked together by van der Waals (vdW) bonds and rotate around an axis parallel to the [001][2] direction at the center and corners of the hexagonal basic unit [3–5]. Each tellurium atom is covalently bonded to its two nearest atoms in the same chain. Early studies have shown that tellurium has a smaller effective mass and higher hole migration due to spin–orbit coupling. The smaller bandgap and higher mobility make tellurium well-suited for fast-response devices [6–8].

Size and shape are two typical parameters that determine nanomaterials [9]. The miniaturization of size gives nanomaterials some unique physical and chemical properties, such as bulk effects and temperature [1]. The properties of tellurium nanocrystals depend mainly on the crystal structure of elemental tellurium, where individual helical chains of tellurium atoms are stacked together by vdW bonds and rotate around an axis parallel to the [001][2] direction at the center and corners of the hexagonal basic unit [3–5] surface and interfacial effects, small size effects, and macroscopic quantum tunneling effects [10–13]. In comparison, shape controls the growth direction of nanocrystals through molecular dynamics and, thus, the surface structure of nanocrystals. Therefore, controlling the size and shape for tuning into the properties of nanocrystalline materials provides excellent flexibility and dramatically extends the range of applications of nanocrystals. As a result, shape control engineering has been a popular field in recent years, and materials with different shapes exhibit different properties and are used in different fields [14–16]. In this chapter, we report on the synthetic routes for tellurium nanomaterials with different morphologies (e.g. nanoparticles [NPs], nanowires [NWs], nanotubes, nanosheets) that require only a change in reaction parameters. These earlier works demonstrate the importance of controlling the morphology of the nanocrystals to maximize their properties. We mainly summarize and highlight

*Single Element Semiconductors: Properties and Devices*, First Edition. Yi Shi and Shancheng Yan.

recent advances in the rational design and morphology-controlled synthesis of tellurium nanostructures and their properties and applications in different fields.

## 5.2 Controlled Synthesis of Tellurium Nanomaterials

Nanomaterials are materials with at least one dimension less than 100 nm in their spatial structure. Since its introduction in the mid-20th century, it has received much attention from the scientific community. Size and shape are two critical parameters of nanomaterials. Size determines the size of the specific surface area and the number of one-dimensional (1D) atoms, and shape controls the structure of the nanocrystals; together, size and shape control the physical and chemical properties of nanomaterials [17–19]. Nanomaterials can be classified into zero-dimensional (0D), 1D, two-dimensional (2D), and three-dimensional (3D) materials according to their dimensionality. When the size of an arbitrary nanomaterial is miniaturized to nanometers, the control of the material's dimensions plays a crucial role in its properties due to the different ways in which electrons interact with the 0D, 1D, 2D, and 3D materials [20, 21]. In the past decades, tellurium nanocrystals of various shapes, including 0D NPs, 1D and 2D nanostructures, and even more complex chiral structures, have been successfully synthesized by different synthesis protocols (e.g. hydrothermal methods, gas-phase deposition, and microbiome-assisted synthesis methods) [22–24].

### 5.2.1 0D Tellurium Nanomaterial Synthesis

0D nanostructures are less than 100 nm nanocrystals in all three dimensions and are widely used in electronics, catalysis, optics, and biology [25–27]. Tellurium NPs are generally used as precursors for the synthesis of 1D nanostructures. At the beginning of the formation of 1D nanostructures, seeds of tellurium NPs are generally formed. Müller [28] and his team prepared single-crystalline tellurium NPs by introducing $H_2Te$ gas into a bromohexadecyl pyridine (CPB) solution of an organic solvent in a room-temperature solution. The synthesized NPs were uniform in size with a diameter of about $22 \pm 3$ nm, and the transmission electron microscopy (TEM) and high-resolution transmission electron microscopy (HRTEM) of the tellurium NPs are shown in Figure 5.1a and b. Tellurium nanorods were finally synthesized by controlling the reaction time and the amount of $H_2Te$ gas. Webster [28] and his team synthesized tellurium NPs using microwave-assisted reaction green using orange, lemon, and lime extracts as reducing and capping agents to expand the application of tellurium NPs in the biological field. Using different extracts to synthesize tellurium NPs TEM, as shown in Figure 5.1c (lemon juice) and d (lime juice), the structure of tellurium NPs with cubic shape was observed. Tellurium NPs showed significant antibacterial activity against gram-negative and positive bacteria in the 5~50 μg/mL concentration range.

### 5.2.2 Controllable Synthesis of 1D Tellurium Nanostructures

1D structures, including NWs, nanotubes, and nanoribbons, have received much attention in the last two decades for their applications in semiconductors and energy storage devices [29–31]. Tellurium nanocrystals are highly anisotropic, with atoms and atoms interlinked

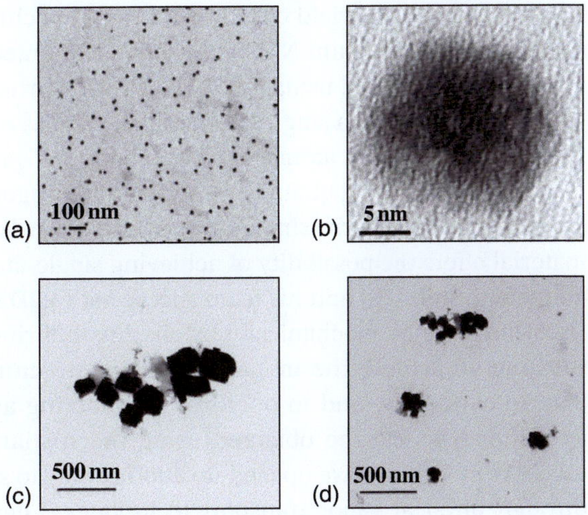


**Figure 5.1** (a) TEM images of Te nanoseeds. (b) HRTEM image of a single seed lemon. *Source:* Reprinted with permission from Ref. [28]. Copyright 2008, WILEY-VCH, (c) and lime (d) juices with 50 mM of metallic salt concentration. *Source:* Reprinted with permission from Ref. [29]. Copyright 2019, Royal Society of Chemistry.

through covalent bonds into a helical chain-like structure with a strong tendency to grow in the c-axis direction, contributing to the prerequisite for the growth of 1D tellurium nanostructures [32].

#### 5.2.2.1 Controlled Synthesis of Tellurium NWs

NWs are typically 1D nanostructures with a wide range of applications in semiconductor nanodevices, such as field effect transistors (FETs) and photodetectors (PDs), due to their miniaturization in two dimensions [33, 34]. To obtain tellurium NWs, controlling both dimensions of the nanocrystals at the nanoscale level is necessary. Ohasi [35] and his team synthesized tellurium nanowhiskers by thermal evaporation and found that the temperature of the substrate is critical to control the synthesis. Komarneni [36] and his team proposed a green chemical starch to assist the large-scale synthesis of tellurium NWs. Where $H_2TeO_4{\cdot}2H_2O$ was used as a tellurium source, starch is an effective reducing agent and a novel morphology guide for tellurium NWs. Figure 5.2a shows the TEM image of the synthesized sample. It can be seen that the obtained tellurium crystals have a homogeneous linear morphology; the average diameter of these NWs is 25 nm, and the length can be up to 10 μm. Figure 5.2b shows the HRTEM image, where the planar spacing of the tellurium NW's long axis is 0.59 nm. This spacing corresponds to the (001) crystallographic plane, indicating a preferential orientation toward the [001] growth direction. Yu [37] and his team demonstrated the first sub-kilogram-scale synthesis of ultrathin tellurium

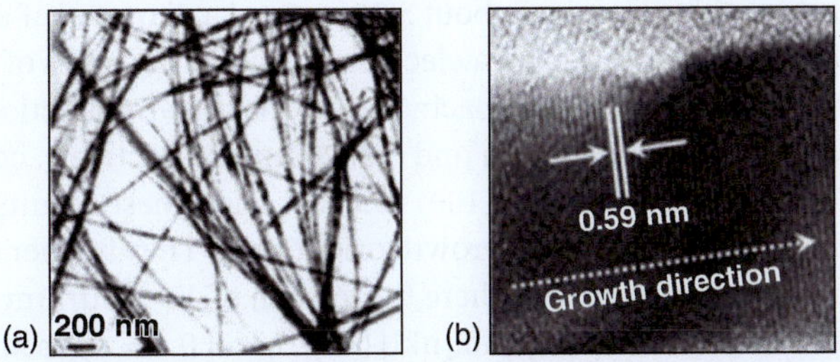


**Figure 5.2** (a) TEM images and (b) HRTEM image of the obtained Te sample. *Source:* Reprinted with permission from Ref. [36]. Copyright 2005, American Chemical Society.

NWs to scale up the yield of tellurium NWs to achieve a cheaper synthesis. They synthesized ultrathin tellurium NWs with uniform diameters of a few nanometers by a one-pot hydrothermal method using $Na_2TeO_3$ as the tellurium source and hydrazine hydrate as the reducing agent, with a single synthesis yielding as much as 150 g. Covalent bonds link the tellurium nanocrystal atoms to form a chain-like structure, with three atoms per helix per turn and neighboring chains arranged in a hexagonal shape. The chains of neighboring atoms combine to form single crystals through vdW forces, and this 1D, weakly bonded material offers the possibility of achieving single-atom chain structures with atomic-scale diameters. Shih [38] and his team succeeded in 2D sheets and 1D NWs 1~2 nm thick and ~100 nm wide by mechanically exfoliating tellurium single crystals directly on a silicon substrate to achieve the monatomic chain structures. Since individual tellurium atoms tend to covalently bind to only two neighboring atoms in the chain, single-atom chain tellurium NWs can be obtained using the spatial confinement of single-walled nanotubes. Ye et al. [39] have opened up another way to synthesize single-chain tellurium NWs through physical vapor transport techniques. They used carbon nanotubes and boron nitride nanotubes filled with tellurium vapor at high temperatures, and by controlling the inner diameter of these nanotubes, they synthesized tellurium NWs with few-atom chains and single-atom chains. High-angle annular dark-field scanning transmission electron microscopy (HAADF-STEM) and HRTEM images show that single-chain tellurium NWs have been successfully synthesized in single-walled carbon nanotubes (SWCNTs) with a diameter of 0.8 nm as a container. As the diameters of the SWCNTs were enlarged to 1.2 and 1.5 nm, the number of tellurium atom chains in the cavities increased to two and three. When the diameter of the SWCNT is expanded to 2.3 nm, tellurium NWs formed by 19 atomic chains are obtained.

#### 5.2.2.2 Controlled Synthesis of Tellurium Nanotubes

As a hollow 1D structure, nanotubes have attracted much attention due to their large specific surface area. Xia et al. [40] synthesized single-crystal nanotubes by a solution-phase approach. They dissolved tellurium powder in ethylene glycol and synthesized single-crystal tellurium nanotubes using condensation reflux. Furthermore, the growth process of tellurium nanotubes was revealed by TEM (Figure 5.3a). At the beginning of the reaction, tellurium atoms grow by uniform nucleation to form seeds with concave ends, and as the reaction continues, tellurium nanotubes are eventually formed due to the anisotropy of their growth (Figure 5.3b). Lu [41] and his team achieved the transformation of tellurium NWs to nanotubes by microwave-assisted synthesis using $TeO_2$ powder as the source of tellurium, ethylene glycol as the solvent and the reductant, and the presence of potassium hydroxide through controlling the temperature. Figure 5.3c shows the synthesized nanotubes with a length of about 5.75 μm and a diameter of about 240 nm. The HRTEM images and SAED (selected area electron diffraction) images of tellurium nanotubes are shown in Figure 5.3d. The facet spacing along the growth direction and perpendicular to the growth direction are 0.5856 nm and 0.3823 nm, respectively, corresponding to the lattice spacing between the (001) and (100) facets of t-Te. These results suggest that tellurium nanotubes have [001] preferential growth orientation. They further explored the growth mechanism of tellurium nanotubes, where the growth of 1D tellurium nanostructures can be viewed as a combination of [001] and [hkl] ($h = 0, k = 0, l = 0$) vectors and at temperatures lower than

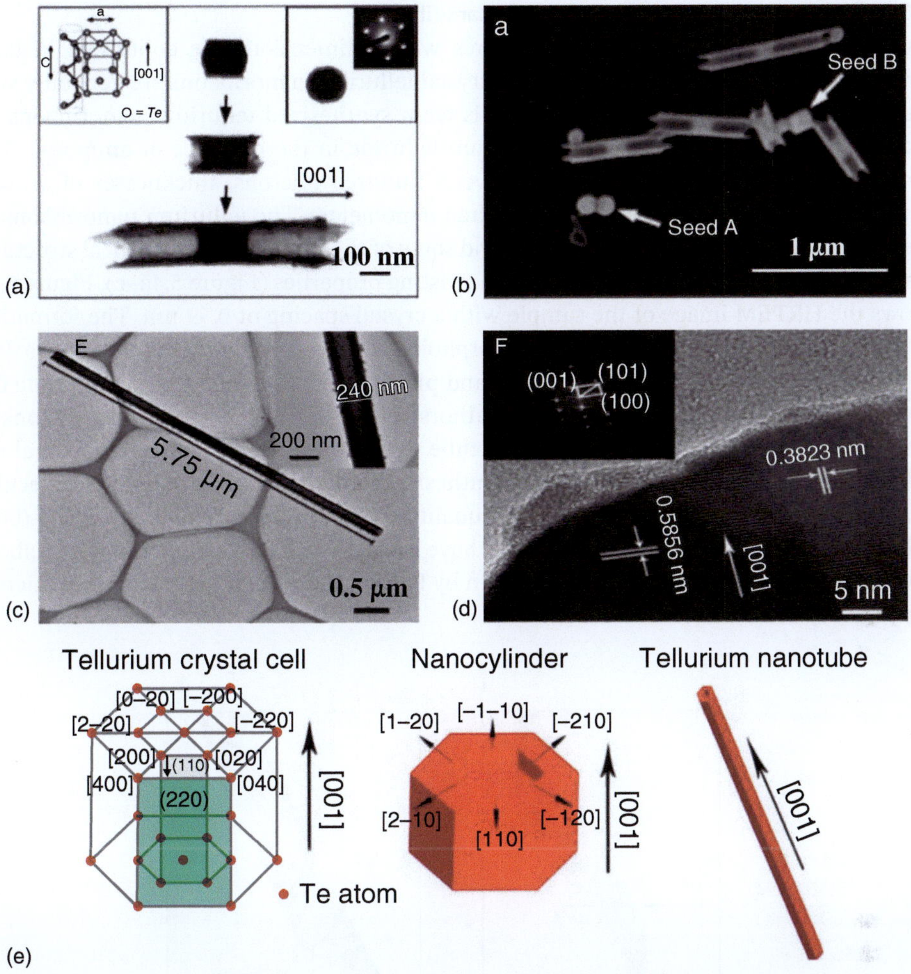


**Figure 5.3** (a) TEM images of tellurium nanotubes at three different stages of growth. (b) SEM images of tellurium nanotubes that were synthesized by refluxing a solution of ortho-telluric acid in ethylene glycol for 4 min. *Source:* Reprinted with permission from Ref. [40]. Copyright 2002, WILEY-VCH. (c) TEM images and (d) HRTEM images of Te nanotubes synthesized in the presence of NaOH (0.3 g) at 200°C for 25 min. (e) Schematic diagram of the growth mechanism of tellurium nanotubes. *Source:* Reprinted with permission from Ref. [41]. Copyright 2013, Royal Society of Chemistry.

155°C, low thermal energies limit the further formation of 1D tellurium nanostructures. As the temperature rises to 170°C, the tellurium nanostructures grow in the extended [001] crystal direction to form tellurium NWs, as the crystal surface energy in the (001) plane is much higher than that in the (110) plane. As the temperature increased to 185°C, the (110) plane, which had been relatively stable at 170°C, became violently involved in the reaction, with the (110) plane and its centrosymmetric plane promoting circumferential growth and the combined effect of [001] crystal-directed and circumferential growth promoting the formation of tellurium nanostructures (Figure 5.3e).

### 5.2.2.3 Controlled Synthesis of Tellurium Nanoribbons

Nanoribbons are 1D ribbon nanostructures whose dimensions are controlled in both dimensions to within 100 nm, and single-crystal tellurium nanoribbons are generally synthesized hydrothermally. Qian [42] and his team synthesized tellurium nanoribbons by disproportionate inverse synthesis of sodium telluride in the presence of ammonia. The synthesized products reach lengths of several hundred microns, thicknesses of around 8 nm, and widths of thirty to several hundred nanometers. The tellurium nanoribbons in the samples have a uniform morphology, and some of the samples show a helical structure, which indicates that they have good helical twisting properties (Figure 5.4a–c). Figure 5.4d shows the HRTEM image of the sample with a crystal spacing of 0.59 nm. The formation of tellurium nanoribbons with uniform morphology depends on their growth kinetics. It is subject to reaction pressure, temperature, and pH, and proper regulation can promote the synthesis of tellurium nanoribbons. The authors believe that using helical nanoribbons as templates allows for the formation of nanotube structures through epitaxial growth along the edges. In addition to hydrothermal synthesis, Li and his team proposed a molecular beam epitaxy method to synthesize high-quality tellurium nanoribbons on $SrTiO_3$(001) substrates. The individual Te nanoribbons have a length of about 1 μm, a width of about 100 nm, and a thickness of 15 nm, as shown by Figure 5.4d SEM and atomic force microscopy (AFM) images.

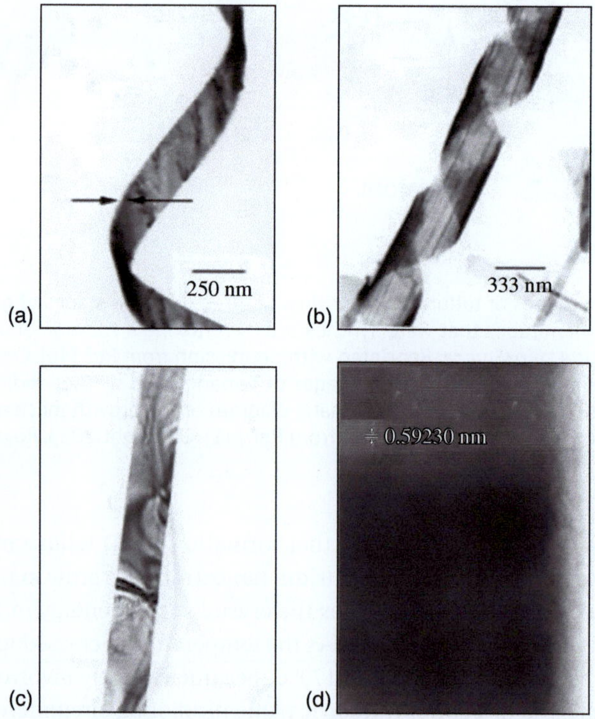


**Figure 5.4** (a–c) TEM images of tellurium nanosheets. (d) HRTEM images. *Source:* Reprinted with permission from Ref. [48]. Copyright 2008, WILEY-VCH.

### 5.2.3 Synthesis of 2D Tellurium Nanostructures

Moore's Law has governed advanced semiconductor manufacturing since its introduction in 1965. With further miniaturization in size, it became clear that this process could not continue forever. As devices approach the 2-nm node, their power consumption, reliability, and mobility degradation become increasingly severe. To continue Moore's Law, attention has turned to 2D material platforms [43–45]. With 2D graphene coming into the limelight, 2D materials have garnered significant interest in micro- and nanodevices. However, the field of 2D materials faces major challenges, notably the zero bandgap of graphene, which limits its application in PDs [46]. Another popular 2D material, black phosphorus (BP), suffers from severe environmental and structural degradation despite its excellent optoelectronic properties, making it difficult to achieve large-scale applications. Tellurium nanomaterials have excellent mobility, adjustable bandgap, and good environmental stability [47]. Their bulk material formation is by their atomic chains through vdW force interactions, giving them the potential to become 2D materials. In 2017, the first large-scale synthesis of 2D tellurium nanosheets was reported, making them a new addition to the list of single-element 2D materials [48]. Compared to the synthesis methods of tellurium nanomaterials for 0D and 1D materials, there are relatively few synthesis strategies for 2D tellurium, including gas-phase deposition, hydrothermal synthesis, vdW epitaxy, and liquid-phase exfoliation.

#### 5.2.3.1 Gas-phase Deposition

Vapor-phase deposition is depositing or reacting a heated substrate on a specific substrate to produce a specified morphology. It is usually classified into physical vapor deposition (PVD) and chemical vapor deposition (CVD). Apte et al. [49] placed bulk tellurium on Si/$SiO_2$ substrate by PVD. They thermally evaporated in an Ar/$H_2$ atmosphere at 650°C to form ultrathin tellurium nanofilms (Figure 5.5a). These tellurium nanofilms can have transverse dimensions up to 50 μm. The synthesized tellurium nanofilms were found to have a uniform thickness of $0.85 \pm 0.1$ nm, corresponding to three atomic layers by AFM. The presence of the α-Te phase was confirmed by scanning tunneling microscopy (Figure 5.5b). Peng et al. [50] placed $SnTe_2$ powder into a ceramic boat placed in the middle of a quartz tube by the CVD method and heated it in an atmosphere where the nitrogen flow rate was 50 standard state cubic centimeters per minute. The decomposed Sn and Te atoms were blown by nitrogen to different regions of the substrate. Tellurium nanosheets with a length of 20–80 μm and a thickness of 200 nm were obtained in the growth region at a temperature of 400°C (Figure 5.5c). TEM and HRTEM images of Te nanosheets and NWs are shown in Figure 5.5d. HRTEM indicates that the synthesized tellurium nanosheets have high crystallinity, and the lattice spacing of the tellurium nanosheets is 0.59 nm, which agrees with the spacing of (001) hexagonal tellurium lattice planes as studied by previous researchers.

#### 5.2.3.2 Hydrothermal Synthesis

Hydrothermal synthesis is considered to be a cost-effective and straightforward synthesis method. Wu [52] and his team synthesized large-area, high-quality 2D tellurium nanocrystals by substrate-less hydrothermal synthesis using $Na_2TeO_3$ as the tellurium source and hydrazine hydrate as the reductant at 180°C. Figure 5.6a shows a typical atomically resolved HRTEM image of a tellurium sheet. The spacing of the samples is 0.222 nm and

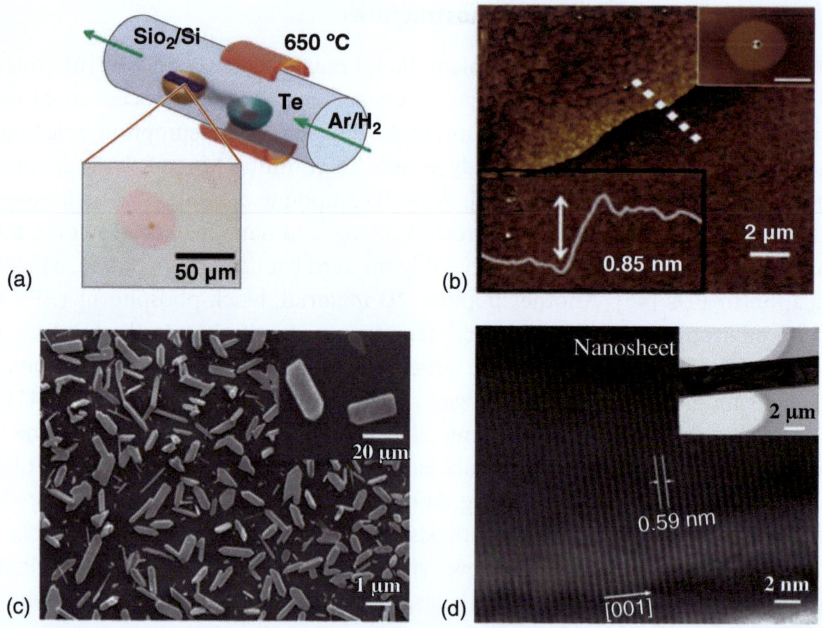


**Figure 5.5** (a) Schematic of vapor deposition of tellurene on $SiO_2$/Si substrates, and optical micrograph of a typical flake. (b) AFM image and height profile (insets: low-mag view and line profile of a 0.85-nm thick flake; scale bar = 10 μm). *Source:* Reprinted with permission from Ref. [49]. Copyright 2019, IOP Publishing. (c) SEM images of Te nanosheets. Insets are corresponding high-magnification SEM images. (d) HRTEM of Te nanosheets. Insets are corresponding low-magnification TEM images. *Source:* Reprinted with permission from Ref. [50]. Copyright 2021, American Association for the Advancement of Science.

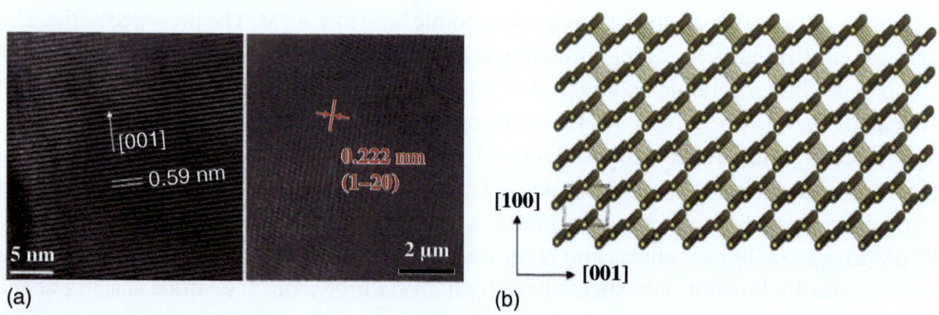


**Figure 5.6** (a) HRTEM images of tellurene. (b) 3D illustration of the structure of tellurene. *Source:* Reprinted with permission from Ref. [52]. Copyright 2023, MDPI.

0.6 nm, which corresponds to the tellurium (0001) and (1210) crystal planes, respectively. By analyzing dozens of samples of different sizes, Figure 5.6b shows a 3D view of the tellurium structure. All samples were grown laterally along the [0001] and [1-210] directions and stacked vertically along the [1010] direction to form 2D tellurium nanosheets. They concluded that the $Na_2TeO_3$/PVP molar ratio is a critical factor in the synthesis of tellurium nanosheets. During the initial growth phase, polyvinylpyrrolidone (PVP) was preferentially adsorbed on the {10-10} facets. This promotes the nanocrystal extension [0001] growth to form a 1D structure. As the reaction continues, due to the continuous growth of nanocrystals, the PVP capping is insufficient, and the {10-10} surface of the formed structure will be partially covered. Since the {10-10} surface has the lowest free energy in telluric crystals, the growth of the {10-10} surface along the [1-210] direction is significantly increased by the thermodynamically driven assembly. Enhanced growth along the [1-210] direction and sustained [0001] growth lead to the formation of 2D tellurium. By controlling the molar ratio of $Na_2TeO_3$/PVP, the size and thickness of the resulting nanosheet products could be controlled. To further obtain ultrathin tellurium nanosheets with a thickness of less than 10 nm, large-area monolayers, bilayers, trilayers, and few layers of 2D tellurium nanocrystals were obtained by placing the 2D tellurium nanosheets into a mixture of acidic salt solution and acetone after a specific period.

#### 5.2.3.3 vdW Epitaxy

The unique vdW epitaxy of 2D materials in recent years has also opened new doors for the epitaxial growth of 2D materials. vdW epitaxy converts the juxtaposition growth mode between the crystal substrate and the material into a vdW low-barrier-induced growth mode, thus allowing a significant lattice mismatch between the epitaxial layer and the substrate, which can be used to grow high-quality 2D thin films. He et al. [51] synthesized 2D Te nanosheets on mica and graphene substrates by vdW epitaxy. The 2D Te nanosheets grown on mica substrates were in the lateral range of 6–10 with about 30–80 nm thicknesses. Single 2D tellurium nanosheet devices were fabricated on flexible mica substrates. The tellurium nanosheet devices on the substrate still showed good optoelectronic performance after 100 flexible mica sheet bending cycles.

#### 5.2.3.4 Liquid-phase Stripping

Liquid-phase exfoliation, a method of obtaining 2D materials by breaking the weak force of 2D laminar structure interactions in 3D precursors, has received much attention due to its low cost and ease of implementation. In the previous paper, Wu et al. [52] successfully exfoliated tellurium nanosheets [53] with few or single layers by placing the synthesized tellurium nanosheets into a mixture of hydrochloric acid and acetone. Recently, Zhang [54] and his team prepared 2D single-crystal tellurium nanosheets by ultrasound-assisted liquid-phase exfoliation using substable 1T'-$MoTe_2$ as the tellurium source and *N*-methylpyrrolidone (NMP) as the solvent, and obtained 2D tellurium nanosheets with lateral lengths ranging from 5 to 100 $\mu$m. They found that the 2D growth of 2D tellurium occurs only in the synthesized mixture (1T'-$MoTe_2$/NMP). Only 1D tellurium nanostructures would eventually appear when 1T'-MoTe2 was absent, and 1T'-MoTe2 was not only a source of tellurium but also a template for the growth of 2D tellurium nanosheets.

### 5.2.4 Synthesis of Chiral Tellurium Nanomaterials

Chirality refers to an object that does not coincide with its mirror image. At the same time, chiral molecules are molecules with a particular configuration or conformation that are not identical to their mirror images and cannot coincide. Their unique structure brings some special properties, such as optical properties. Chiral structures can include circular dichroism and rotational linear polarization of light, which makes chiral materials play a significant role in molecular detection, polarized light detection, and broadband polarization. Tellurium nanostructures are formed by chains of tellurium atoms covalently bonded into a helical chain structure that naturally has a quasi-1D chiral helix, which can be described by either a right-handed (P3121) space group or a left-handed (P3221) space group [55] (Figure 5.7a). Despite the natural in-plane chirality of tellurium nanocrystals, controlling them to possess chiral shapes remains an exciting direction.

To obtain tellurium nanostructures with chiral shapes, Markovich [56] and his team reduced $TeO_2$ by thiolated chiral biomolecules, in which the sulfhydryl group reacts strongly with Te. Shape control of the chiral crystals was achieved by varying the biomolecule type, the reagent addition order, reaction time, and temperature. Five different tellurium nanocrystals were obtained by controlling the reaction conditions to explore the chiral tellurium nanocrystal synthesis scheme. Figure 5.7b shows the CD absorption

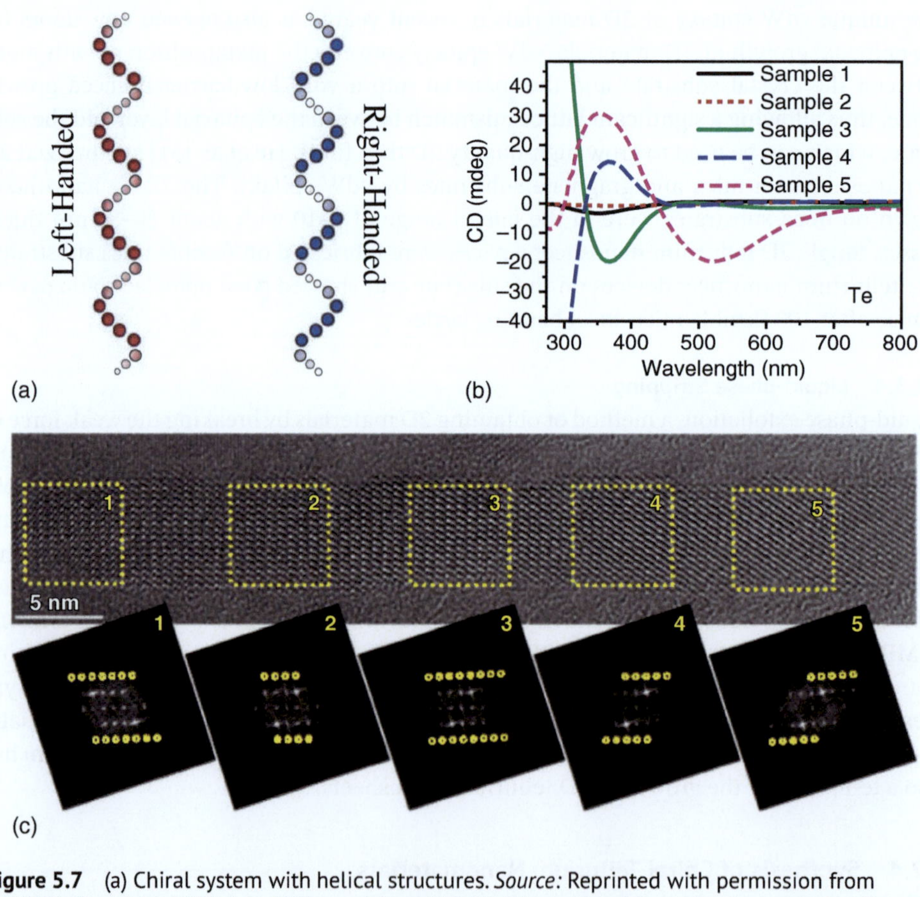


**Figure 5.7** (a) Chiral system with helical structures. *Source:* Reprinted with permission from Ref. [55]. Copyright 2020, John Wiley and Sons Ltd. (b) CD spectra of five tellurium samples. *Source:* Reprinted with permission from Ref. [56]. Copyright 2014, SpringerLink. (c) TEM image of a 5.7-nm diameter Te NW and binned FFTs from different areas along the NW's length showing differences. *Source:* Reprinted with permission from Ref. [57]. Copyright 2021, Royal Society of Chemistry.

spectra of the five samples, and it can be seen that the peak asymmetry value of sample 5 is more than 0.003, which is much larger than the other Te samples. The strong chiral activity of sample 5 compared to other Te samples has to exclude another factor other than the effect of chiral lattice symmetry: shape chirality. Since the shape of Te nanostructures is usually quite complex, the authors speculate that the filling of the nanocrystal surface with chiral surfactants may be a key factor leading to the formation of their chiral deformation.

Chiral molecule synthesis pathways have been a controversial topic since their discovery. According to one view, Ben-Moshe and his team observed by experimentally reducing tellurium dioxide with a chiral thiocyanine ligand and in the presence of hydrazine that left- and right-handed ligands affect the left- and right-handedness of tellurium chiral crystals. However, when the reaction was carried out with a nonchiral mercaptopropionic acid as a ligand, the tellurium NPs still formed a chiral shape similar to the penicillamine ligand. This suggests that the chiral ligand is not a determining factor in forming chiral crystals. They experimentally found that only crystal cores containing helical dislocations, responsible for the formation of twisted NWs and possibly the pathway for forming chiral crystals with different chiral polyhedral shapes, could continue to grow under shallow supersaturation conditions. In the same year, Londoño-Calderon et al. [57] synthesized ultrathin tellurium NWs with helical twists and chirality by hydrothermal reduction of $Na_2TeO_3$ by hydrazine hydrate in the presence of ammonia. TEM images of tellurium NWs and merged fast fourier transform (FFT) pictures of different regions along the NWs are shown in Figure 5.7c. Successive FFTs show apparent orientation differences, and tellurium NWs were calculated to have a torsion rate of 0.2–0.3 $nm^{-1}$ out of 0.2–0.3 $nm^{-1}$ by the rotational symmetry of the crystals.

## 5.3 Properties and Applications of Tellurium Nanostructures

Tellurium nanostructures exhibit unique physicochemical properties due to their miniaturized size compared to bulk tellurium nanomaterials [58, 59]. Their superior properties include a tunable indirect bandgap, high carrier mobility, excellent thermoelectric properties, and stability. These properties give them great potential for high-performance electronic and optoelectronic devices [60–62]. After this chapter, we summarize the tellurium nanomaterial properties and their applications.

### 5.3.1 Electrical Property

Tellurium belongs to p-type narrowband semiconductors, and due to its strong spin–orbit coupling, bulk Te has a small effective mass and high hole mobility, which makes it widely used in the field of fast devices, especially as field effect transistors [63, 64]. Wu [52] and his team synthesized tellurium nanosheets on a large scale by liquid-phase method and fabricated tellurium nanosheet transistor devices on high-dielectric substrates by electron-beam lithography technique. Because Pd has a relatively high Fermi energy level, it can be p-type transistor-matched to obtain small contact resistance, using Pd as the metal contact. The authors investigated the electrical properties of the devices at room temperature. Figure 5.8a shows the transfer curves of a 2D Te field effect tube with a thickness of 7.5 nm measured at different bias voltages. Due to its narrow bandgap characteristics, the device exhibits p-type characteristics with slight bipolar transport behavior and a switching ratio as high as ~$1 \times 10^5$. To further investigate the thickness dependence of the transistor device performance on the 2D tellurium nanochannels, the authors fabricated pairs of more than 50 2D tellurium nanosheet long-channel devices, with a range of nanosheet thicknesses from monolayer (~0.5 nm) to 35 nm. As can be seen from Figure 5.8b, as the thickness of the nanosheets gets progressively thinner, the mobility slowly rises until it reaches a peak of 700 $cm^2V^{-1}$ $s^{-1}$ at a thickness of 16 nm. As the thickness decreases to at least layers and monolayers, the mobility decreases to single digits. The mobility of the FET is affected by the contribution of the carrier layer near the semiconductor-oxide interface. As the thickness of the nanosheet decreases, it is affected more by charge impurities at the interface and surface scattering, which reveals the reason for the decrease in device mobility. When the thickness of the nanosheets is reduced to a few layers, the decrease in the thickness of the

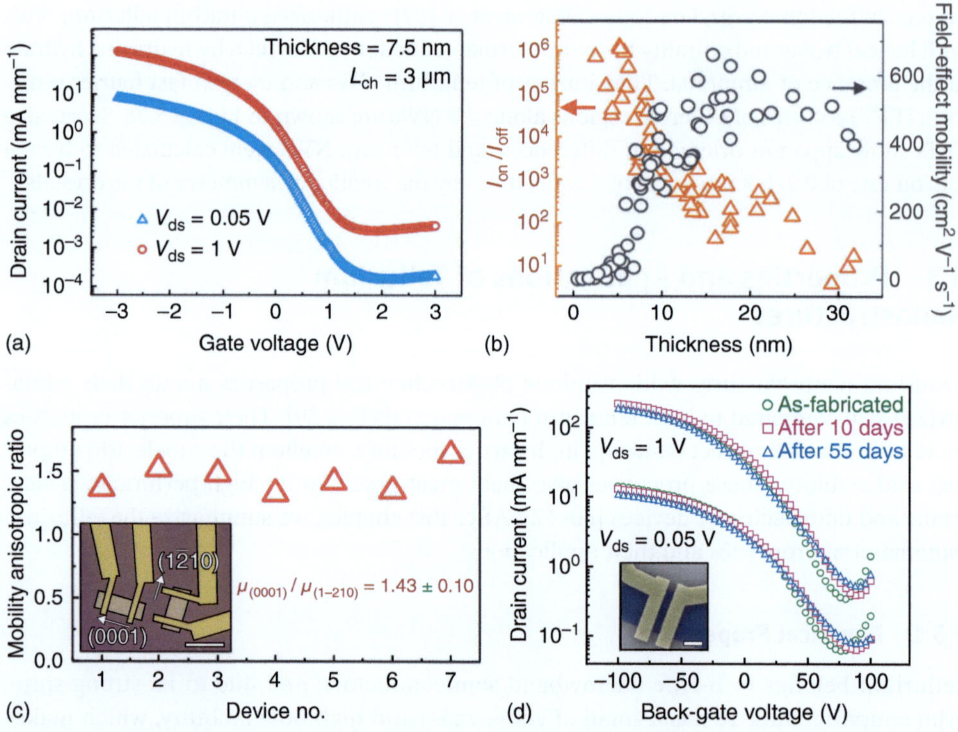


**Figure 5.8** 2D tellurene FET device performance. (a) Transfer curve of a typical long-channel 2D tellurene transistor with a thickness of 7.5 nm. (b) Thickness-dependent on/off ratio (orange triangles) and field-effect mobility (gray circles) for 2D tellurene transistors. (c) Anisotropic mobility measurements along [0001] and [1-210] directions. Inset: Optical image of a typical device for anisotropic transport measurement. Scale bar, 10 μm. (d) Transfer curves of a typical 2D tellurene transistor with a thickness of 15 nm, with stability measured for up to 55 days. Inset: False-colored SEM image of a tellurene transistor. The scale bar is 10 μm. *Source:* Reprinted with permission from Ref. [52]. Copyright 2018, SpringerLink.

tellurium nanosheets with fewer layers induces a broadening of their bandgap, which causes a drastic increase in the contact resistance at the metal-semiconductor interface, leading to a large error in the mobility extraction. BP, another high-quality p-type narrowband semiconductor, has a solution-synthesized 2D Te with about 2–3 times higher mobility than BP for the same device structure, geometry, and mobility extraction method. The gate control voltage for thicker samples prevents the device from being fully depleted. Once the crystal thickness exceeds the maximum depletion width of the film, the thickness dependence/off ratio (Supplementary Note) drops dramatically from ~$1 \times 10^6$ to less than 10. To investigate the anisotropy of the electron transport in 2D tellurium nanosheets for field effect transistors, they used a dry etching method to trim the 2D Te sheet from the same two identical rectangles (Figure 5.8c). One was made into a field effect transistor aligned along the [0001] direction of the 1D chain of atoms. The other rectangle was aligned along the [1-210] direction (inset). Mobility along the [0001] and [1-210] directions was extracted from seven 2D Te samples, and the average anisotropic mobility ratio among the seven devices was $1.43 \pm 0.1$. To test the stability of the devices, the electrical properties of the 15-nm-thick transistors were tested without any encapsulation after 10 days and two months of exposure to air. The samples were stable, the transfer curves overlapped well, and there was no significant degradation in the electrical properties of the devices (Figure 5.8d).

### 5.3.2 Photoconductive Properties

When light strikes a semiconductor material, photons hit electrons in the semiconductor material, causing them to jump to the conduction band, forming a hole and electron pair. This payload pair (electron-hole pair) moves through the semiconductor and forms a current when a bias voltage is applied externally [65–67]. Tellurium, an important narrow bandgap p-type elemental semiconductor with a bulk bandgap of ~0.35 eV, possesses a high hole mobility. The narrower bandgap and excellent electrical properties make the optoelectronic properties of low-dimensional Te nanomaterials very suitable for use as PDs.

To investigate the optoelectronic properties of low-dimensional tellurium nanomaterials, Peng et al. [50] synthesized low-dimensional tellurium NWs and tellurium nanosheets by CVD. Figure 5.9a shows the photoresponse spectra of Te NW devices under laser irradiation at a wavelength of 500~2500 nm and a power intensity of 0.01 mW mm$^{-2}$. The tellurium NWs show excellent response in the wavelength range of 1100~1500 nm, and the response reaches the highest at 1550 nm. Therefore, the 1550 nm laser is mainly used to characterize the optoelectronic performance of the devices. The dependence of photocurrent and gain on laser power is plotted in Figure 5.9b. Where the photocurrent is positively correlated with the power, which can be fitted by the nonlinear function $I_{ph} = P^{0.79}$, indicating that the device is subjected to the photogate effect, electron-hole generation, capture and complexation processes under laser irradiation. The responsivity is a crucial parameter to measure the efficiency of the PD, and Figure 5.9c shows that the tellurium nanomaterial exhibits a responsivity of up to 6650 A W$^{-1}$ under the irradiation of a 1550 nm laser, which represents one of the most advanced levels of low-dimensional devices in the 1550 nm band. Detectivity (D*) serves as another essential performance metric for PDs. The telluric NW PD has a specific detectivity of up to $1.23 \times 10^{12}$ Jones at a bias voltage of 0.1 V. The

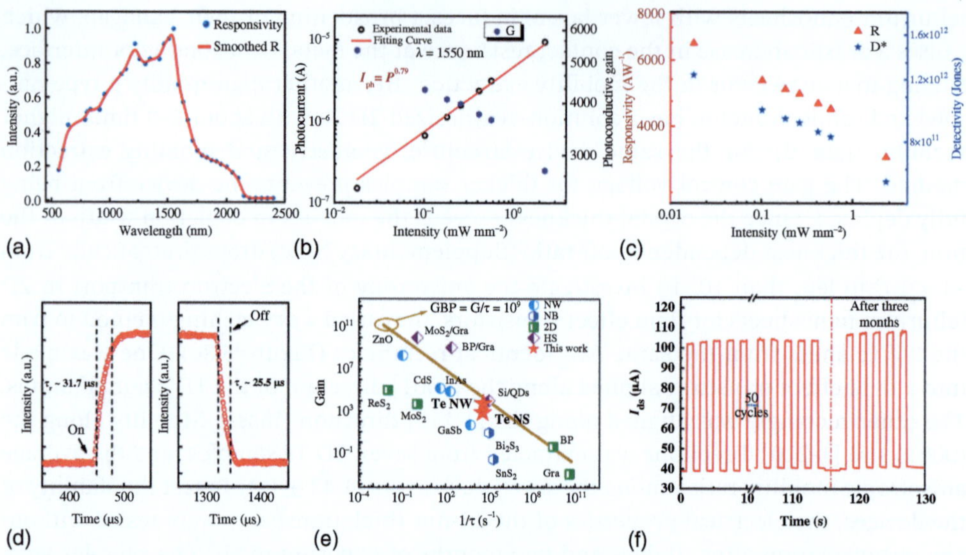


**Figure 5.9** (a) Photoresponse spectrum of Te NW device under the laser illumination with the wavelength from 500 to 2500 nm and a power intensity of 0.01 mW mm$^{-2}$. The device operated at a bias of 0.1 V. (b) Photocurrent and photoconductive gain versus power intensity under 1550 nm laser illumination. (c) Responsivity and detectivity versus power intensity under 1550 nm laser illumination. (d) Time-resolved photoresponse of Te NW PD under 1550 nm laser illumination. (e) GBP of low-dimensional PDs, exhibiting the high GBP of low-dimensional Te PDs. The yellow line represents a typical magnitude order of GBP for traditional thin-film PDs. (f) Stability of Te NWs device. The device was measured at the $V_{ds}$ = 0.1 V. *Source:* Reprinted with permission from Ref. [50]. Copyright 2021, American Association for the Advancement of Science.

dependence of the responsivity and detectivity as a function of the incident laser intensity is shown in Figure 5.9d. Both the response rate and detection rate decrease significantly with increasing laser intensity. To test the response rate of the device, the rise and fall times of the NW device were 31.7 and 25.5 μs, respectively. Compared to other fast response devices, they have been one of the fastest among all the photoconductive infrared detectors reported in the literature. It is well known that there is a competitive relationship between the gain and response time of PDs. Low-dimensional PDs, especially those in photoconductive mode, usually have high optical gain. However, high gain is always accompanied by a slow frequency response, limited by the average lifetime of the photogenerated carriers in the material. Without considering gain attenuation, we refer to the joint effect of gain and response time as the gain bandwidth product (GBP), which is inversely related to the response time, and the GBP is constant for each device. A larger value of GBP for a PD means that the device's performance is closer to the theoretical limit.

Figure 5.9e shows an at-a-glance plot of the GBP values of low-dimensional photoconductive detectors. The GBP values of Te nanosheet and NW PDs are significantly higher than those of other photoconductive PDs. In addition, the GBP of low-dimensional Te PDs is comparable to that of conventional thin-film PDs. Air stability is also essential for low-dimensional PDs. The photovoltaic performance of the low-dimensional tellurium nanodevices was measured at room temperature and atmospheric pressure. It showed good stability after running 50 cycles (Figure 5.9f). Exposing the devices to air for three months showed no significant degradation in the photoelectric response of the devices. The air stability of the tellurium nanodevices is excellent compared to other infrared detectors, such as BP and III–V semiconductor NWs. In conclusion, the low-dimensional tellurium PDs' responsiveness, response time, and detection capability are impressive.

### 5.3.3 Piezoelectric Thermoelectric Properties

Tellurium is a popular candidate as a novel semiconductor device with low thermal conductivity (1.6 W $m^{-1}$ $K^{-1}$), excellent transport properties (up to ~1000 $cm^2$ $V^{-1}$ $s^{-1}$), and anisotropic atomic structure [68]. Single-crystal tellurium nanocrystals consist of ternary helical chains of tellurium atoms bonded to the hexagonal system by vdW forces. Te has a high electron transport rate and conductivity within the molecular chain. As a result, it has good thermoelectric properties and conductivity along the z-axis [6]. Shen [69] and his team systematically investigated tellurium nanomaterials' piezoelectric and thermoelectric properties and applied them to sensor devices. When tellurium is subjected to pressure, the deformation causes its energy band structure to change. The authors performed first-principles calculations of the energy bands of tellurium through density flooding (Figure 5.10a). There is a significant difference in the energy band structure of Te when an

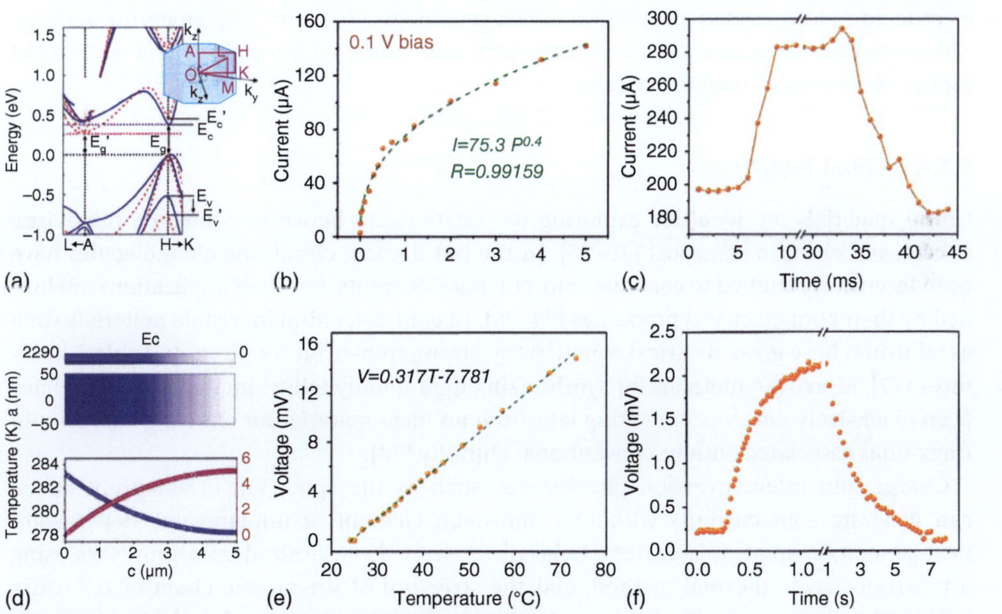


**Figure 5.10** (a) Band structure of Te. (b) The curve of the current to the pressure at 0.1 V. (c) The piezoresistive response time of the BTS. (d) When the Te NW with a temperature of 283.15 K at one end is in the air of 273.15 K, the temperature, electric potential, and z-axis components of electric field intensity of the Te NW along the z-axis. (e) The curve of the voltage to bottom temperature. (f) The thermoelectric response time of the BTS. *Source:* Reprinted with permission from Ref. [69]. Copyright 2022, SpringerLink.

external pressure of 0 kPa (blue solid line) and 0.1 MPa (red dashed line) is applied to the [001] crystal phase. Stimulated by pressure, the energy bands broaden along the z-axis, the energy band at the H-point becomes steeper, the effective mass decreases, carrier mobility increases, and resistance is reduced. To assess the performance of tellurium NW devices, the team meticulously studied the piezoelectric properties of the NW devices. Figure 5.10b shows current versus pressure at a bias voltage of 0.1 V. When the external pressure reaches 5 kPa, the current increases from an initial 2.82 μA to 141.82 μA. The current-voltage relationship can be fitted using I/M a = 75.3 × (P/kPa) · 0.4. Response time is a critical factor for the sensor device, and Figure 5.10c shows this sensor's piezoresistive response time relationship curve. The rise time of the device in response to pressure is $<$5 ms, and the relaxation time is $<$10 ms. When one end of the tellurium nanodevice is stimulated by temperature, the temperature difference leads to the unilateral accumulation of carriers, and the low-temperature side of the tellurium nanodevice leads to the accumulation of hole carriers with high potential. To investigate the thermoelectric properties of the device, a heat source of 283.15 K was applied to the bottom of the tellurium NW device, and the variation of the voltage and electric field with temperature is shown in Figure 5.10d, where the temperature only affects the output voltage over a small range. At an external temperature of 23°C, the measured output voltage was linearly related to the bottom temperature with a coefficient of 0.317 mV/K (Figure 5.10e). The output voltage versus time curve is shown in Figure 5.9f. The measured temperature response time is $<$0.75 s and the relaxation time is about 5 s. The above tests show that the tellurium nanodevices have excellent thermoelectric and piezoresistive properties, which provide the basis for implementing sensors. Moreover, the two signals do not interfere with each other, which makes them well-suited for the realization of bimodal sensors.

### 5.3.4 Chiral Properties

Chiral materials are ideal for exploring the relationship between symmetry, relativistic effects, and electron transport [70–73]. In the last decade, chiral organic molecules have been intensively studied to generate spin-polarized currents, but their applications are limited by their poor electrical properties [74–76]. In contrast, chiral inorganic materials such as tellurium have good electrical conductivity, strong spin–orbit coupling, and chiral structures [77]. Moreover, methods for synthesizing high-quality tellurium nanomaterials have been extensively developed, making tellurium an ideal material for studying charge-spin transitions associated with unconventional chirality [78].

Charge-spin interconversion phenomena, such as the spin Hall or Edelstein effect, can generate spin currents without a magnetic element, a fundamental step toward next-generation spintronic devices. Calavalle et al. [79] synthesized tellurium NWs using a passivated hydrothermal method, and the structure of the atomic chain of tellurium NWs with two opposing chiralities, as observed by STEM images matches the space groups P3121 and P3221 models. To detect the current-induced spin polarization associated with the tellurium NW chirality, a four-probe device for testing the magnetic transport properties was constructed using standard lithography with Pt as the contact (Figure 5.11a). The resistance of the device is plotted to decrease monotonically with decreasing temperature (Figure 5.11b). The figure illustrates the magnetoresistance curves for magnetic fields

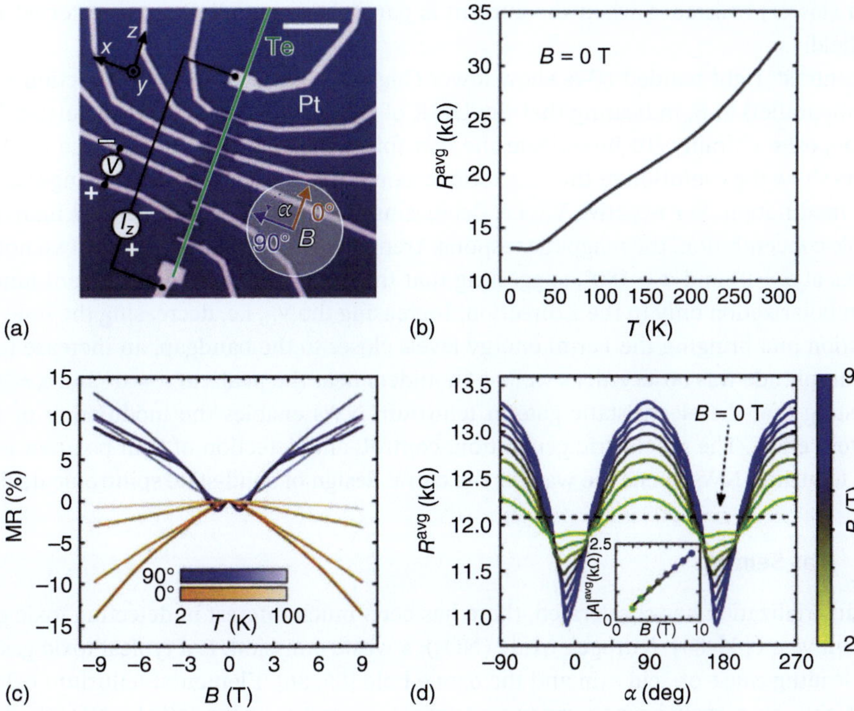


**Figure 5.11** (a) Optical image of a typical Pt contact touching Te NW. (b) Temperature dependence of the four-probe resistance at zero magnetic field. (c) Te magnetoresistance measured along the chiral z-axis (α = 0°) and in the transverse direction (α = 90°) at different temperatures (2, 10, 50, and 100 K). (d) Angular dependence of Ravg at different magnetic fields and T = 10 K. *Source:* Reprinted with permission from Ref. [79]. Copyright 2022, SpringerLink.

applied perpendicular or parallel to the direction of the current in the face at temperatures ranging from 2 to 100 K. When a field parallel to the current is applied, the device exhibits negative magnetoresistance characteristics, reaching a maximum value of – 15% at T = 2 K and B = 9 T (Figure 5.11c). When a perpendicular magnetic field is applied, the magnetoresistance changes from negative to positive with magnetic field strength at low temperatures, and this phenomenon disappears when T = 50 K. The figure exhibits the angular dependence curves of the magnetoresistance of the device, where the transverse and longitudinal magnetoresistance are almost linear to the magnetic field but with opposite slopes. The angular dependence difference increases linearly with the applied magnetic field strength enhancement (Figure 5.11d). To reveal the net spin polarization of telluric chiral devices, Figure 5.10e and f show the measured device resistance as a function of the angle α between the NW and the magnetic field (B = 9 T) for two telluric NW devices of opposite chirality at currents of opposite directions ($+I_z$ and $-I_z$). For both right-handed and left-handed tellurium NWs, the resistance was found to strongly depend on the relative alignment between $I_z$ and B, which is the hallmark of unidirectional magnetoresistance (UMR), and this dependence demonstrates the presence of a net spin polarization in the tellurium crystal. The figure indicates that left-handed tellurium NWs show significantly higher (lower) resistance when the current is parallel (antiparallel) to the external magnetic field.

In contrast, right-handed NWs show lower (higher) resistance in the z-direction parallel (antiparallel) to B, indicating that the UMR of tellurium is reflected in tellurium NWs with opposite chirality. To investigate the control of the UMR by gate voltage ($V_G$), the authors show the evolution of the magnetic response under carrier density changes driven by $V_G$ modulation. For negative $V_G$, i.e. decreasing the Fermi energy level and increasing the hole concentration, the magnetic response trend is relatively flat, except for two notable features at $\alpha = 0°$ and $\alpha = 180°$, suggesting that the current induces a significant amount of spin polarization only in the z direction. Increasing the $V_G$, i.e. decreasing the hole concentration and bringing the Fermi energy levels closer to the bandgap, an increase in the UMR amplitude was observed, as well as shoulders near the peaks at $\alpha = 0°$ and $\alpha = 180°$, suggesting that the electrostatic gate in tellurium NWs enables the modulation of their Edelstein effect. The all-electric generation, control, and detection of spin polarization in chiral tellurium NWs opens the way for the chiral design of solid-state spintronic devices.

### 5.3.5 Gas Sensing

As industrialization has accelerated, there has been much interest in detecting toxic gases in recent years [80–84]. Nitrogen oxide ($NO_2$), a waste emission, is a typical toxic gas and is the leading cause of acid rain and the ozone hole [85, 86]. Elemental tellurium exhibits remarkable gas-responsive performance at room temperature, especially for $NO_2$ gas detection. Tellurium nanostructures can provide more reactive sites, and adsorption of atomic oxygen and nitrogen affects the chemical activity of tellurium nanomaterials, which ultimately leads to a significant change in their electrical properties. Chen [87] and his team synthesized ultrathin tellurium nanosheets by liquid-phase exfoliation and then prepared tellurium nanosheet solutions, which were sprayed onto prepared electrodes to fabricate a tellurium nanosheet gas sensor. To evaluate the performance of the sensor, the authors tested the transient response and resistance response of the device at different concentrations of $NO_2$ (concentrations of 25~150 ppb) under dry conditions (Figure 5.12a). The curves of the resistance are very smooth, demonstrating the excellent signal-to-noise ratio of the device. At a $NO_2$ concentration of 150 ppb, the sensor was found to have a response factor of 264.3%. It is noteworthy that the tellurium sensor had a response factor of 201.8% for 25 ppb $NO_2$, which is a surprising value for low $NO_2$ concentration detection (Figure 5.12b). Figure 5.12c shows the linear fitting curve of the response value with increasing $NO_2$ concentration. The excellent linearity of the tellurium-based gas sensor for $NO_2$ in the range of 25~150 ppb allows it to be used for precision testing of gases. Response time and recovery time are crucial for the continuous operation of the sensor, as can be seen in Figure 5.12d. The response time and recovery time of the tellurium sensor for 100 ppb $NO_2$ are 26 and 290 s, respectively. Figure 5.12e Response of tellurium-based sensors to 40 ppm $CO_2$, 20 ppm $SO_2$ and $NH_3$, and 100 ppb $NO_2$ at room temperature. Compared to other hazardous (ppm concentration) gases, the electrical signal change of the $NO_2$ sensor at the ultralow concentration (ppb) level is outstanding, demonstrating the excellent selectivity of tellurium nanosheet sensors for $NO_2$. The long-term stability of the tellurium sensor was tested in a 100 ppb $NO_2$ environment at room temperature to test the stability of the device.

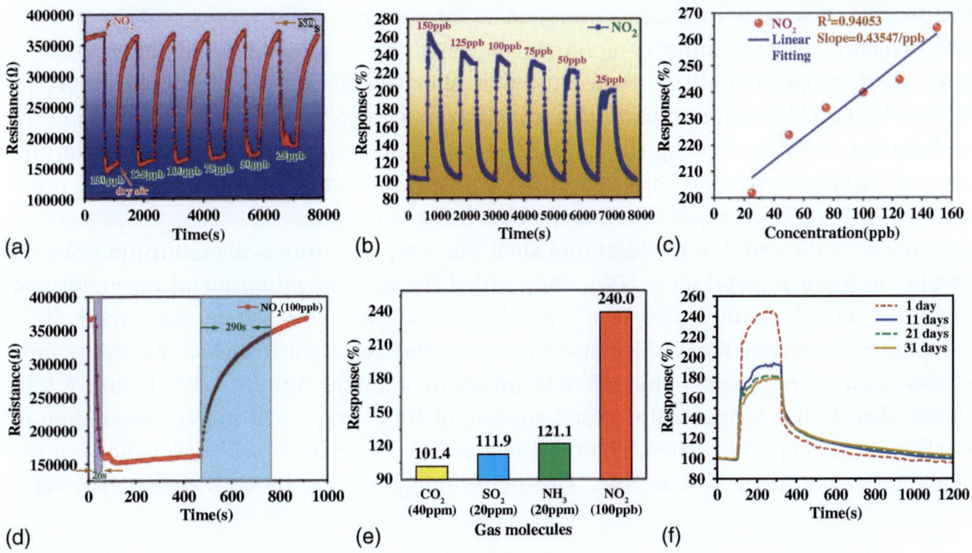

**Figure 5.12** Resistance responses of tellurene sensors at room temperature. The transient (a) resistance and (b) response of tellurene sensors detecting $NO_2$ under different $NO_2$ concentrations of 25–150 ppb level in dry air. (c) Fitted curve of response factor with the $NO_2$ concentration in dry air. (d) Response and recovery time of the tellurene sensor under 100 ppb $NO_2$. (e) Responses of the tellurene-based sensor to 40 ppm $CO_2$, 20 ppm $SO_2$ and $NH_3$, and 100 ppb $NO_2$ at room temperature. (f) Long-term stability of tellurene sensors for 100 ppb $NO_2$ under the atmospheric environment at room temperature. *Source:* Reprinted with permission from Ref. [87]. Copyright 2020, American Chemical Society.

As shown in Figure 5.12f, the tellurium sensor maintains a reasonably high $NO_2$ response after 31 days of exposure to a humid atmosphere. The fast response time and excellent response coefficient in the ppb concentration range make the tellurium nanostructures well-suited for low-concentration gas sensor applications.

### 5.3.6 Chemical Templates

The template method regulates the morphology and size of the synthetic target materials by designing specific elemental nano-shape dimensions as structural templates. As a simple, efficient, and versatile method, the template method is widely used to prepare unique nanostructures (e.g. NW nanotube nanosheets). An excellent template material should have good dispersion, flexible morphology, and excellent reactivity when used as a template [88, 89].

Elements of the same group have similar crystal structures and physicochemical properties. Se and Te have similar crystal structures, and hydrothermal methods can synthesize both. Therefore, hydrothermal methods can reduce selenite and tellurite, and Se/Te alloy nanostructures with a single-crystal nature can be obtained. Cao [90] and his team synthesized Se/Te alloy NWs by the liquid-phase method by controlling the molar ratio of selenium and tellurium, and they found that controlling the surfactant can inhibit the growth direction of Se/Te alloy NWs, which has an enormous impact on the structure of the product. His study provides a prerequisite for constructing complex multilevel Te-Se nanostructures. Metal–sulfur compounds and their two or more alloying nanostructures have a wide range of applications because they offer several new physicochemical properties. Yu [91] and his team prepared TexSey@Se core–shell NWs with different Te/Se ratios and diameters by controlling the amount of selenium source using ultrathin tellurium NWs of 7 nm as templates. They found that the average diameter of the synthesized product was $18.7 \pm 1.5$ nm when the ratio of selenium to tellurium was 1 : 4, and the increase in diameter indicated that the selenium shell was wrapped around the tellurium NWs. To obtain TexSey@Se metal alloy NWs, they added the corresponding metal precursor and hydrazine directly to the TexSey@Se NW solution using a chemical conversion method.

Figure 5.13a shows the TEM image of the synthesized AgSeTe NWs, and the sample shows good homogeneity. The HRTEM image of a single AgSeTe NW (Figure 5.13b) shows clear lattice stripes with crystal spacing of 0.435 and 0.527 nm corresponding to (100) and (011) crystal planes. The inset is a SAED image of AgSeTe NWs showing the single-phase nature of AgSeTe NWs. The image in Figure 5.13c shows the elemental maps

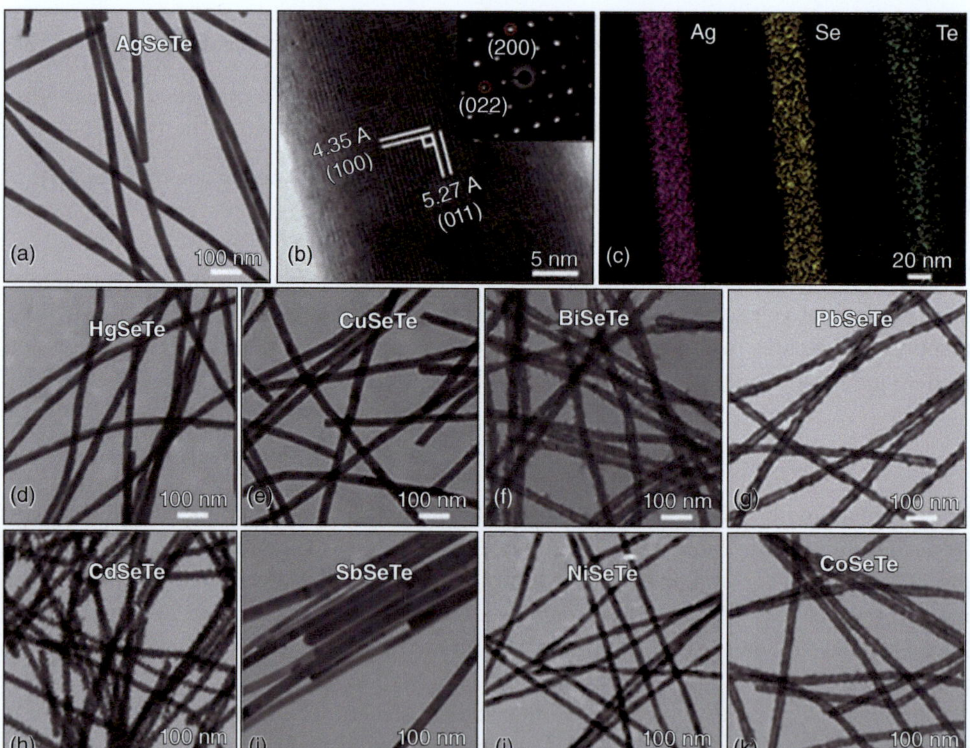


**Figure 5.13** (a) TEM image of AgSeTe NWs. (b) HRTEM image of AgSeTe NWs. Inset: A SAED pattern. (c) Elemental maps of Ag (magenta), Se (yellow), and Te (green). (d–k) TEM images of the remaining eight types of MSeTe NWs: (d) HgSeTe, (e) CuSeTe, (f) BiSeTe, (g) PbSeTe, (h) CdSeTe, (i) SbSeTe, (j) NiSeTe, and (k) CoSeTe. *Source:* Reprinted with permission from Ref. [91]. Copyright 2015, American Association for the Advancement of Science.

of Ag (magenta), Se (yellow), and Te (green), indicating that the three elements have a relatively uniform distribution, which proves that the AgSeTe NWs are ternary alloy structures. Figure 5.13d–k shows the nanostructured TEM images of eight MSeTe (M is a metallic element) alloy NWs (HgSeTe, CuSeTe, BiSeTe, PbSeTe, CdSeTe, SbSeTe, NiSeTe, and CoSeTe). Alloy sulfur NWs with two or more MSeTe-phase integrations were prepared by adding two or more metal precursors in a single batch. 36 bimetallic sulfur NWs of every 2 MSeTe combinations of 9 MSeTe were successfully obtained. The use of tellurium nanostructures as a template strategy shows excellent potential for synthesizing a new generation of 1D metal–sulfide nanostructures with controllable compositions.

### 5.3.7 Batteries

With the rapid development of consumer electronics and new energy vehicles, rechargeable batteries have attracted widespread attention [92–94]. Excellent energy storage devices possess superior energy density, cycling, and stability. In the past decades, lithium-ion batteries have been the mainstream of research. It has been found that elemental selenium and tellurium can undergo two-electron redox reactions with Li [95–97]. At the same time, tellurium also has a high theoretical specific capacity, the highest electronic conductivity of a nonmetallic element, and a large bulk capacity density, which makes it very suitable for constructing cathode materials for rechargeable batteries. The scheme is as follows:

$$Te_n + 2nLi^+ + 2ne^- \Leftrightarrow Li_{2n}Te_n \quad (5.1)$$

However, the attenuation problem of pure tellurium nanomaterials has not been solved. Yu [98] and his team constructed a high-energy tellurium lithium-ion battery using hydrothermal carbonization. They used SWCNT aerogel as an intermediate layer, allowing rapid $Li^+$ transport during charging and discharging due to its hollow structure and excellent electrical properties. They tested the stability of the device, with 81.5% capacity retention after 100 cycles. To explore energy storage materials beyond lithium-ion batteries, Fang [99] and his team proposed a tellurium-aluminum battery (TAB) with a large capacity. In the design of the battery structure, tellurium NWs were synthesized as the battery cathode material using the hydrothermal method. There are two critical factors for the capacity degradation of TAB (i) Tellurium dissolves in the electrolyte, which leads to the loss of the tellurium NW cathode material. (ii) $Te_4(AlCl_4)_2$, a product of electrochemical redox reactions, can also dissolve in the electrolyte during repeated charging and discharging, and these chemicals can pass through the separator, leading to a loss of active material and affecting the battery capacity. To solve these problems, the authors proposed modifying the devices' positive electrode and separator. The authors fabricated three devices: Type I: the original tellurium NW positive device; Type II: the introduction of graphene oxide nanosheets to modify the tellurium nanosheets to form a hybrid positive electrode; and Type III: in addition to Type II, functionalized SWCNTs were also used to deposit on the separator, which is expected to enhance the inhibition of the tellurium-based compounds (Figure 5.14a). Cyclic voltammetry characteristic curves of the three cell types have three pairs of similar redox peaks, mainly due to tellurium NW contributions (Figure 5.14b). Charge-discharge curves (Figure 5.14c) of type III devices exhibit a pronounced piezoelectric plateau and

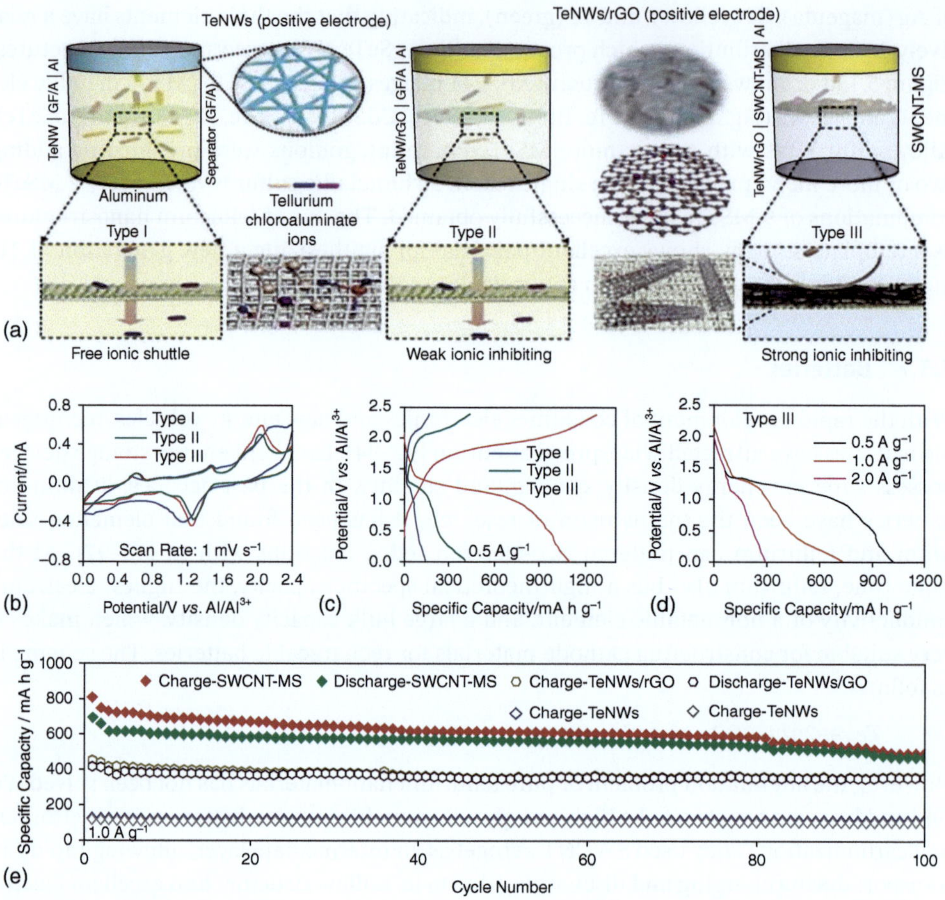


**Figure 5.14** (a) Schemes for illustrating various cell configurations of TABs. Type I: pristine Te NWs (positive electrode); Type II: modification using Te NWs/rGO as positive electrodes (middle); Type III: modification using Te NWs/rGO as positive electrodes and SWCNT-MS as separator (right). (b) Cycling voltammetry profiles of various positive electrodes as marked. (c) The corresponding charge/discharge curves at 0.5 $Ag^{-1}$ are marked. (d) The specific discharge capabilities of Type III TAB at various current densities are marked. (e) Cycling performance of the three types of TABs at a specific current of 1 A $g^{-1}$. *Source:* Reprinted with permission from Ref. [98]. Copyright 2019, Royal Society of Chemistry.

high specific discharge energy (excluding the contribution of graphene oxide and carbon nanotubes). By calculating the capacity contribution of the type III device, it is estimated that the tellurium NWs provide 1026 mA h $g^{-1}$, which is close to the theoretical limit. The discharge curves at different current densities (Figure 5.14d) show that the tellurium NWs of the type III device can provide ultrahigh specific discharge capacities of 734 and 300 mA h $g^{-1}$ at 1.0 and 2.0 A $g^{-1}$, respectively. Figure 5.14e demonstrates that the type III cell presents a greater rechargeability at specific currents, with more than 100 cycles showing good stability. The Li-Te and Te-Al cells' success of Li-Te and Te-Al batteries shows tellurium's potential in high-performance charge/discharge energy storage.

## 5.4 Conclusion

With the development of micro- and nanotechnology and integrated circuits in recent years, tellurium nanomaterials have opened up new opportunities. Tellurium exhibits excellent semiconducting and chiral structural properties, and these studies have led to significant advances in the synthesis of tellurium nanostructures. This chapter summarizes the synthetic routes to tellurium nanostructures (including NWs, nanotubes, nanoribbons, 2D nanosheets, and chiral nanostructures), mainly based on the liquid phase. It also shows the platform of tellurium nanostructure applications for electronics, photoconductivity, sensing, batteries, and spin quantum devices.

## References


**1** Wu, B.Z., Liu, X.H., Yin, J.R., Lee, H. Bulk β-Te to few layered β-tellurenes: indirect to direct band-Gap transitions showing semiconducting property. *Mater. Res. Express* 2017, 4 (9).

**2** Liu, Y.Z., Xiao, J.W., Koo, J., Yan, B.H. Chirality-driven topological electronic structure of DNA-like materials. *Nat. Mater*. 2021, 20 (5): 638–644.

**3** Liang, Z.F., Wang, Y., Hua, C.Q., Xiao, C.C., Chen, M.G., Jiang, Z., Tai, R.Z., Lu, Y.H., Song, F. Electronic structures of ultra-thin tellurium nanoribbons. *Nanoscale* 2019, 11 (30): 14134–14140.

**4** Sapkota, K.R., Lu, P., Medlin, D.L., Wang, G.T., High temperature synthesis and characterization of ultrathin tellurium nanostructures. *APL Mater*. 2019, 7 (8).

**5** Kim, M.S., Ma, X.H., Cho, K.H., Jeon, S.Y., Hur, K., Sung, Y.M. A generalized crystallographic description of all tellurium nanostructures. *Adv. Mater*. 2018, 30 (6): 1702701.

**6** Yang, Y.F., Xu, M.K., Jia, S.J., Wang, B.L., Xu, L.J., Wang, X.X., Liu, H., Liu, Y.S., Guo, Y.Z., Wang, L.D., Duan, S.K., Liu, K., Zhu, M., Pei, J., Duan, W.R., Liu, D.M., Li, H.L. A new opportunity for the emerging tellurium semiconductor: making resistive switching devices. *Nat. Commun*. 2021, 12 (1): 6081.

**7** Yan, J.H., Pang, H., Xu, L., Yang, J., Quhe, R.G., Zhang, X.Y., Pan, Y.Y., Shi, B.W., Liu, S.Q., Xu, L.Q., Yang, J.B., Pan, F., Zhang, Z.Y., Lu, J. Excellent device performance of Sub-5-nm monolayer tellurene transistors. *Adv. Electron. Mater*. 2019, 5 (7): 1900226.

**8** Nguyen, D.A., Park, D.Y., Lee, J., Duong, N.T., Park, C., Nguyen, D.H., Le, T.S., Suh, D., Yang, H., Jeong, M.S. Patterning of type-II dirac semimetal PtTe2 for optimized interface of tellurene optoelectronic device. *Nano Energy* 2021, 86: 106049.

**9** Zhu, H.L., Fan, L., Wang, K.L., Liu, H., Zhang, J.W., Yan, S.C. Progress in the synthesis and application of tellurium nanomaterials. *Nanomaterials* 2023, 13 (14): 2057.

**10** Campos, A., Troc, N., Cottancin, E., Pellarin, M., Weissker, H.C., Lermé, J., Kociak, M., Hillenkamp, M. Plasmonic quantum size effects in silver nanoparticles are dominated by interfaces and local environments. *Nat. Phys*. 2019, 15 (3): 275–280.

**11** Wang, C.X., Chen, J., Yang, G.W., Xu, N.S. Thermodynamic stability and ultrasmall-size effect of nanodiamonds. *Angew. Chem. Int. Ed*. 2005, 44 (45): 7414–7418.

**12** Song, X.X., Liu, F.W., Qiu, C.J., Coy, E., Liu, H., Aperador, W., Zaleski, K., Li, J.J., Song, W., Lu, Z.F., Pan, H.B., Kong, L., Wang, G.C. Nanosurfacing Ti alloy by weak alkalinity-activated solid-state dewetting (AAD) and its biointerfacial enhancement effect. *Mater. Horizons* 2021, 8 (3): 912–924.

**13** Kim, Y., Kim, C., Kim, S.Y., Lee, B.C., Seo, Y., Cho, H., Kim, G.T., Joo, M.K. Emergence of quantum tunneling in ambipolar black phosphorus multilayers without heterojunctions. *Adv. Funct. Mater.* 2022, 32 (13): 2110391.

**14** Xiao, L., An, T.T., Wang, L., Xu, X.L., Sun, H.D. Novel properties and applications of chiral inorganic nanostructures. *Nano Today* 2020, 30: 100824.

**15** Qian, L., Xie, Y., Zou, M.Z., Zhang, J. Building a bridge for carbon nanotubes from nanoscale structure to macroscopic application. *J. Am. Chem. Soc.* 2021, 143 (45): 18805–18819.

**16** Zhang, L., Biesold, G.M., Zhao, C.Y., Xu, H., Lin, Z.Q. Necklace-like nanostructures: from fabrication, properties to applications. *Adv. Mater.* 2022, 34 (40).

**17** Li, Y., Li, Z.W., Chi, C., Shan, H.Y., Zheng, L.H., Fang, Z.Y. Plasmonics of 2D nanomaterials: properties and applications. *Adv. Sci.* 2017, 4 (8): 1600430.

**18** Chu, K.L., Wang, Z.B., Zhou, J.J., Jiang, H. Transport properties in monolayer-bilayer-monolayer graphene planar junctions. *Chin. Phys. B* 2017, 26 (6): 067202.

**19** Hu, Z.Y., Breeze, B., Kashtiban, R.J., Sloan, J., Lloyd-Hughes, J. Zigzag HgTe nanowires modify the electron-phonon interaction in chirality-refined single-walled carbon nanotubes. *ACS Nano* 2022, 16 (4): 6789–6800.

**20** Qi, F.L., Jeong, K.J., Gong, J.X., Tang, Z.Y. Modulation of nano-superstructures and their optical properties. *Acc. Chem. Res.* 2022, 55 (17).

**21** Wang, J., Song, J., Mu, X., Sun, M. Optoelectronic and photoelectric properties and applications of graphene-based nanostructures. *Mater. Today Phys.* 2020, 13: 100196.

**22** Shangguan, L., He, L.B., Ran, Y.T., Hong, H., Zhu, J.H., Gao, Y.T., Sun, L.T., Hydrothermal synthesis of Te nanosheets: growth mechanism and electrical property. *ACS Appl. Mater. Interfaces* 2023, 15 (32): 38707–38715.

**23** Ge, Z.H., Nolas, G.S. Purification and crystal growth of single-crystalline tellurium tubes and rods. *Mater. Lett.* 2017, 194: 20–22.

**24** Hao, G.L., Xiao, J.B., Hao, Y.L., Zhou, G.L., Zhu, H., Gao, H., Xu, Z.Q., Zhao, Z. K., Miao, L.L., Li, J., Sun, H.T., Zhao, C.J. Van der Waals epitaxial growth of mixed-dimensional 1D/2D heterostructures with tellurium nanowires and transition metal dichalcogenide nanosheets for nonlinear optical applications. *Mater. Today Phys.* 2023, 34: 101069.

**25** Liu, Y.W., Chen, Q.L., Cullen, D.A., Xie, Z.X., Lian, T.Q. Efficient hot electron transfer from small Au nanoparticles. *Nano Lett.* 2020, 20 (6): 4322–4329.

**26** Pedone, D., Moglianetti, M., De Luca, E., Bardi, G., Pompa, P.P. Platinum nanoparticles in nanobiomedicine. *Chem. Soc. Rev.* 2017, 46 (16): 4951–4975.

**27** Cao, L., Mueller, T. Catalytic activity maps for alloy nanoparticles. *J. Am. Chem. Soc.* 2023, 145 (13): 7352–7360.

**28** Yuan, J., Schmalz, H., Xu, Y., Miyajima, N., Drechsler, M., Moeller, M.W., Schacher, F., Müller, A.H.E. Room-temperature growth of uniform tellurium nanorods and the assembly of tellurium or Fe3O4 nanoparticles on the nanorods. *Adv. Mater.* 2008, 20 (5): 947.

**29** Bhol, P., Swain, S., Jena, S., Bhatte, K., Rout, C.S., Saxena, M., Jadhav, A.H., Samal, A.K. Co-decorated tellurium nanotubes for energy storage applications. *ACS Appl. Nano Mater.* 2021, 4 (9): 9008–9021.

**30** Lee, S., Je, Y., Seok, B., Kim, H.T., Jo, Y.R., Oh, S.J., Im, B., Kim, D.G., Chee, S.S., An, G.H. Anode surface engineering of zinc-ion batteries using tellurium nanobelt as a protective layer for enhancing energy storage performance. *J. Energy Chem.* 2024, 92: 113–123.

**31** Yin, Y.H., Zhang, Z.F., Zhong, H.X., Shao, C., Wan, X.H., Zhang, C., Robertson, J., Guo, Y.Z. Tellurium nanowire gate-all-around MOSFETs for Sub-5 nm applications. *ACS Appl. Mater. Interfaces* 2021, 13 (2): 3387–3396.

**32** Zhang, B., Hou, W.Y., Ye, X.C., Fu, S.Q., Xie, Y. 1D tellurium nanostructures: photothermally assisted morphology-controlled synthesis and applications in preparing functional nanoscale materials. *Adv. Funct. Mater.* 2007, 17 (3): 486–492.

**33** Dasika, P., Samantaray, D., Murali, K., Abraham, N., Watanbe, K., Taniguchi, T., Ravishankar, N., Majumdar, K., Contact-barrier free, high mobility, dual-gated junctionless transistor using tellurium nanowire. *Adv. Funct. Mater.* 2021, 31 (13): 2006278.

**34** Wang, R., Wang, J.L., Liu, T., He, Z., Wang, H., Liu, J.W., Yu, S.H. Controllable inverse photoconductance in semiconducting nanowire films. *Adv. Mater.* 2022, 34 (36): 2204698.

**35** Furuta, N., Itinose, H., Maruyama, N., Ohasi, Y. Morphology and dislocation structure of tellurium whiskers grown from the vapor. *Jpn. J. Appl. Phys.* 1972, 11 (8): 1113–11181118.

**36** Lu, Q.Y., Gao, F., Komarneni, S. A green chemical approach to the synthesis of Te nanowires. *Langmuir* 2005, 21 (13): 6002–6005.

**37** Wang, K., Yang, Y., Liang, H.W., Liu, J.W., Yu, S.H. First sub-kilogram-scale synthesis of high quality ultrathin tellurium nanowires. *Mater. Horizons* 2014, 1 (3): 338–343.

**38** Churchill, H.O.H., Salamo, G.J., Yu, S.Q., Hironaka, T., Hu, X., Stacy, J., Shih, I. Toward single atom chains with exfoliated tellurium. *Nanoscale Res. Lett.* 2017, 12: 1–6.

**39** Qin, J.K., Liao, P.Y., Si, M.W., Gao, S.Y., Qiu, G., Jian, J., Wang, Q.X., Zhang, S.Q., Huang, S.Y., Charnas, A., Wang, Y.X., Kim, M.J., Wu, W.Z., Xu, X.F., Wang, H.Y., Yang, L., Yap, Y.K., Ye, P.D.D. Raman response and transport properties of tellurium atomic chains encapsulated in nanotubes. *Nat. Electron.* 2020, 3 (3): 141–147.

**40** Mayers, B., Xia, Y.N. Formation of tellurium nanotubes through concentration depletion at the surfaces of seeds. *Adv. Mater.* 2002, 14 (4): 279.

**41** Li, Z.L., Zheng, S.Q., Zhang, Y.Z., Teng, R.Y., Huang, T., Chen, C.F., Lu, G.W. Controlled synthesis of tellurium nanowires and nanotubes via a facile, efficient, and relatively green solution phase method. *J. Mater. Chem. A* 2013, 1 (47): 15046–15052.

**42** Mo, M.S., Zeng, J.H., Liu, X.M., Yu, W.C., Zhang, S.Y., Qian, Y.T. Controlled hydrothermal synthesis of thin single-crystal tellurium nanobelts and nanotubes. *Adv. Mater.* 2002, 14 (22): 1658–1662.

**43** Yu, M., Tan, C., Yin, Y., Tang, J., Gao, X., Liu, H., Ding, F., Peng, H. Integrated 2D multi-fin field-effect transistors. *Nat. Commun.* 2024, 15 (1): 3622.

**44** Jayachandran, D., Pendurthi, R., Sadaf, M.U., Sakib, N.U., Pannone, A., Chen, C., Han, Y., Trainor, N., Kumari, S., Mc Knight, T.V., Redwing, J.M., Yang, Y., Das, S. Three-dimensional integration of two-dimensional field-effect transistors. *Nature* 2024, 625 (7994): 276–281.

45 Jiang, S.W., Li, L.Z., Wang, Z.F., Shan, J., Mak, K.F. Spin tunnel field-effect transistors based on two-dimensional van der Waals heterostructures. *Nat. Electron.* 2019, 2 (4): 159–163.

46 Niu, W.H., Ma, J., Feng, X.L. Precise structural regulation and band-gap engineering of curved graphene nanoribbons br. *Acc. Chem. Res.* 2022, 55 (23): 3322–3333.

47 Wu, D.Z., Peng, Z.J., Jin, C.H., Zhang, Z.Y. Effective passivation of black phosphorus transistor against ambient degradation by an ultra-thin tin oxide film. *Sci. Bull.* 2019, 64 (9): 570–574.

48 Londoño-Calderon, A., Williams, D.J., Ophus, C., Pettes, M.T. 1D to 2D transition in tellurium observed by 4D electron microscopy. *Small* 2020, 16 (49).

49 Apte, A., Bianco, E., Krishnamoorthy, A., Yazdi, S., Rao, R., Glavin, N., Kumazoe, H., Varshney, V., Roy, A., Shimojo, F., Ringe, E., Kalia, R.K., Nakano, A., Tiwari, C.S., Vashishta, P., Kochat, V., Ajayan, P.M. Polytypism in ultrathin tellurium. *2D Mater.* 2019, 6 (1).

50 Peng, M., Xie, R.Z., Wang, Z., Wang, P., Wang, F., Ge, H.N., Wang, Y., Zhong, F., Wu, P.S., Ye, J.F., Li, Q., Zhang, L.L., Ge, X., Ye, Y., Lei, Y.C., Jiang, W., Hu, Z.G., Wu, F., Zhou, X.H., Miao, J.S., Wang, J.L., Yan, H.G., Shan, C.X., Dai, J.N., Chen, C.Q., Chen, X.S., Lu, W., Hu, W.D. Blackbody-sensitive room-temperature infrared photodetectors based on low-dimensional tellurium grown by chemical vapor deposition. *Sci. Adv.* 2021, 7 (16).

51 Wang, Q.S., Safdar, M., Xu, K., Mirza, M., Wang, Z.X., He, J. Van der Waals epitaxy and photoresponse of hexagonal tellurium nanoplates on flexible mica sheets. *ACS Nano* 2014, 8 (7): 7497–7505.

52 Wang, Y.X., Qiu, G., Wang, R.X., Huang, S.Y., Wang, Q.X., Liu, Y.Y., Du, Y.C., Goddard, W.A., Kim, M.J., Xu, X.F., Ye, P.D., Wu, W.Z. Field-effect transistors made from solution-grown two-dimensional tellurene. *Nat. Electron.* 2018, 1 (4): 228–236.

53 Zhu, H., Fan, L., Wang, Y., Wang, K., Liu, H., Zhang, J., Yan, S. Birefringence after tellurium nanosheets and copper intercalation. *Symmetry* 2023, 15 (10): 1885.

54 Xie, Z.J., Xing, C.Y., Huang, W.C., Fan, T.J., Li, Z.J., Zhao, J.L., Xiang, Y.J., Guo, Z.N., Li, J.Q., Yang, Z.G., Dong, B.Q., Qu, J.L., Fan, D.Y., Zhang, H. Ultrathin 2D nonlayered tellurium nanosheets: facile liquid-phase exfoliation, characterization, and photoresponse with high performance and enhanced stability. *Adv. Funct. Mater.* 2018, 28 (16): 1705833.

55 Mokashi-Punekar, S., Zhou, Y., Brooks, S.C., Rosi, N.L. Construction of chiral, helical nanoparticle superstructures: progress and prospects. *Adv. Mater.* 2020, 32 (41): 1905975.

56 Ben-Moshe, A., Wolf, S.G., Bar Sadan, M., Houben, L., Fan, Z.Y., Govorov, A.O., Markovich, G. Enantioselective control of lattice and shape chirality in inorganic nanostructures using chiral biomolecules. *Nat. Commun.* 2014, 5: 4302.

57 Londoño-Calderon, A., Williams, D.J., Schneider, M.M., Savitzky, B.H., Ophus, C., Ma, S.J., Zhu, H.Y., Pettes, M.T. Intrinsic helical twist and chirality in ultrathin tellurium nanowires. *Nanoscale* 2021, 13 (21): 9606–9614.

58 Wang, Y.X., Liu, S., Wu, Z., Liu, G.L., Yang, X., Wei, T.Q., Wang, Q.J., Ye, Y., Li, D.Y., Zhu, J. Enhanced thermoelectric performance of van der Waals tellurium via vacancy engineering. *Mater. Today Phys.* 2021, 18: 100379.

59 Huang, W.C., Zhang, Y., You, Q., Huang, P., Wang, Y.Z., Huang, Z.Y.N., Ge, Y.Q., Wu, L.M., Dong, Z.J., Dai, X.Y., Xiang, Y.J., Li, J.Q., Zhang, X.W., Zhang, H. Enhanced photodetection properties of Tellurium@Selenium roll-to-roll nanotube heterojunctions. *Small* 2019, 15 (23): 1900902.

**60** Li, L.L., Xu, H., Li, Z.X., Liu, L.C., Lou, Z., Wang, L.L. CMOS-compatible Te/silicon ultrafast near-infrared photodetector. *Small* 2023, 19 (42): 2303114.

**61** Ma, W.L., Gao, Y.Q., Shang, L.Y., Zhou, W., Yao, N.J., Jiang, L., Qiu, Q.X., Li, J.B., Shi, Y., Hu, Z.G., Huang, Z.M. Ultrabroadband tellurium photoelectric detector from visible to millimeter wave. *Adv. Sci.* 2022, 9 (5): 2103873.

**62** Zhang, W.M., Wang, B., Li, K.Q., Sun, Y.Q., Zhou, J., Sun, Z.M. Towards ultralow-power and high-speed electronics: tunnel transistor based on single-chain Tellurium. *Mater. Today Phys.* 2024, 40: 101313.

**63** Yang, P., Zha, J.J., Gao, G.Y., Zheng, L., Huang, H.X., Xia, Y.P., Xu, S.C., Xiong, T.F., Zhang, Z.M., Yang, Z.B., Chen, Y., Ki, D.K., Liou, J.J., Liao, W.G., Tan, C.L. Growth of tellurium nanobelts on h-BN for p-type transistors with ultrahigh hole mobility. *Nano-Micro Lett.* 2022, 14 (1): 109.

**64** Chen, J.W., Zhou, Y., Yan, J.M., Liu, J.D., Xu, L., Wang, J.L., Wan, T.Q., He, Y.H., Zhang, W.J., Chai, Y. Room-temperature valley transistors for low-power neuromorphic computing. *Nat. Commun.* 2022, 13 (1): 7758.

**65** Teng, F., Hu, K., Ouyang, W.X., Fang, X.S. Photoelectric detectors based on inorganic p-type semiconductor materials. *Adv. Mater.* 2018, 30 (35): 1706262.

**66** Xie, C., Mak, C., Tao, X.M., Yan, F. Photodetectors based on two-dimensional layered materials beyond graphene. *Adv. Funct. Mater.* 2017, 27 (19): 1603886.

**67** Zou, Y.X., Zhang, Z.K., Yan, J.W., Lin, L.H., Huang, G.Y., Tan, Y.D., You, Z., Li, P. High-temperature flexible WSe2 photo-detectors with ultrahigh photoresponsivity. *Nat. Commun.* 2022, 13 (1): 4372.

**68** Naqi, M., Choi, K.H., Yoo, H., Chae, S., Kim, B.J., Oh, S., Jeon, J., Wang, C., Liu, N., Kim, S., Choi, J.Y. Nanonet: low-temperature-processed tellurium nanowire network for scalable p-type field-effect transistors and a highly sensitive phototransistor array. *NPG Asia Mater.* 2021, 13 (1): 46.

**69** Li, L.L., Zhao, S.F., Ran, W.H., Li, Z.X., Yan, Y.X., Zhong, B.W., Lou, Z., Wang, L.L., Shen, G.Z. Dual sensing signal decoupling based on tellurium anisotropy for VR interaction and neuro-reflex system application. *Nat. Commun.* 2022, 13 (1): 5975.

**70** Pop, F., Auban-Senzier, P., Canadell, E., Rikken, G., Avarvari, N. Electrical magnetochiral anisotropy in a bulk chiral molecular conductor. *Nat. Commun.* 2014, 5: 3757.

**71** Yoda, T., Yokoyama, T., Murakami, S. Current-induced orbital and spin magnetizations in crystals with helical structure. *Sci. Rep.* 2015, 5: 12024.

**72** Inui, A., Aoki, R., Nishiue, Y., Shiota, K., Kousaka, Y., Shishido, H., Hirobe, D., Suda, M., Ohe, J., Kishine, J., Yamamoto, H.M., Togawa, Y. Chirality-induced spin-polarized state of a chiral crystal CrNb3S6. *Phys. Rev. Lett.* 2020, 124 (16): 166602.

**73** Shiota, K., Inui, A., Hosaka, Y., Amano, R., Onuki, Y., Hedo, M., Nakama, T., Hirobe, D., Ohe, J., Kishine, J., Yamamoto, H.M., Shishido, H., Togawa, Y. Chirality-induced spin polarization over macroscopic distances in chiral disilicide crystals. *Phys. Rev. Lett.* 2021, 127 (12): 126602.

**74** Ray, K., Ananthavel, S.P., Waldeck, D.H., Naaman, R. Asymmetric scattering of polarized electrons by organized organic films of chiral molecules. *Science* 1999, 283 (5403): 814–816.

**75** Yang, S.H. Spintronics on chiral objects. *Appl. Phys. Lett.* 2020, 116 (12).

**76** Suda, M., Thathong, Y., Promarak, V., Kojima, H., Nakamura, M., Shiraogawa, T., Ehara, M., Yamamoto, H.M. Light-driven molecular switch for reconfigurable spin filters. *Nat. Commun.* 2019, 10: 2455.

**77** Song, T.C., Anderson, E., Tu, M.W.Y., Seyler, K., Taniguchi, T., Watanabe, K., McGuire, M.A., Li, X.S., Cao, T., Xiao, D., Yao, W., Xu, X.D. Spin photovoltaic effect in magnetic van der Waals heterostructures. *Sci. Adv.* 2021, 7 (36): eabg8094.

**78** Dong, Z.Y., Ma, Y.H. Atomic-level handedness determination of chiral crystals using aberration-corrected scanning transmission electron microscopy. *Nat. Commun.* 2020, 11 (1): 1588

**79** Calavalle, F., Suárez-Rodríguez, M., Martín-García, B., Johansson, A., Vaz, D.C., Yang, H.Z., Maznichenko, I., Ostanin, S., Mateo-Alonso, A., Chuvilin, A., Mertig, I., Gobbi, M., Casanova, F., Hueso, L.E. Gate-tuneable and chirality-dependent charge-to-spin conversion in Te nanowires. *Nat. Mater.* 2022, 21 (5): 526–532.

**80** Fahad, H.M., Gupta, N., Han, R., Desai, S.B., Javey, A. Highly sensitive bulk silicon chemical sensors with Sub-5 nm thin charge inversion layers. *ACS Nano* 2018, 12 (3): 2948–2954.

**81** Guo, S.Q., Yang, D., Zhang, S., Dong, Q., Li, B.C., Tran, N., Li, Z.Y., Xiong, Y.J., Zaghloul, M.E. Development of a cloud-based epidermal MoSe 2 device for hazardous gas sensing. *Adv. Funct. Mater.* 2019, 29 (18): 1900138.

**82** van den Broek, J., Abegg, S., Pratsinis, S.E., Güntner, A.T. Highly selective detection of methanol over ethanol by a handheld gas sensor. *Nat. Commun.* 2019, 10: 4220.

**83** Woellner, M., Hausdorf, S., Klein, N., Mueller, P., Smith, M.W., Kaskel, S. Adsorption and detection of hazardous trace gases by metal-organic frameworks. *Adv. Mater.* 2018, 30 (37): 1704679.

**84** Dincer, C., Bruch, R., Costa-Rama, E., Fernandez-Abedu, M.T., Merkoçi, A., Manz, A., Urban, G.A., Güder, F. Disposable sensors in diagnostics, food, and environmental monitoring. *Adv. Mater.* 2019, 31 (30).

**85** Wang, Z., Huang, L.Z., Zhu, X.F., Zhou, X., Chi, L.F. An ultrasensitive organic semiconductor $NO_2$ sensor based on crystalline TIPS-pentacene films. *Adv. Mater.* 2017, 29 (38): 1703192.

**86** Kim, S.G., Jun, J., Lee, J.S., Jang, J. A highly sensitive wireless nitrogen dioxide gas sensor based on an organic conductive nanocomposite paste. *J. Mater. Chem. A* 2019, 7 (14): 8451–8459.

**87** Cui, H.P., Zheng, K., Xie, Z.J., Yu, J.B., Zhu, X.Y., Ren, H., Wang, Z.P., Zhang, F., Li, X.D., Tao, L.Q., Zhang, H., Chen, X.P. Tellurene nanoflake-based $NO_2$ sensors with superior sensitivity and a sub-parts-per-billion detection limit. *ACS Appl. Mater. Interfaces* 2020, 12 (42): 47704–47713.

**88** Moon, G.D., Ko, S., Min, Y., Zeng, J., Xia, Y.N., Jeong, U. Chemical transformations of nanostructured materials. *Nano Today* 2011, 6 (2): 186–203.

**89** Buck, M.R., Schaak, R.E. Emerging strategies for the total synthesis of inorganic nanostructures. *Angew. Chem. Int. Ed.* 2013, 52 (24): 6154–6178.

**90** Tao, H., Shan, X.D., Yu, D.P., Liu, H.M., Qin, D.H., Cao, Y. Solution grown Se/Te nanowires: nucleation, evolution, and the role of triganol Te seeds. *Nanoscale Res. Lett.* 2009, 4 (9): 963–970.

**91** Yang, Y., Wang, K., Liang, H.W., Liu, G.Q., Feng, M., Xu, L., Liu, J.W., Wang, J.L., Yu, S.H. A new generation of alloyed/multimetal chalcogenide nanowires by chemical transformation. *Sci. Adv.* 2015, 1 (10): e1500714.

**92** Armand, M., Tarascon, J.M. Building better batteries. *Nature* 2008, 451 (7179): 652–657.

**93** Chen, Z., Zhang, D.Q., Wang, X.L., Jia, X.L., Wei, F., Li, H.X., Lu, Y.F. High-performance energy-storage architectures from carbon nanotubes and nanocrystal building blocks. *Adv. Mater.* 2012, 24 (15): 2030–2036.

**94** Dong, S.M., Chen, X., Gu, L., Zhou, X.H., Li, L.F., Liu, Z.H., Han, P.X., Xu, H.X., Yao, J.H., Wang, H.B., Zhang, X.Y., Shang, C.Q., Cui, G.L., Chen, L.Q. One dimensional $MnO_2$ titanium nitride nanotube coaxial arrays for high performance electrochemical capacitive energy storage. *Energy Environ. Sci.* 2011, 4 (9): 3502–3508.

**95** Chiang, Y.M. Building a better battery. *Science* 2010, 330 (6010): 1485–1486.

**96** Goodenough, J.B. Evolution of strategies for modern rechargeable batteries. *Acc. Chem. Res.* 2013, 46 (5): 1053–1061.

**97** Liu, N., Lu, Z.D., Zhao, J., McDowell, M.T., Lee, H.W., Zhao, W.T., Cui, Y. A pomegranate-inspired nanoscale design for large-volume-change lithium battery anodes. *Nat. Nanotechnol.* 2014, 9 (3): 187–192.

**98** Xu, J., Xin, S., Liu, J.W., Wang, J.L., Lei, Y., Yu, S.H. Elastic carbon nanotube aerogel meets tellurium nanowires: a binder- and collector-free electrode for Li-Te batteries. *Adv. Funct. Mater.* 2016, 26 (21): 3580–3588.

**99** Zhang, X.F., Jiao, S.Q., Tu, J.G., Song, W.L., Xiao, X., Li, S.J., Wang, M.Y., Lei, H.P., Tian, D.H., Chen, H.S., Fang, D.N. Rechargeable ultrahigh-capacity tellurium-aluminum batteries. *Energy Environ. Sci.* 2019, 12 (6): 1918–1927.

# VI

# Selenium

*Shancheng Yan*

## 6.1 Introduction

Selenium [1–3], a nonmetallic element, is crucial as a group IV semiconductor. It has been widely studied due to its unique properties, such as high photoconductivity, high-voltage electrical and thermoelectric response, thermal conductivity, and anisotropy. The electrical conductivity of selenium can be increased by a factor of 1000 when exposed to light. This is mainly due to its electrical structure of $5s^25p^4$ configuration with six outer electrons [4, 5]. Selenium has several allotropes, i.e. structural variants with different atomic arrangements. α-Selenium exhibits a monoclinic crystalline structure, distinguished by its red color and helically arranged atomic configuration. Similar to a-selenium [6, 7], β-selenium also possesses a monoclinic structure and remains relatively stable at higher temperatures; however, it undergoes a phase transition to a grayish-white color. The most stable of all the isomers is *t*-Se, where each selenium atom shares two electrons with two other selenium atoms, forming a selenium chain that resembles a DNA helical structure with a gray color. This is why many selenium nanostructures are naturally self-contained and chiral. Since 1818, when Betzelius of Sweden identified the element selenium, a long quest for selenium has been undertaken. It was not until 1957 that selenium was discovered as an essential trace element for plants and animals. Low-dose selenium intake can enhance immunity and prevent diabetes, cataracts, and cardiovascular diseases. Even selenium can fight cancer to a certain extent. However, excessive selenium intake can cause gastrointestinal disorders and irritate the respiratory and mucous membranes.

Nanotechnology, as an up-and-coming emerging technology, has attracted much attention. Nanomaterials have many exceptional properties due to their close coherence length to electrons and their proximity to the wavelength of light on the scale. For example, small size effect [8, 9], surface and interface effect [10], quantum size effect [11–13], enhancement effect, etc. Meanwhile, different shapes of nanomaterials also present different properties, and the purpose of tuning device performance can be realized by adjusting the nanostructure of selenium. After much research, the methods for synthesizing selenium nanostructures can be mainly categorized into bottom-up and top-down methods, and the growth mechanisms are vapor–liquid–solid (VLS) and solid–liquid–solid (SLS), with standard methods such as hydrothermal [14–16], solvent-thermal, vapor-phase deposition [17–20], laser [21–24], and liquid-phase separation [25–28]. The same synthesis method

*Single Element Semiconductors: Properties and Devices*, First Edition. Yi Shi and Shancheng Yan.

can sometimes simultaneously produce two or more selenium nanostructures. Due to the multiple unique properties of selenium nanostructures, they have been applied in optoelectronic devices [29–31], sensing [32–34], biology [35–38], and energy storage.

This chapter discusses the synthesis of selenium nanomaterials, summarizing the various synthesis methods. The properties and applications of selenium nanomaterials with different morphologies are discussed in depth. The outline of this chapter is as follows: Section 6.2 discusses the synthesis methods for different nano-morphologies of selenium, including zero-dimensional (0D) nanostructures (Section 6.2.1), one-dimensional (1D) nanostructures (Section 6.2.2), and two-dimensional (2D) nanostructures (Section 6.2.3). In Section 6.3, we discuss the properties and applications of selenium nanomaterials, including optical properties (Section 6.3.1), Raman spectroscopy (Section 6.3.2), and photoluminescence (PL) spectroscopy (Section 6.3.3); and the applications include photovoltaic applications (Section 6.3.4), energy storage applications (Section 6.3.5), and biological applications (Section 6.3.6). The conclusion is given in Section 6.4.

## 6.2 Synthesis of Selenium Nanomaterials

When the size is controlled to the nanoscale, the material will exhibit many unique properties. Nanomaterials often exhibit different optical, electrical, mechanical, and thermal properties than when they are in bulk. Different nanostructures often possess different properties as well. So far, much research has been done in synthesizing selenium nanomaterials with impressive results. The major nanostructures include 0D selenium nanoparticles (NPs) and nanorods, 1D selenium nanowires (NWs), nanotubes, and nanorods, and 2D selenium nanosheets and films.

### 6.2.1 Synthesis of 0D Selenium Nanostructures

NPs [39–41], as typical 0D nanostructures, mostly have particle sizes between 1 and 100 nm. Selenium NPs are among the most extensively studied selenium nanostructures. Various chemical methods exist for synthesizing these NPs. For instance, Boroumand et al. [42] produced selenium NPs using a chemical reduction method. They combined sodium selenite (50 mM) with a stabilizer (0.1% polyvinyl alcohol (PVA) or 1% CS) and stirred the mixture for 5 min. After that, 50 mM ascorbic acid aqueous solution was added with vigorous stirring to reduce the selenium NPs quickly. These transmission electron microscopy (TEM) images confirm that the selenium particles obtained are on the nanometer scale, with a shape close to spherical and more uniform in size. The average diameters of the PVA-SeNPs and CS-SeNPs are $66.6 \pm 8.5$ and $48.5 \pm 2.8$ nm, respectively. This synthesis method uses no harmful reducing agent and is a low-cost and green synthesis. Shah et al. [43] have used a facile and feasible wet-chemistry method to synthesize selenium NPs. They successfully prepared $Na_2SeSO_3$ solution by reacting 120 mL of $Na_2SO_3$ aqueous solution (17%) with 2 g of Se powder and refluxing it at 70°C for 7 h. The solution of $Na_2SeSO_3$ was used as a reducing agent for synthesizing selenium NPs. The $Na_2SeSO_3$ solution was used as a precursor and mixed with 1 g of PVA, an appropriate acrylonitrile, in 100 mL of water. The stabilized selenium NPs were obtained after stirring. The atomic force microscopy (AFM) images of the NPs are shown in Figure 6.1a,b, from which we can see that all the NPs are spherical with a particle size between 50 and 100 nm. The synthesis method can be carried out at room temperature, a completely new synthesis method. Unlike the purely chemical synthesis method, many have utilized biomolecular mediation to synthesize selenium NPs. Fardsadegh et al. [44] proposed a hydrothermal synthesis method using aloe vera leaf extract. They prepared sun-dried aloe vera as a dried aloe vera powder, after which deionized water was added, boiled, and filtered to obtain the extract. Afterward, 0.263 g of sodium selenite was added to 100 mL of deionized water, mixed with the aloe leaf extract, and placed in a high-pressure facultative. The final selenium NPs were obtained at 121°C and 1.5 bar for 15 min. The average particle size of this NP was 50 nm, and the morphology was spherical. The zeta potential value of selenium NPs was found to be −18 mV by dynamic light scattering analysis. This data proves that the NPs synthesized by this method are highly stable. The polydispersity index value of 0.344 implies that the NPs are monodisperse. The selenium NPs synthesized through aloe vera leaf extract have higher zeta potential value and smaller particle size for higher stability than those synthesized from geranium leaf extract. This indicates that the stabilizer used in the hydrothermal synthesis affects the morphology of selenium NPs to some extent. Mellinas et al. [45] synthesized selenium NPs using a microwave-assisted method using cacao bean shell extract, similar to the hydrothermal synthesis of aloe vera leaf extract, as mentioned previously. They placed 6 g of cocoa bean shell extract (CBSE) into 150 mL of deionized water. Afterward, 1 mol of hydrochloric acid was added to adjust the pH to 2, and the mixture was heated in a microwave oven at 100°C for 5 min. While heating and stirring was done at a rate of 400 rpm. The CBSE obtained after centrifugation was placed at −20°C for storage. Sodium selenite served as the selenium source. Fifty milliliters of CBSE were mixed with it and then heated and stirred in a microwave oven. Subsequently, the solution was centrifuged to obtain a lyophilized solution of selenium NPs. The TEM images of the NPs obtained under optimal conditions are shown in Figure 6.1c,d, with a 1–3 nm particle size and uniform distribution. It can also be seen from the figure that crystals with good stability dominate the structure of the NPs. It is worth mentioning that the biological activity of selenium NPs increases with the decrease in particle size, indicating that the NPs prepared by this method are promising for biological applications. There are many similar methods of synthesizing selenium NPs using plant extracts, such as Kokila et al. [46] synthesized selenium NPs with particle sizes of 4–16 nm using Diospyros montana leaf extract, Sharma et al. [47] synthesized selenium NPs using raisin extract in green with particle size of 3–18 nm, Anu et al. [48] synthesized selenium NPs using Allium sativum extract green synthesized selenium NPs with particle size ranging from 40 to 100 nm, Krishnan et al. [49] used leaf extract of Dillenia green synthesized selenium NPs with particle sizes ranging from 50 to 900 nm, and so on. From these examples, we can see that the natural extracts largely influence the size and morphology of selenium NPs synthesized by biomolecule-mediated synthesis. Different extracts have different reducing and encapsulating abilities, which also determine the application potential of the synthesized selenium NPs.

To synthesize relatively pure NPs without surface contamination is challenging, and NPs synthesized by physical methods will be purer than chemically synthesized versus biosynthesized NPs. Selenium NPs were prepared by Geoffrion et al. [50] using pulsed laser ablation, which is a new and green synthesis method. They used a Q-switched Nd:YAG

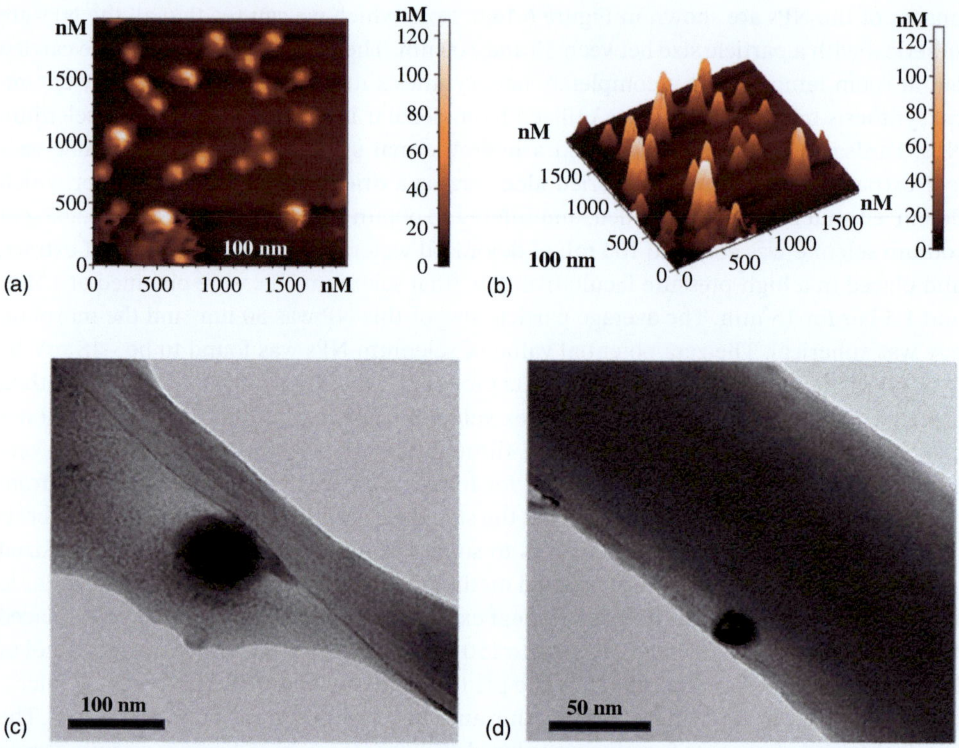


**Figure 6.1** (a) 2D AFM image of selenium NPs formed by the reaction of $5 \times 10^{-4}$ mol dm$^{-3}$ sodium selenosulphate with $9.0 \times 10^{-2}$ mol dm$^{-3}$ acrylonitrile, in the presence of 0.05% PVA. (b) The corresponding 3D AFM image. *Source:* Reprinted with permission from Ref. [43]. Copyright 2008, American Chemical Society). (c, d) TEM image of Se NPs obtained under optimal conditions. *Source:* Reprinted with permission from Ref. [45]. Copyright 2019, MDPI.

laser with a wavelength of 1064 nm, which continuously emitted pulses with an energy of 16.5 mJ pulse$^{-1}$ at 1000 Hz. Many selenium NPs escaped from the selenium block irradiated by the pulses and were waiting to be collected in deionized water. The production of NPs was maximized when the frequency reached 3000 Hz at a concentration of about 40 ppm. The TEM image of the NPs is illustrated in Figure 6.2a, where it can be seen that the particles are uniformly spherical and have a size of 20–60 nm. Meanwhile, the electron diffraction image (Figure 6.2b) proved the NPs to be amorphous selenium. Stabilized selenium NPs were prepared by Guisbiers et al. [51] using a laser ablation technique. They placed excess selenium powder in deionized water, after which a laser beam was focused in the middle of the solution. Stirring at 1000 rpm was maintained during irradiation to ensure the selenium powder was uniformly dispersed in the water. An Nd:YAG laser was used with a pulse duration of 3.6 ns and a repetition frequency of 10 Hz. The laser's energy was monitored during the experiments and kept constant at ~$23 \pm 1$ mJ pulse$^{-1}$. After focusing, the pulse energy corresponded to an effect of ~2 J cm$^{-2}$. The size of the particles is between 100 and 200 nm, and the particles have an amorphous structure. This indicates that the selenium NPs grow very fast after the irradiation pulse, the surrounding liquid

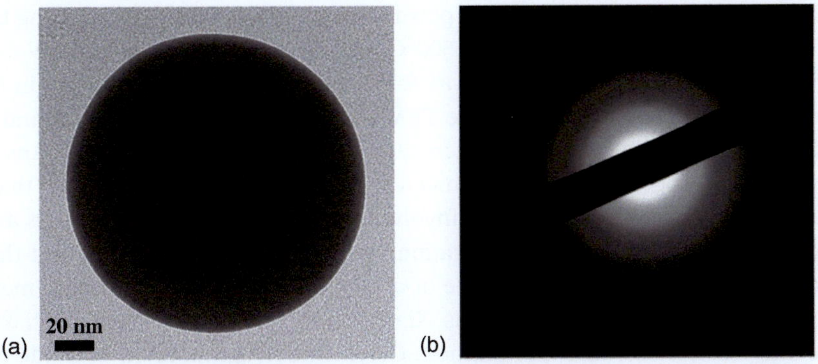


**Figure 6.2** (a) TEM image of a representative selenium NP with its size ~85 nm. (b) Diffraction pattern revealing the amorphous structure of the selenium NP. *Source:* Reprinted with permission from Ref. [50]. Copyright 2020, American Chemical Society.

medium quenches rapidly, and the vast majority of selenium NPs do not have enough time to crystallize. The selenium NPs synthesized by this class of physical methods are relatively pure and do not pollute the environment.

## 6.2.2 Synthesis of 1D Selenium Nanostructures

1D selenium nanostructures are generally considered to include NWs [52, 53], nanotubes [54, 55], nanorods, and nanobelts, which have been widely studied due to their unique morphology and properties.

### 6.2.2.1 Synthesis of Selenium NWs

As the more common 1D nanostructures, NWs have been widely studied due to their unique properties. As early as 2002, Gates et al. [56] proposed an acoustochemical method to synthesize selenium NWs on a large scale. They used excess hydrazine to reduce selenite.

$$H_2SeO_3 + H_2N_2H_2 \rightarrow \alpha\text{-Se}(\downarrow) + N_2(\uparrow) + 3H_2O \quad (6.1)$$

Colloids were prepared at room temperature to avoid premature nucleation, and after sonication of the colloids, the colloidal suspension was stored in the dark. Selenium NWs of 20 μm length were observed after 1 h, and the growth was complete after 5 h. The prepared NWs have controllable lateral scales and even lengths of hundreds of micrometers. Almost all of the a-Se was converted into *t*-Se NWs after 5 h of growth. Control over the length of the NWs can be achieved by varying the concentration of the a-Se colloid, and the length of the selenium NWs increased from 40 to 100 nm when the concentration of selenium was increased by a factor of 3. The method can be carried out at room temperature and atmospheric pressure, which means that it is possible to scale up to mass production of selenium NWs, and experimentally, they have synthesized selenium NWs in a single run of 1 g. The method can be used to produce selenium NWs in a single run. In 2005, Zhang et al. [57] proposed a carbon-thermal chemical vapor deposition method to synthesize selenium NWs, which involves heating a mixture of activated carbon and selenium, followed by thermal evaporation and decomposition into elemental selenium. They mixed different ratios of Se powder and activated carbon powder into a quartz boat. After heating them in a tube furnace with nitrogen gas at 400–500°C, they formed gray selenium NW samples. When the ratio of selenium to carbon was 4:1, the samples with high crystallinity were obtained by heating at 500°C for 2 h. The TEM image is shown in Figure 6.3, and all the products were NWs with uniform diameters ranging from 20 to 60 nm and lengths up to a few micrometers. A portion of the NWs had a seed at the front end, indicating that some NWs were grown from NPs. The method involves three steps: the wire undergoes a chemical reaction, followed by thermal evaporation and, finally, decomposition, and the NWs obtained by this method are virtually free of crystal defects. The solvothermal method is also typical in chemical synthesis methods. Xia et al. [58] synthesized selenium NWs using a simple and feasible solvothermal method. Cetyltrimethylammonium bromide (CTAB) and $SeO_2$ were mixed into 30 mL of distilled water for stirring, and then sodium formaldehyde sulfite was put into distilled water to get its aqueous solution. Selenium NWs were obtained by combining the two mixed solutions and heating them for 4 h at 50°C with stirring. The chemical reaction of the method is as follows:

$$SeO_2 + H_2O \rightarrow H_2SeO_3 \rightarrow SeO_3^{2-} + 2H^+ \tag{6.2}$$

$$NaHSO_2CH_2O2H_2O + H^+ \rightarrow Na^+ + CH_2OHSOOH + 2H_2O \tag{6.3}$$

$$CH_2OHSOOH \rightarrow HSO_2^- + HCHO + H^+ \tag{6.4}$$

$$HSO_2^- \rightarrow SO_2^{2-} + H^+ \tag{6.5}$$

The reduction reaction is as follows:

$$SO_2^{2-} + SeO_3^{2-} + H^+ \rightarrow Se\downarrow + SO_3^{2-} + H_2O \tag{6.6}$$

The diameter of the NW is uniform, about 100 nm. The corresponding selected area electron diffraction shows that the reflections in the (001), (111), and (110) planes are apparent,

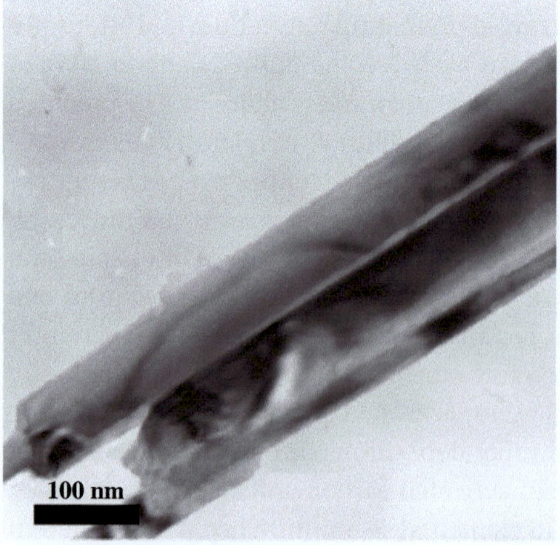


**Figure 6.3** TEM image of several Se NWs assembled parallel to each other. *Source:* Reprinted with permission from Ref. [57]. Copyright 2005, American Chemical Society.

and the NW is a single crystal. The above syntheses are typical chemical syntheses with simple and efficient reactions and low cost.

A representative physical method to synthesize selenium NWs was proposed by Ren et al. [59] by a gas-phase growth method. They used Si powder as a catalyst and Se powder as a Se source. The quartz boat covered with Si powder was placed in the center of the tube furnace to ensure a temperature of 950°C in its vicinity, and the Se powder was placed at the inlet end of the tube furnace at a temperature of about 680°C. The final selenium NWs obtained were collected using copper sheets. Figure 6.4a demonstrates the TEM image of the NWs, which can be up to a few micrometers long, with diameters greater than 20 nm and uniform diameters of the individual NWs. The NWs synthesized by this method were well crystallized with an amorphous layer of selenium wrapped around the exterior of the NWs. Yan et al. [60] reported a novel physical synthesis method to form the NWs by immersing an amorphous selenium film in a propanol solution without any chemical reaction directly through a phase transition. The TEM image of the NWs is shown in Figure 6.4b; the NWs have a diameter of about 100 nm, are uniform in thickness and linear shape, and preferentially grow along the [001] direction. The presence of acetone exacerbates the anisotropy between the (001) crystallographic and prismatic planes of Se,

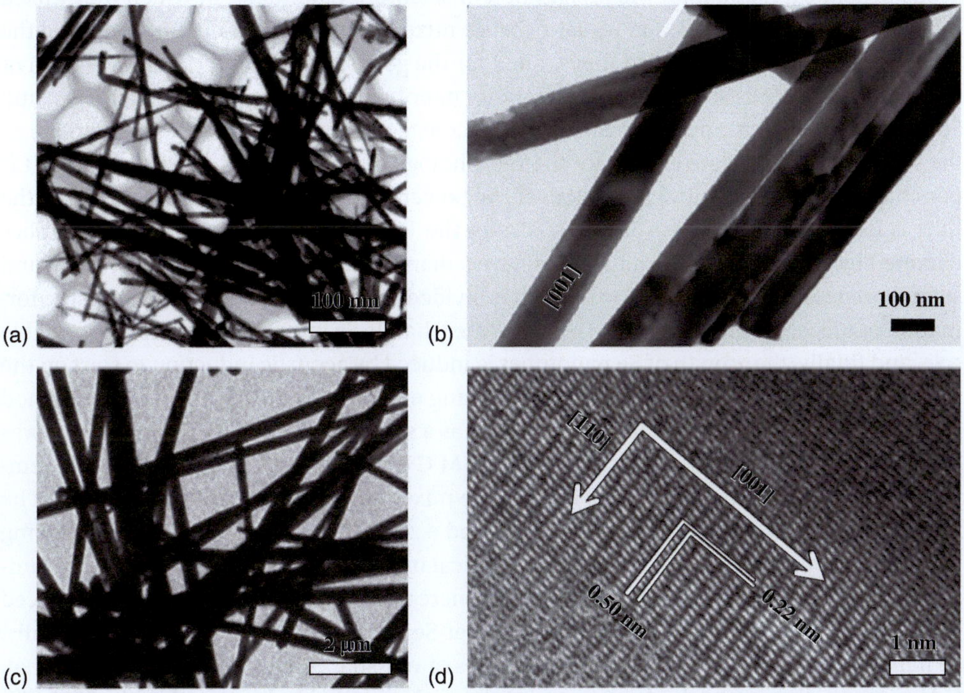


**Figure 6.4** (a) TEM image of the Se NWs. The diameters of the Se NWs vary in the range between 20 and 300 nm. *Source:* Reprinted with permission from Ref. [59]. Copyright 2004, American Chemical Society). (b) Bright-field TEM image of NWs. *Source:* Reprinted with permission from Ref. [60]. Copyright 2018, American Chemical Society. (c) TEM image of *t-Se* NWs. (d) HRTEM image revealed the growth of Se NWs was along [001]. *Source:* Reprinted with permission from Ref. [61]. Copyright 2020, NATL INST R&D MATERIALS PHYSICS.

favoring the growth of 1D nanostructures. The most significant advantage of this method is that it can be carried out on any solid substrate, facilitating large-scale synthesis.

The liquid-phase method for preparing selenium NWs was proposed by Lei et al. [61]. They prepared an aqueous solution of $CH_4N_2Se$, which was vigorously stirred at room temperature to make amorphous a-Se colloids. Afterward, the colloid was centrifuged, and the brick-red precipitate was kept in the dark to obtain the selenium NWs. TEM images and high-resolution transmission electron microscopy (HRTEM) images are shown in Figure 6.4c,d. The spacing of the crystal faces perpendicular to the direction of the long axis of the *t*-Se NWs and parallel to the *t*-Se NWs is about 0.50 and 0.22 nm, which corresponds to the (001) face and the (110) face of the *t*-Se, respectively. In addition, the successive streaks show that the NWs have few defects and very excellent crystal quality. This method reveals that the transformation of a-Se to *t*-Se does not necessarily need to go through three steps of solid–liquid–solid but can also be performed directly from solid to solid state, which inspires new synthesis pathways.

#### 6.2.2.2 Synthesis of Selenium Nanotubes

Unlike NWs, nanotubes have received much attention as hollow 1D structures, and the most common nanotube material is carbon nanotubes. Back in 2004, Li et al. [62] proposed an acoustochemical method for the preparation of selenium nanotubes, in which they mixed hydrazine hydrate, ethanol, and water to form a mixed solution and then ultrasonicated the solution for 0.5–1 h, after which they waited for the generation of nanotubes. At 30 min of sonication, *t*-Se seeds started to transform into nanotubes, and then sonicated for 1 h, and Se nanotubes were formed. For this type of acoustochemical method, the parameters of the sonication process are crucial for the morphology of the selenium nanostructures. It is possible to prepare selenium nanotubes if the power of sonication is strong enough or the time is long enough. Figure 6.5a demonstrates the TEM image of the selenium nanotubes obtained by this method. The nanotubes have a diameter of about 200 nm and a spiral and semi-closed tubular structure. The method is divided into three stages, starting with the formation of unstable spherical R-Se particles. After that, R-Se is transformed into more stable *t*-Se, and finally, the growth of Se nanotubes is induced by acoustic induction to start on the surface of spherical *t*-Se. In the same year, Zhang et al. [63] proposed a chemical method to synthesize selenium nanotubes using CTAB as a soft template. 2 mL of 1 M selenite was used as the selenium source, and 20 mL of 0.6 M CTAB solution was used as the soft template, after which ascorbic acid was added. The mixed solution was sonicated at 80°C. The nanotubes have a diameter of about 100 nm and a smooth surface. It is worth mentioning that the concentration of CTAB is the most critical in this method, which enhances the anisotropy of *t*-Se as the concentration of CTAB increases. And, since CTAB largely reduced the surface tension of the mixed solution, larger Se seeds were generated, and eventually, Se nanotubes were formed on the soft templates of rod-shaped CTAB micelles. Zhang et al. [64] successfully prepared selenium nanotubes using a liquid-phase method in 2009. An appropriate amount of Se powder and 10 mL of hydrazine, 30 mL of ammonia, and 60 mL of alcohol were added to water to form a mixed solution, which was kept in a Teflon-lined stainless steel autoclave at 200°C for 20 h. The mixture of Se powder and methanol was used as an acoustic solvent. When methanol was used as the sonochemical solvent, a tubular product was obtained, and the field-emission scanning electron microscopy (FESEM)

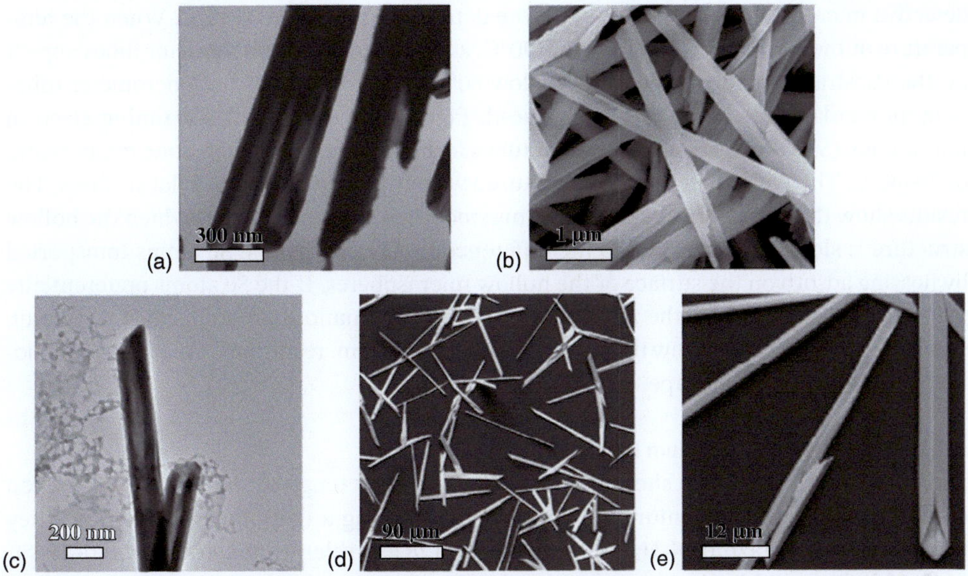


**Figure 6.5** (a) TEM image of selenium nanotubes prepared in ethylene glycol by Bransonic B5-E ultrasonic cleaner. *Source:* Reprinted with permission from Ref. [62]. Copyright 2005, American Chemical Society. (b) FESEM image of Se nanostructures prepared by the hydrothermal process and subsequent sonication process using methanol as a solvent. *Source:* Reprinted with permission from Ref. [64]. Copyright 2006, American Chemical Society. (c) TEM image of the as-synthesized Se nanotubes. *Source:* Reprinted with permission from Ref. [66]. Copyright 2012, Elsevier B.V. (d, e) Low- and high-magnification SEM images of the Se microstructures synthesized at the temperature deposition zones of 70–80°C. *Source:* Reprinted with permission from Ref. [67]. Copyright 2010, American Chemical Society.

image is shown in Figure 6.5b. The diameter of the nanotubes was about 300–500 nm, and the wall thickness was about 20 nm, which was a uniform diameter. In this method, ethanol and acetone can also be used as sonochemical solvents to prepare nanotubes. In 2010, Chen et al. [65] synthesized nanotubes on a large scale in an aqueous solution at 85°C without using any template or surfactant. The method is based on the "solid–solution–solid" growth mechanism, where amorphous Se NPs are first generated and then transformed into crystal seeds to form nanotubes after a high-temperature phase transition. Ren et al. [66] devised a low-temperature method for preparing selenium nanotubes by reducing selenite with hydrazine hydrate. The TEM images of the nanotubes are shown in Figure 6.5c, and the length of the tubes can be up to a few micrometers. The preparation method still follows the "solid–solution–solid" growth mechanism, in which a-Se nanorods are dissolved into amorphous Se colloids, which are then crystallized into selenium nanotubes. Unlike chemical synthesis, the most typical method for physically synthesizing selenium nanotubes is gas-phase deposition, in which Filippo et al. [67] used this method to prepare selenium nanotubes by heating and evaporating Se powders at 300°C, then cooling and depositing them on quartz substrates at temperatures lower than 140°C. The growth temperature is critical for the formation of the nanostructures. The growth temperature had a great influence on the morphology of the nanostructures formed, with a large number of defective micrometer spheres present in the deposition zone at 90–140°C. When the temperature of the deposition zone is below 90°C, a large number of micrometer tubes appear on the substrate, and at temperatures below 80°C, the structure of the micrometer tubes is more regular and the defects are reduced. Figure 6.5d,e shows the scanning electron microscopy (SEM) images of the micron tubes formed at the deposition zone temperature of 70–80°C. The nanotubes are well structured with triangular edges and flat surfaces. The results show that the deposited Se first forms smooth microspheres, after which the hollow structure is slowly formed. Due to the difference in free energy, the Se atoms transported by heating adsorb on the surface of the hollow microspheres. If the Se atoms preferentially move toward the edges of the microspheres, defect-free nanotubes are formed. However, some Se atoms will still follow the spiral growth mechanism, resulting in an irregular nanotube with a semi-closed shape.

#### 6.2.2.3 Synthesis of Selenium Nanorods

Unlike NWs, nanorods are shorter and have a straighter longitudinal morphology. Chen et al. [68] synthesized selenium nanorods (Se-NRs) using a hydrothermal method. They mixed 1 mL of hydrazine ($N_2H_4$, 50%, v/v) with 0.128 g of selenite and kept it at 110–180°C for 2–12 h. The resulting black product was Se-NRs. Different synthesis temperatures determine the size of the diameter of the nanorods, and at a temperature of 130°C, the rods have a more regular shape with a diameter of about 2 μm and a length of about 5 μm, and the SEM image is shown in Figure 6.6a. Although *t-Se* is anisotropic, nanorods can only be grown when the growth rate is fast enough, and higher temperatures can accelerate the enrichment of selenium, so the nanorods at high temperatures have better morphology and larger diameters. Se-NRs were prepared using a modified hydrothermal method by Chen et al. [69]. Without surfactants and templates, the reaction was carried out in water and ethanol using green and nonpolluting glucose as the reducing agent. The method, similar to most hydrothermal methods, follows the solid–solution–solid growth mechanism, and the ratio of water and ethanol has an essential effect on the morphology of the nanorods. Biomolecule-assisted synthesis methods are more common in synthesizing Se-NRs than chemical synthesis. Kannan [70] reported for the first time a simple precipitation method for the synthesis of Se-NRs utilizing l-arginine and ascorbic acid, two biomolecule substances with attractive structures and self-assembling effects, for assisted synthesis. The FESEM image of the nanorods is shown in Figure 6.6b, with a large number of nanorods attached to the surface with uniform distribution. The diameter was about 40 nm, and the length could be up to a few micrometers. In 2021, Velayati et al. [71] synthesized Se-NRs by a green and simple biological method using ascorbic acid as a reducing agent, gum arabic as a stabilizer, and sodium selenite ($Na_2SeO_3$) as a source of selenium. The FESEM image is shown in Figure 6.6c, and the rods have diameters of about 20–100 nm and lengths of up to hundreds of nanometers. The following year, Velayati et al. [72] proposed another simple and green method to prepare Se-NRs using chitosan (CS) as a stabilizer. They added selenite as a source of selenium to a mixture of CS solution and ascorbic acid, stirred it, and finally centrifuged it to obtain Se-NRs. These biosynthesis methods mentioned above utilize natural biomolecules, which are environmentally safe and less costly. However, to pursue higher quality nanorods, there is always an effort to explore the physical method of synthesis. Jiang et al. [73] prepared extremely pure nanorods using laser ablation. They

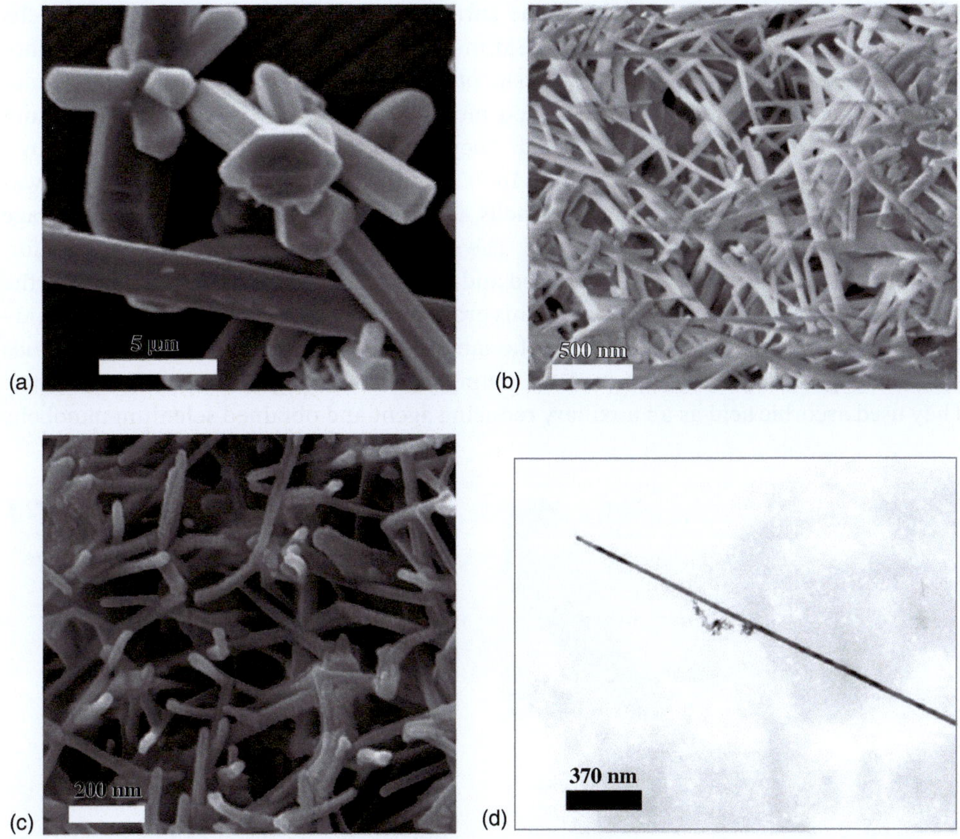


**Figure 6.6** (a) SEM image of selenium rods prepared at 130°C. *Source:* Reprinted with permission from Ref. [68]. Copyright 2006, Elsevier B.V. (b) FESEM image of Se-NRs. *Source:* Reprinted with permission from Ref. [70]. Copyright 2014, Indian Academy of Sciences. (c) FESEM image of Se-NRs. *Source:* Reprinted with permission from Ref. [71]. Copyright 2021, Elsevier B.V. (d) TEM image of one as-prepared Se-NR. *Source:* Reprinted with permission from Ref. [73]. Copyright 2003, Elsevier B.V.

put selenium powder into the bottom of a quartz tube and focused the laser on Se powder. After cooling and deposition, a gray-black product appeared on the wall of the quartz tube, i.e. the Se-NRs. The TEM images of Se-NRs grown at temperatures of 170–120°C are shown in Figure 6.6d. The nanorods have diameters of about 30–80 nm and lengths of several micrometers. The nanorods synthesized by this method are single crystals and remain structurally stable after months of storage. Chen et al. [74] synthesized Se-NRs by ultrasonication using a mixed ethylenediamine, selenium powder, and water solution. The diameter of the nanorods can be controlled by adjusting the concentration of selenium added, while the length of the nanorods can be determined by controlling the reaction time.

#### 6.2.2.4 Synthesis of Selenium Nanobelts

There are few studies on selenium nanobelts. Ma et al. [75] synthesized single-crystal selenium nanobelts using a liquid-phase method. They chose poly(oxyethylene octadecyl ether) ($C_{18}EO_{20}$) micellar solution as the surfactant and prepared selenium nanobelts with a thickness of about 30 nm. The TEM images are shown in Figure 6.7a. The nanobelts have a 100–300 nm width and very smooth edges. They speculated that these nanobelts are not flexible enough to bend too much. Xie et al. [76] synthesized selenium nanobelts using a hydrothermal method. They mixed $SeO_2$, glucose, and polyvinylpyrrolidone into a reactor to be heated at 160°C for 20 h. After drying, the product was obtained. The TEM images of the nanobelts are shown in Figure 6.7b, with an average width of 85 nm and a thickness of 25 nm. The whole growth process started with the formation of R-Se NPs, which were dissolved and then crystallized into *t-Se*, and finally, the nanoribbons were grown on the seeds. This growth can be attributed to the solid–liquid–solid phase transition mechanism. Not the same as the two chemical synthesis methods above, Wang et al. [77] synthesized selenium nanobelts using biomolecule assistance. They used ascorbic acid as an auxiliary reducing agent and obtained selenium nanobelts

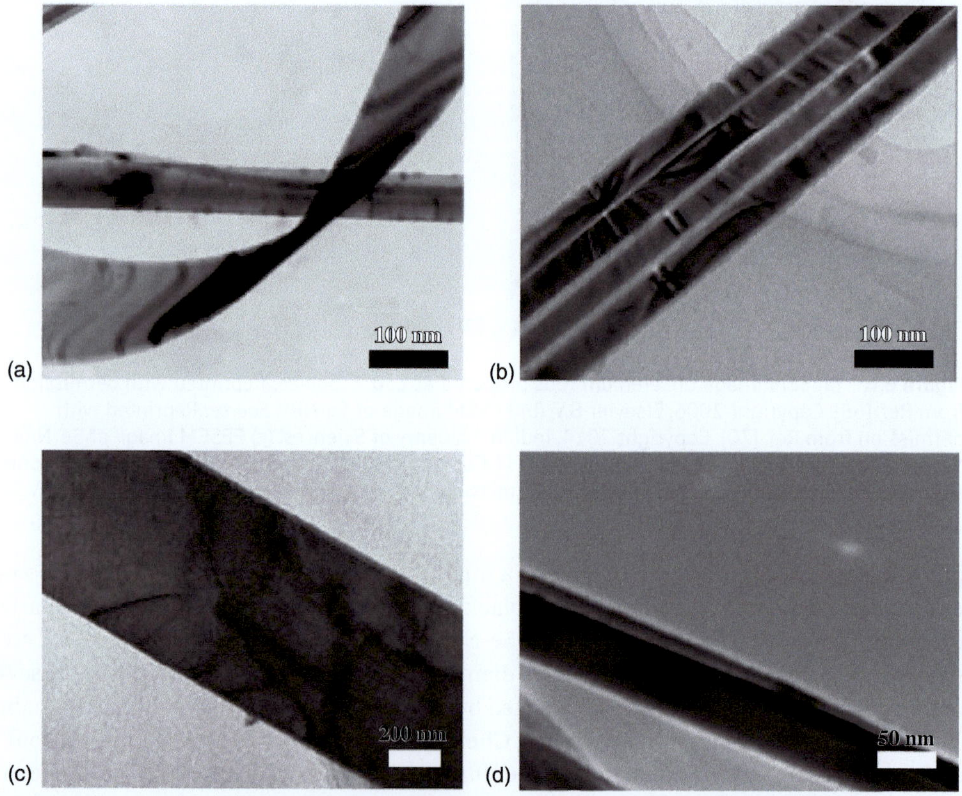


**Figure 6.7** (a) TEM image of Se nanobelts obtained in $C_{18}EO_{20}$ micellar solution. *Source:* Reprinted with permission from Ref. [75]. Copyright 2005, American Chemical Society. (b) TEM image of the *t*-Se nanobelts synthesized. *Source:* Reprinted with permission from Ref. [76]. Copyright 2006, American Chemical Society. (c) TEM image of a single Se nanobelt. *Source:* Reprinted with permission from Ref. [77]. Copyright 2018, Elsevier B.V. (d) TEM image of one as-prepared Se nanobelt. *Source:* Reprinted with permission from Ref. [78]. Copyright 2007, American Chemical Society.

with a length of about 800 nm after a hydrothermal process. Figure 6.7c demonstrates the TEM images of the nanobelts, all of which have a uniform and smooth morphology, and the nanobelt crystallinity is as high as 96.1%. In pursuit of high-quality nanobelts, Wang et al. [78] prepared selenium nanobelts by vacuum vapor deposition. The TEM images of the nanobelts, with individual nanobelt widths of about 500 nm, are illustrated in Figure 6.7d. The study results show that the morphology of the generated nanostructures is highly dependent on the growth temperature. Se sea urchin-like nanostructures consisting of nanorods and open nanorods were formed as intermediates during nanobelt preparation. As evaporation proceeds, these intermediates undergo agglomeration through particle-nanorod collisions, and eventually Se nanobelts grow from the roots formed by agglomeration. *t*-Se nanobelts grow following a gas–solid mechanism mediated by agglomeration.

### 6.2.3 Synthesis of 2DSelenium Nanostructures

2D materials are layered structures consisting of single or few layers of atoms or molecules, where electrons can only move freely on a 2D scale and usually have some exciting properties. Common 2D materials are graphene, black phosphorus nanosheets, and several van der Waals (vdW) layered semiconductors with single or few layers. The liquid-phase exfoliation method is the most common means of preparing selenium nanosheets. The first preparation of selenium nanosheets using the liquid-phase exfoliation method was reported by Xing et al. [79]. They combined probe and water bath sonication to strip selenium solution (isopropanol) with multiple sonication cycles, resulting in selenium nanosheets. The power of probe sonication was 600 W for 9 h, and the energy power of water bath sonication was 400 W for 6 h. The TEM images are shown in Figure 6.8a, and the nanosheets have transverse dimensions of about 40–110 nm and thicknesses of about 3–6 nm. This study demonstrated that bulk selenium has a typical chain structure. When the weak vdW forces within the chain are broken by sonication, the inner linkages in the bulk selenium break to form nanosheets. Fan et al. [80] also prepared selenium nanosheets using a liquid-phase exfoliation method similar to that of Xing et al. [79]. High-quality selenium nanosheets were obtained using probe sonication followed by water bath sonication. The energy power of probe sonication was 200 W, and the energy power of water bath sonication was 360 W for 10 h. The TEM images of the nanosheets are shown in Figure 6.8b, with transverse dimensions of about 50–120 nm. This method is also effective for materials such as black phosphorus. Vapor-phase deposition is also a standard method for preparing selenium nanosheets. Qin et al. [81] used physical vapor deposition to prepare large, high-quality selenium nanosheets. The AFM image of a single nanosheet is shown in Figure 6.8c, with an irregular tetragonal shape, jagged edges, a flat surface, and a thickness of about 5 nm. As shown in Figure 6.8d, some of the nanosheets have a twinned structure with an average thickness of 15 nm. The growth mechanism of the method is similar to that of most sheet structures. The first one is formed into nanorods, which, due to the anisotropy of selenium, have a binding energy in the direction of the c-axis higher than the binding energy in the direction of the $< 1\bar{2}10 >$, causing the nanorods to grow along the $< 1\bar{2}10 >$ parallel direction to form nanosheets.

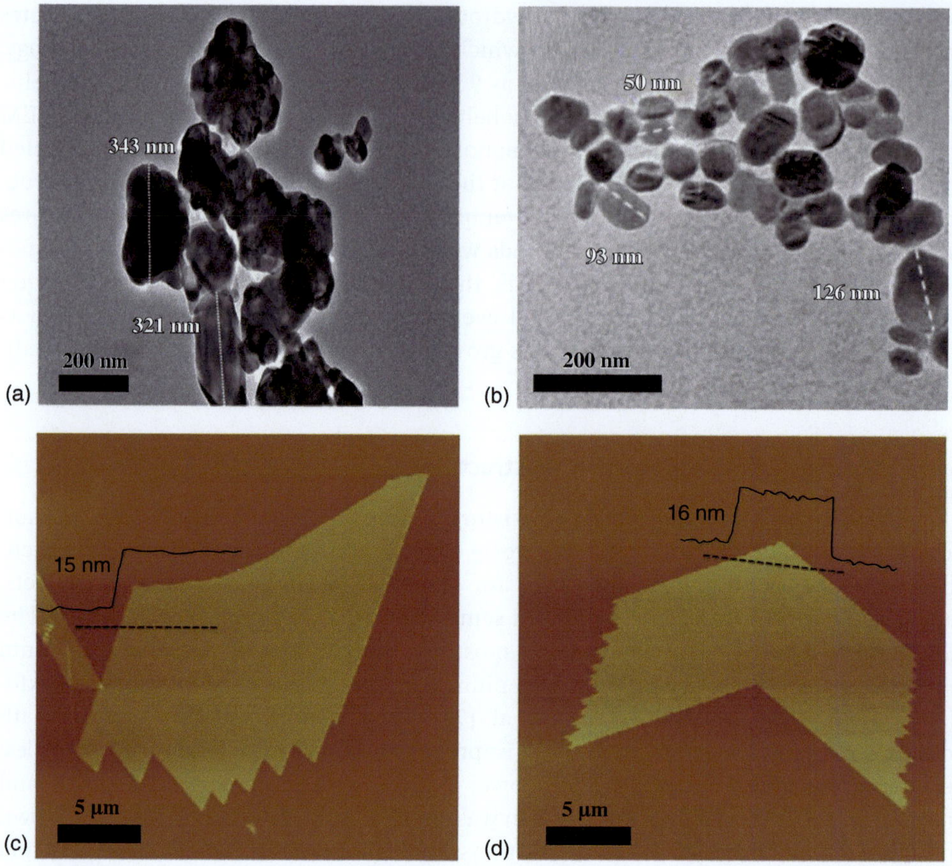


**Figure 6.8** (a) TEM image of the as-prepared 2D nonlayered Se nanosheets obtained by liquid-phase exfoliation at various centrifugation speeds of 1000 rpm. *Source:* Reprinted with permission from Ref. [79]. Copyright 2017, WILEY-VCH. (b) TEM image of the Se nanosheets. *Source:* Reprinted with permission from Ref. [80]. Copyright 2019, IOP. (c, d) AFM height profile of saw-like and twin-structure nanosheets. *Source:* Reprinted with permission from Ref. [81]. Copyright 2017, American Chemical Society.

## 6.3 Properties and Applications of Selenium Nanostructures

Compared with bulk selenium, a wide variety of selenium nanostructures have many interesting properties, such as quantum confinement effect, surface effect, volume effect, etc. Studying the properties of selenium nanostructures and applying them in various fields has far-reaching implications for nanotechnology research. This section discusses the optical properties of selenium nanostructures, Raman and PL spectroscopy, and the applications of selenium nanostructures in optoelectronic devices, energy storage, and biology.

### 6.3.1 Optical Properties

Selenium is a typical p-type exogenous semiconductor with a relatively narrow bandgap (1.8–2.2 eV) and a strong response to light [82]. *t*-Se nanobelts and NWs prepared by Ma et al. [75] were characterized by UV-vis absorption spectroscopy, and the results are shown in Figure 6.9a. The Se NWs obtained in $C_{12}EO_{10}$ micelles showed three absorption peaks at 570, 450, and 346 nm. Due to interchain interactions perpendicular to the c-axis within the *t*-Se crystals, peaks higher than 530 nm can be used to characterize the degree of perfection of the crystals and the interactions between the atomic chains. However, the *t*-Se nanobelts obtained in $C_{18}EO_{20}$ micelles have a low-energy peak at 660 nm, which is redshifted by 90 nm compared to the NW. The results indicate that the interchain interactions in the nanobelts are much weaker than those in the NW. The nanosheets have a rectangular cross-section, while the NW cross-section is equilateral hexagonal, which makes the NW triangular phase more perfect. The experimental results show that the optical absorption properties of selenium depend on the nanostructural morphology of selenium. The synthesized selenium nanobelts were also characterized by UV-vis absorption spectra by Xie et al. [76] The results are shown in Figure 6.9b. The nanoribbon has three prominent absorption peaks located at 589 nm, 450 nm, and 360 nm, and the absorption peak at 589 nm can be attributed to molecular leaps between the *t*-Se helical chains. Xing et al. [79] carried out UV-vis absorption spectroscopy measurements on the nanosheets. They found that single-probe sonication of the nanosheets for 1.5 h had almost the same effect as sonication using a water bath for 20 h, thus demonstrating that the anisotropy of each of the probe sonications can be used to prepare selenium nanostructures with anisotropy more efficiently. Moreover, the absorbance of the solution was enhanced with the increase of sonication time, indicating that the number of nanosheets in the solution increased, and the further stripping of selenium from the bulk could be achieved by continuing the water bath sonication after the probe sonication. Using UV-vis absorption spectroscopy,

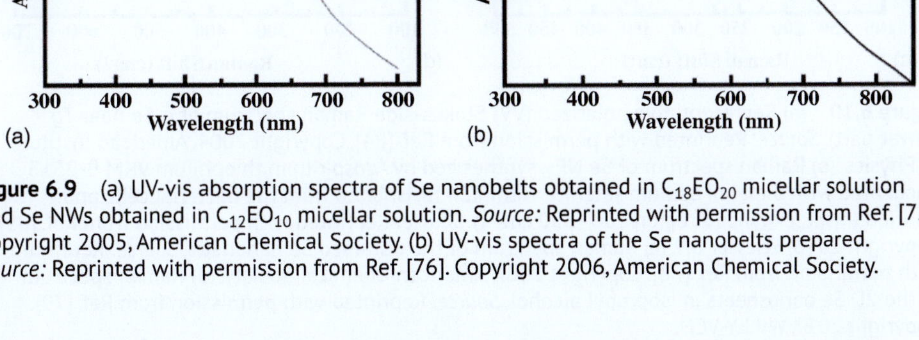


**Figure 6.9** (a) UV-vis absorption spectra of Se nanobelts obtained in $C_{18}EO_{20}$ micellar solution and Se NWs obtained in $C_{12}EO_{10}$ micellar solution. *Source:* Reprinted with permission from Ref. [75]. Copyright 2005, American Chemical Society. (b) UV-vis spectra of the Se nanobelts prepared. *Source:* Reprinted with permission from Ref. [76]. Copyright 2006, American Chemical Society.

they demonstrated that the type of sonication affected the efficiency of selenium nanosheet stripping and confirmed that prolonged water bath sonication favors nanosheet stripping.

### 6.3.2 Raman Spectroscopy

Raman spectroscopy [83] is a fundamental means of analyzing the phase, crystallinity, and intermolecular interactions of selenium nanostructures based on the interaction of light and chemical bonds within the material. The Stokes side Raman spectrum of Se in the amorphous state [84] is shown in Figure 6.10a, and the spectrum can be roughly divided

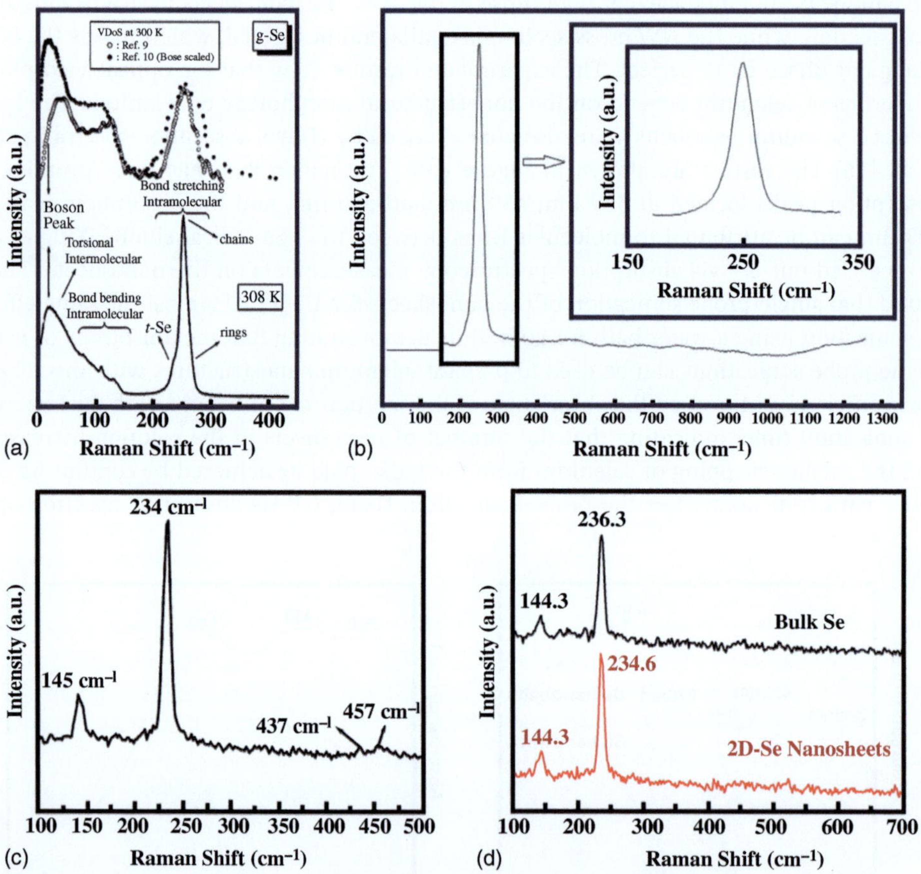


**Figure 6.10** (a) Representative polarized (VV) Stokes-side Raman spectrum of g-Se near *Tg* (lower part). *Source:* Reprinted with permission from Ref. [84]. Copyright 2004, American Institute of Physics. (b) Raman spectrum of Se NPs, synthesized by Azospirillum thiophilum VKM B-2513 (incubated with 5 mM of sodium selenite within 24 h), isolated from the bacterial cells and purified (inset: expanded region 150–350 cm−$^{1}$). *Source:* Reprinted with permission from Ref. [85]. Copyright 2018, Elsevier B.V. (c) Raman spectrum of the obtained Se nanobelts. *Source:* Reprinted with permission from Ref. [78]. Copyright 2007, American Chemical Society. (d) Raman spectrum of the 2D Se nanosheets in isopropyl alcohol. *Source:* Reprinted with permission from Ref. [79]. Copyright 2017, WILEY-VCH.

into three regions: the high-frequency region of 200–300 $cm^{-1}$, the mid-frequency region of 70–150 $cm^{-1}$, and the low-frequency region of 0–70 $cm^{-1}$. There is a 150–200 $cm^{-1}$ gap between the high-frequency region and the mid-frequency region, which is consistent with the Se-Se bond-stretching vibrational mode characteristics. Three peaks of 82, 112, and 138 $cm^{-1}$ are present in the mid-frequency region, where the two sharp peaks of 112 and 138 $cm^{-1}$ are due to the bond-bending vibration of the ring and helix materials. Whereas, among the three peaks in the high-frequency region, the bond stretching vibrational modes of both chains and rings are located at 250 $cm^{-1}$ attachment, while the mode of the peak at 235 $cm^{-1}$ belongs to the triangular structure chains. The Raman spectrum of the 0D amorphous selenium NPs synthesized by Tugarova et al. [85] is illustrated in Figure 6.10b, from which it is found that only one peak at 250 $cm^{-1}$ is present. The selenium nanobelts prepared by Wang et al. [78] are typical 1D nanostructures with a Raman spectrum, as shown in Figure 6.10c. Only one strong resonance peak at 234 $cm^{-1}$ is present in the spectrum attributed to the stretching mode of the helical selenium chains of the *t*-Se phase, and no peaks of amorphous selenium and monoclinic selenium appear. The peak at 145 $cm^{-1}$ belongs to the transverse photophotonic mode, while the two peaks near 437 and 457 $cm^{-1}$ belong to the second-order mode of *t*-Se, and the results show that this *t*-Se nanoribbon is well-crystallized. Xing et al. [79] compared the Raman spectra of bulk Se and 2D Se nanosheets as shown in Figure 6.10d. There are two resonance peaks of 144.3 and 236.3 $cm^{-1}$ for bulk Se, which represent the transverse optical phonon mode and the stretching mode of the chain structure of *t*-Se, respectively. The Raman spectroscopy results of the 2D Se nanosheet prove that the nanosheet is a *t*-Se crystalline phase, but the peak at 234.6 $cm^{-1}$ belonging to the chain structure stretching mode of *t*-Se is blueshifted by 1.7 $cm^{-1}$ compared to that of the bulk. Almost all *t*-Se nanostructures will have the two peaks around 144 and 234 $cm^{-1}$, and it is just that the difference in crystallinity will lead to some shifts of the resonance peaks.

### 6.3.3 Photoluminescence

Since selenium is an essential trace element for people and a fluorescent semiconductor material, small-sized selenium NPs can be regarded as fluorescent quantum dots, which have great potential in bioimaging. Prasad and Selvaraj [86] synthesized selenium NPs using Terminalia arjuna leaf extract, and their PL spectra are shown in Figure 6.11a. The excitation peak is 390 nm and the emission peak is 595 nm. The PL spectrum of selenium NPs synthesized by Hassanien et al. [87] using Olea Europaea leaf extract is shown in Figure 6.11b, with an excitation peak located at around 399 nm and an emission peak at 599 nm, which follows the results of the previous studies. Khalid et al. [88] characterized using a near-field confocal fluorescence microscope with an excitation wavelength of 532 nm. A reddish fluorescence was collected, with a near-infrared detection window of 560~750 nm. A representative confocal image of Se NPs deposited on a Si substrate with a size of 50 × 50 μm is shown in Figure 6.11c. It can be seen from the figure that the background fluorescence is very low, and the fluorescence of individual NPs is distinct, with small particles centered in the red color and large particles appearing in the shape of a white speckle. Figure 6.11d shows the background minus emission data for the Se NP selected for the inset in Figure 6.11c, with counts recorded over 20 min averaging about

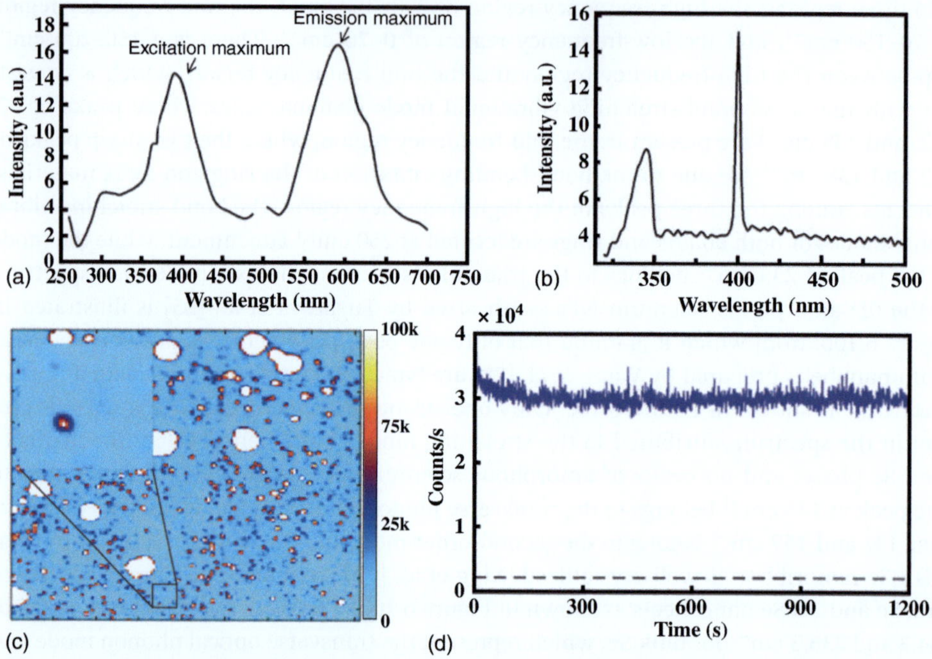


**Figure 6.11** (a) PL spectra of selenium NPs synthesized using leaf extract. *Source:* Reprinted with permission from Ref. [86]. Copyright 2014, SpringerLink. (b) PL spectra of Se-NPs synthesized using M. Oleifera leaf extract. The excitation peak is around 399 nm. *Source:* Reprinted with permission from Ref. [87]. Copyright 2019, Wiley-VCH). (c) A 50 × 50 μm$^2$ near-field confocal fluorescence scan (coarse) for Se NPs on the Si substrate. The excitation wavelength of 532 nm is used at an input power of 100 μW. The intensity scale bar is shown to the right of the scan window. The inset shows a fine scan of 5 × 5 μm$^2$ dimensions. The scan shows distinctively bright Se NP fluorescing with respect to the background. (d) Background-subtracted emission for the selected Se NP is shown in the inset of (c). The dashed line indicates the low-background counts from the Si substrate. *Source:* Reprinted with permission from Ref. [88]. Copyright 2016, Royal Society of Chemistry.

30 counts/s. By calculating the emission counts for 10 selenium NPs, it was found that the average stable emission counts for the distribution of Se NPs existed in the range of 10–30 kcnts/s. They also found that the average lifetime of five selenium NPs immersed in phosphate buffer saline was 4.53 ± 0.77 ns, and this short fluorescence lifetime allows selenium NPs to be used as biomarkers for fluorescence lifetime microscopy.

### 6.3.4 Electronics/Optoelectronics

Selenium, as a narrow bandgap semiconductor, has excellent application prospects in optoelectronics, and selenium photodetectors (PDs) are the more common optoelectronic devices today. For PDs, dark current is an essential measure of their performance, which reduces the dynamic range of detection and increases the effect of noise. In 2011, Abbaszadeh et al. [89] proposed a transverse PD based on a-Se, which uses polyimide as the blocking contact. The dark current test shows that the device uses a blocking layer, and its dark current is significantly reduced. The test results show that the peak simulated electric field at 1 μm depth of the nonblocking contact device is 4 and 16 V/μm at bias voltages of 25 and 100 V, respectively. a-Se device using a blocking layer has a bias voltage of 300 V and a simulated electric field of 33 V/um at 1 μm depth. It can be seen that the blocking contact not only reduces the dark current but also maintains a high electric field in the a-Se layer. The light-to-dark current ratio of the a-Se device with a blocking layer is much larger than that of the device without a blocking layer. This simple lateral device structure can be integrated into an X-ray detector, providing the feasibility of low-cost lateral a-Se devices for indirect conversion digital X-ray imaging. Only two years later, Abbaszadeh et al. [90] also proposed for the first time that a-Se semiconductor detectors could be used to detect ultraviolet photons. The detector's spectral response in the UV (and visible) spectral region was measured by scanning a monochromator equipped with an a-Se detector with a microplasma selected for the spectral lamp, which was found to have a significant UV response in the 200–400 nm band. The dark current density is still less than 3 $pA/mm^2$ even at a maximum voltage of 43 V/μm.

In 2012, Bubon et al. [91] optimized a resistive interfacial layer deposition technique that avoids dielectric breakdown of the edges even if the electric field is locally enhanced due to avalanche multiplication and also ensures the transport properties, which is impossible with the previously known a-Se structures. Analysis by the time-of-flight technique shows that for an a-se layer with a thickness of 15 μm, a stabilized gain of 200 can be obtained at 104 V/μm, the maximum theoretical gain at this thickness. Fan et al. [80] used an electrochemical method to prepare nanosheets for photodetection, resulting in a photocurrent density of up to 1.28 $\mu A/cm^2$ as shown in Figure 6.12a. The calculated photoresponsivity was 10.45 μA/W, indicating the good performance of the detector. After that, a stability test was carried out, and the photocurrent intensity was kept around 0.46 $\mu A/cm^2$ at the $51^{st}$~$60^{th}$, $501^{st}$~$510^{th}$, and $991^{st}$~$1000^{th}$ cycles at an interval of every 5 s. The photocurrent intensity of the $991^{st}-100^{th}$ cycle is shown in Figure 6.12b, and the results indicate excellent cyclic stability.

Due to the narrow bandgap property of selenium, its low-dimensional nanostructures are also suitable for applying field effect transistors (FETs). In 2008, Qin et al. [92] prepared a single selenium NW FET with Ni/Au as the source-drain contact for the first time.

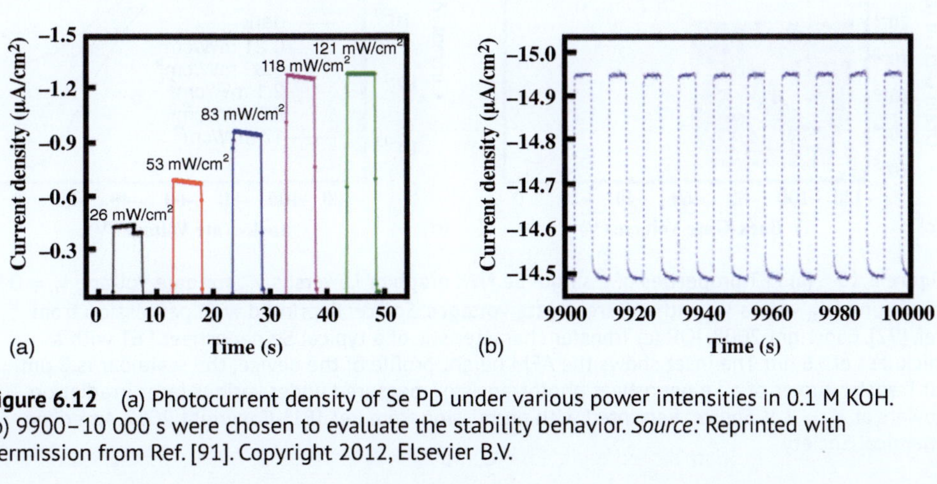


**Figure 6.12** (a) Photocurrent density of Se PD under various power intensities in 0.1 M KOH. (b) 9900–10 000 s were chosen to evaluate the stability behavior. *Source:* Reprinted with permission from Ref. [91]. Copyright 2012, Elsevier B.V.

Figure 6.13a demonstrates the FET characteristics of a single selenium NW, with a resistance of about $7.327 \times 10^{-3}\Omega$ when $V_g = 0$ V. Figure 6.13b illustrates the $I_{ds}$ versus $V_{ds}$ plot at different gate voltages, where the current is suppressed with a positive increase in $V_g$. This gate voltage dependence is a typical feature of p-type FETs. Estimation of the study results showed that the hole mobility of the device was about 30.7 $cm^2\,V^{-1}\,s^{-1}$, and subsequent tests showed that the switching ratio of this single selenium NW device was about 1000. These showed that selenium NWs could be used as FETs for applications in other electronic devices. Qin et al. [81] used the physical vapor deposition method to prepare selenium nanosheets up to a thickness of 5 nm. Figure 6.13c shows the electrical characteristics of a Se nanosheet FET device with a channel length of 16 nm, which is a typical p-type transport mode with a current switching ratio of greater than 106. The selenium nanosheet FETs have a relatively low on-state current, with a maximum value of about 20 mA/mm at $V_{ds} = 3$ V. Figure 6.13d shows the curves of $I_{ds}$ versus $V_g$ for different illumination power densities, and the photocurrent response of the Se nanosheet phototransistor has significant gate tunability. When the nanosheet phototransistor was under very low

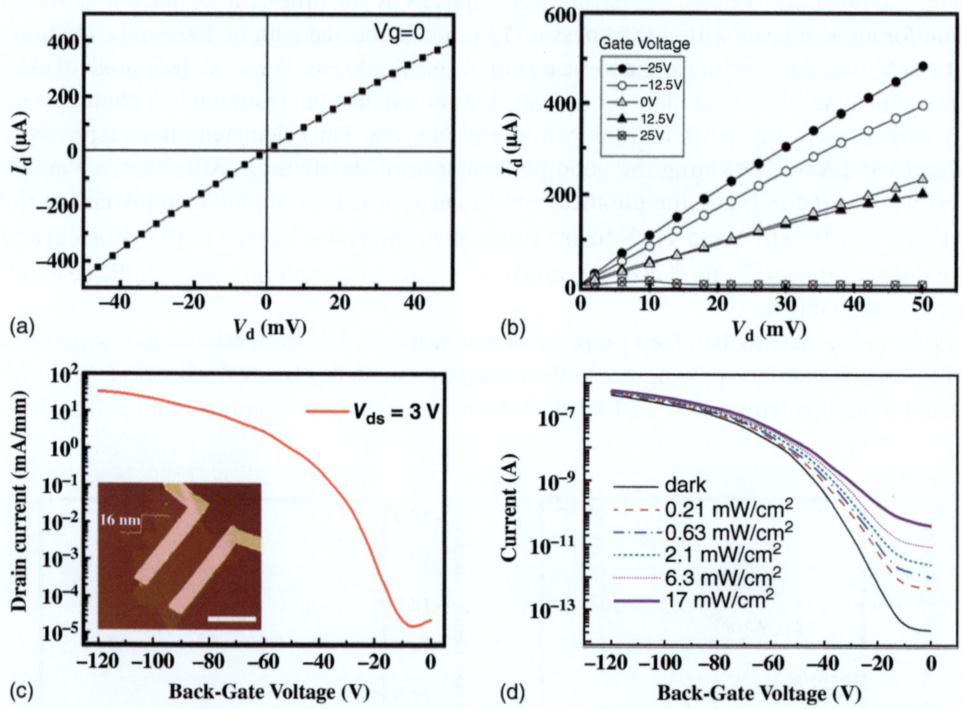


**Figure 6.13** (a) FET properties of a single Se NW: graph of $I_{ds}$ versus $V_{ds}$ for gate voltage $V_g = 0$ V. (b) Graph of $I_{ds}$ versus $V_{ds}$ with different gate voltages. *Source:* Reprinted with permission from Ref. [92]. Copyright 2008, IOP. (c) Transfer characteristic of a typical Se nanosheet FET with a thickness of 16 nm. The inset shows the AFM height profile of the device; the scale bar is 5 μm. (d) Transfer curves of a Se nanosheet phototransistor measured under various laser irradiation powers at $V_{ds} = 3$ V. *Source:* Reprinted with permission from Ref. [81]. Copyright 2017, American Chemical Society.

illumination power (0.21 mW/$cm^2$), it still showed good photoresponse, and the final calculated photoresponsivity was 263 A/W, which was one of the highest values at that time.

### 6.3.5 Energy Storage and Conversion

Solar cells have attracted much attention due to their environmental friendliness and renewable energy [93]. Power conversion efficiency is an essential criterion for evaluating the performance of solar cells, and short circuit current ($J_{sc}$), open-circuit voltage ($V_{oc}$), and fill factor are the decisive parameters affecting power conversion efficiency. These parameters correlate highly with the semiconductor's band gap, carrier lifetime, carrier diffusion length, and radiation complex. For solar cells, the bandgap of Se is large enough to make Se widely used in solar cells. As early as 1982, Ito et al. [94] deposited layered CdSe between glass and crystalline Se and obtained well-crystallized Se films on glass with a film thickness of about 1 μm, where CdSe was used as a bonding medium with a thickness of about 20–30 nm. The prepared selenium films had good integrity and degree of crystallization. After that, Se-based solar cells with an $SnO_2$/CdSe/Se/Au structure were obtained, and the conversion efficiency was up to 4.6% at 60°C and 100 mW · $cm^{-2}$ irradiation intensity. After 20 days, the performance of the device became better due to the stress release. The initial crystalline Se tensile stress was about $10^{11}$ Pa, but the stress became an order of magnitude smaller after 40 days, and the J-V curve of this cell is shown in Figure 6.14a. Later, in 1985, Nakada and Kunioka [95] innovatively proposed $TiO_2$ layers and sputter-deposited them as in solar cells. This cell has an ITO/$n$-$TiO_2$/p-Se/Au structure, which is shown schematically in Figure 6.14b, and the efficiency of this cell reached 5.01%, a record that would stand for more than 30 years. The Se deposited at high temperatures will have voids, and because of its poor adhesion to indium tin oxide, the cell efficiency is not high without the $TiO_2$ layer. The $TiO_2$ layer can effectively avoid the creation of voids on the surface of the Se film, thus improving the efficiency of the cell. The optimal thickness of the $TiO_2$ film is 500 nm. Todorov et al. [96] broke the solar-cell efficiency record designed by Nakada and Kunioka [95] in 2017 by redesigning a new device structure with an efficiency of 6.5%. They designed a glass/FTO/$n$-type buffer (ZnMgO or $TiO_2$)/Se/$MoO_x$/Au structured solar cell, which, for the first time, introduced a $MoO_x$ layer between the selenium and gold back-catalysts to reduce recombination and improve collection. Meanwhile, the thickness of the 100 nm selenium absorber allowed a significant reduction in fabrication time, reduced material miniaturization, and improved surface field efficiency. Figure 6.14c demonstrates the femtosecond pump-probe ultraviolet photoelectron spectroscopy (fs-UPS) spectral superposition of Se and ZnMgO, where a new record $V_{oc}$ of 969 mV was obtained by optimizing the electron-transferring layer using a bandgap-tunable $Zn_xMg_{1-x}O$ buffer. Di et al. [97] used crystalline selenium as a hole-transporting layer in planar chalcogenide solar cells with a fluorine-doped tin oxide (FTO)/$SnO_2$/Calcite/Selenium/Au structure. Figure 6.14d compares the J-V curves of Perovskite solar cells with crystalline Se and 2,2′,7,7′-tetrakis($N$, $N$-p-dimethoxyphenylamino)-9,9′-spirobifluorene (spiroO-MeTAD) as the thermal medium, and the best crystalline The power conversion efficiency of the best crystalline Se device is 8.3%, and it is demonstrated that there is a good carrier transport capability within the crystalline Se. It is worth mentioning that the crystalline Se device still maintains 80% of the original efficiency after 40 days of placement, indicating the remarkable stability of the crystalline Se device.

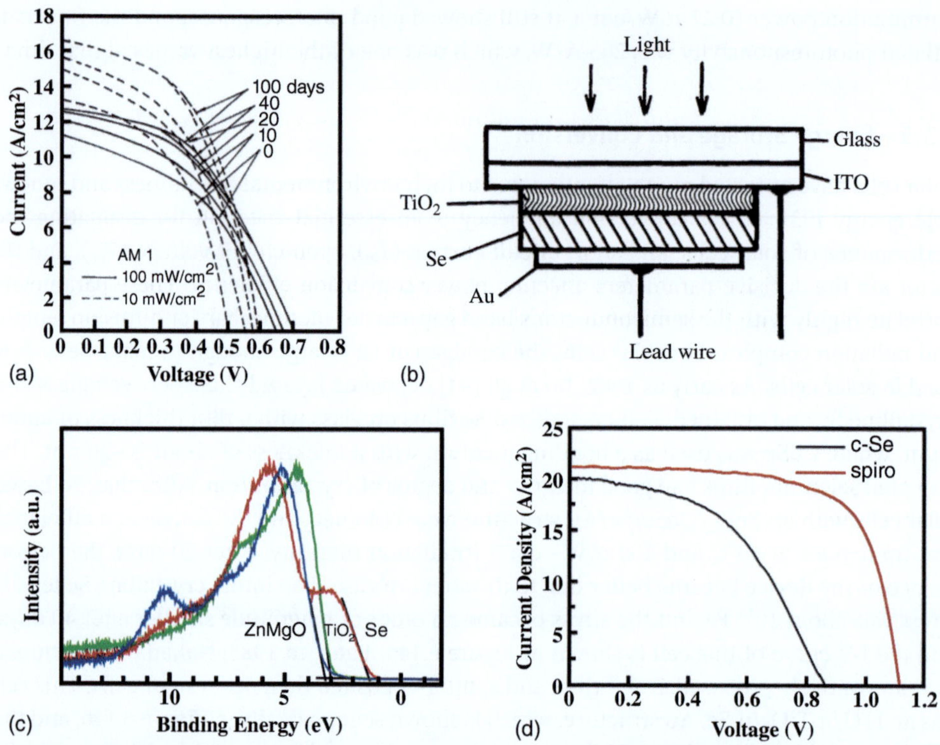


**Figure 6.14** (a) J–V curves of Se solar cells. *Source:* Reprinted with permission from Ref. [94]. Copyright 1982, IOP. (b) Cross-section of a $TiO_2$/Se solar cell. *Source:* Reprinted with permission from Ref. [95]. Copyright 1985, IOP. (c) fs-UPS spectra of Se (red), $TiO_2$ (green), and ZnMgO (blue) under flat-band conditions. *Source:* Reprinted with permission from Ref. [96]. Copyright 2017, SpringerLink. (d) J-V curves with reverse scan mode for the best-performing devices based on spiroOMeTAD and c-Se: normalized external quantum efficiency of perovskite solar cells based on spiroOMeTAD and c-Se. *Source:* Reprinted with permission from Ref. [97]. Copyright 2018, Elsevier B.V.

### 6.3.6 Biological Applications

Selenium, an essential trace element for human beings, has been widely studied for its excellent bioactivity and low toxicity. It has also been shown to possess some degree of anticancer efficacy. The anticancer properties of the synthesized Se-NRs were evaluated against Hep-G2 and MCF-7 cell lines by Sholkamy et al. [98]. The nanorods exhibited dose-dependent growth inhibition on both cancer cells when the nanorod concentration was 10–100 μg/mL. The higher the concentration, the stronger the inhibition of the cancer cells. The experimental results are shown in Figure 6.15a. However, comparing the two types of cells, the inhibitory effect of Se-NRs on MCF-7 cells at the same concentration was significantly higher than that on Hep-G2 cells. These results suggest that biosynthesized Se-NRs have anticancer effects and can be used as potential anticancer drugs. Selenium NPs prepared by Vahdati and Moghadam [99] were used in antimicrobial experiments, and the synergistic antimicrobial activity of selenium NPs with lysozyme was reported for the first time. They carried out an independent bacterial inhibition test against Staphylococcus aureus with different concentrations of Se NPs and lysozyme, the results of which are shown in Figure 6.15b. After 20 h of incubation, all concentrations of selenium NPs were effective in inhibiting *S. aureus* with a minimum inhibitory concentration of 82 μg/mL. After stability studies, it was found that the nanohybridized system retained its appearance and characteristic absorption bands even after several weeks of storage under refrigerated conditions. It was shown that selenium NPs synergize the antimicrobial properties of *Staphylococcus aureus* strains when interacting with biomolecules with good stability. The selenium NPs prepared by Hashem et al. [100] using a biosynthetic approach are used for various biomedical applications. The bacteriostatic activity of such selenium NPs

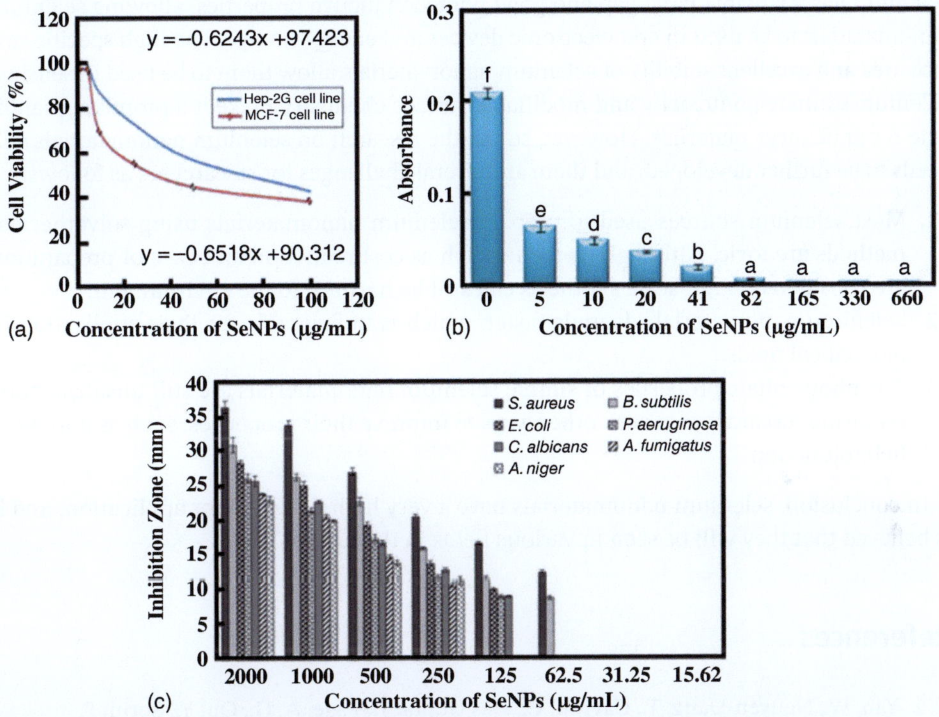


**Figure 6.15** (a) Effect of Se-NRs on the viability of Hep-G2 and MCF-7 human cancer cells under in vitro conditions. Human cancer cells were treated with various concentrations (10 μg/mL, 25 μg/mL, 50 μg/mL, and 100 μg/mL) of Se-NRs, and the MTT assay revealed that the biosynthesized Se-NRs induced cell death of Hep-G2 and MCF-7 human cancer cells. *Source:* Reprinted with permission from Ref. [98]. Copyright 2015, Dove Medical Press Ltd. (b) Growth inhibition of *S. aureus* under treatment with different concentrations of Se NPs after 20 h. Absorbance was recorded at 630 nm. Each bar indicates the mean ± standard deviations of three replications. Bars not labeled by the same letter represent statistical significance at $P \leqslant 0.05$ using ANOVA followed by Tukey's HSD test. *Source:* Reprinted with permission from Ref. [99]. Copyright 2020, SpringerLink. (c) Antimicrobial activity of mycosynthesized Se-NPs by *P. expansum* and minimum inhibitory concentration for Se-NPs at various concentrations (2000, 1000, 500, 250, 125, 62.5, 31.25, and 15.62 μg/mL) against different pathogenic bacteria and fungi. *Source:* Reprinted with permission from Ref. [100]. Copyright 2021, SpringerLink.

against different bacterial strains is shown in Figure 6.15c. The results showed that selenium NPs exhibited bacteriostatic activity and antioxidant activity against gram-positive, gram-negative, *Staphylococcus aureus, Candida albicans, Aspergillus niger*, and *Aspergillus fumigatus*. They also tested the anticancer activity of selenium NPs, and the results showed that they exhibited good anticancer activity against PC3 cell lines, with an inhibition rate of 67.37% against PC3 cells and no toxicity to normal human cells. They showed nonhemolytic activity against human erythrocytes at concentrations up to 250 μg/mL.

## 6.4 Conclusion

Various nanostructures of selenium have many interesting properties and are widely studied. Selenium has a suitable band gap and good photoconductive properties, allowing selenium nanomaterials to be used in optoelectronic devices and energy storage. The high specific surface area and excellent stability of selenium nanomaterials allow them to be used in sensing. Selenium's innate nontoxicity and modifiable surface chemistry make it a promising candidate for anticancer materials. However, so far, the research on selenium nanomaterials still needs to be further developed, and there are several challenges for researchers as follows:

1. Most selenium sources used to prepare selenium nanomaterials using solvothermal methods are toxic. Although elements such as cost reduction and ease of preparation are considered, large-scale synthesis can still be harmful to the environment.
2. Selenium nanomaterials degrade faster, which is unfavorable for their application in biomedical fields.
3. The photovoltaic properties of simple selenium nanomaterials are still unsatisfactory, requiring researchers to seek other ways to improve their properties, such as doping or heterojunction.

In conclusion, selenium nanomaterials have a very high potential for application, and it is believed that they will be seen in various fields in the future.

## References

1 Yan, W., Nguyen-Dang, T., Cayron, C., Das Gupta, T., Page, A. G., Qu, Y., Sorin, F. Microstructure tailoring of selenium-core multimaterial optoelectronic fibers. *Opt. Mater. Express* 2017, 7 (4): 1388–1397.

2 Liu, H., Fan, L., Wei, Q., Song, H., Zhu, H., Hu, M., Yan, S. Preparation of single crystal triangular selenium nanostructures and photoelectric properties. *Mater. Express* 2022, 12 (4): 609–615.

3 Rahman, A., Krause, B., Hoang, T.B., Guisbiers, G. Tailoring the optical properties of selenium nanoneedles by pulsed laser ablation in liquids: implications for solar cells and photocells. *ACS Appl. Nano Mater*. 2023, 6 (3): 2258–2265.

4 Huang, W., Wang, M., Hu, L., Wang, C., Xie, Z., Zhang, H. Recent advances in semiconducting monoelemental selenium nanostructures for device applications. *Adv. Funct. Mater*. 2020, 30 (42): 2003301.

**5** Husen, A., Siddiqi, K.S. Plants and microbes assisted selenium nanoparticles: characterization and application. *J. Nanobiotechnol.* 2014, 12 (28): 1–10.

**6** Ramoutar, R.R., Brumaghim, J.L. Antioxidant and anticancer properties and mechanisms of inorganic selenium, oxo-sulfur, and oxo-selenium compounds. *Cell Biochem. Biophys.* 2010, 58 (1): 1–23.

**7** Cui, D., Yan, C., Miao, J., Zhang, X., Chen, J., Sun, L., Meng, L., Liang, T., Li, Q. Synthesis, characterization and antitumor properties of selenium nanoparticles coupling with ferulic acid. *Mater. Sci. Eng. C* 2018, 90: 104–112.

**8** Liu, S., Yu, T., Lich, L.V., Yin, S., Bui, T.Q. Size and surface effects on mechanical behavior of thin nanoplates incorporating microstructures using isogeometric analysis. *Comput. Struct.* 2019, 212: 173–187.

**9** Journy, N., Mansouri, I., Allodji, R.S., Demoor-Goldschmidt, C., Ghazi, D., Haddy, N., Rubino, C., Veres, C., Zrafi, W.S., Rivera, S., Diallo, I., De Vathaire, F. Volume effects of radiotherapy on the risk of second primary cancers: a systematic review of clinical and epidemiological studies. *Radiot. Oncol.* 2019, 131: 150–159.

**10** Abo-Dahab, S.M., Abdelhafez, M.A., Mebarek-Oudina, F., Bilal, S.M. MHD Casson nanofluid flow over nonlinearly heated porous medium in presence of extending surface effect with suction/injection. *Indian J. Phys.* 2021, 95 (12): 2703–2717.

**11** Li, S., Yuan, M., Zhuang, W., Zhao, X., Tie, S., Xiang, J., Lan, S. Optically-controlled quantum size effect in a hybrid nanocavity composed of a perovskite nanoparticle and a thin gold film. *Laser Photonics Rev.* 2021, 15 (3): 2000480.

**12** Vorontsov, A.V., Valdés, H. Quantum size effect and visible light activity of anatase nanosheet quantum dots. *J. Photochem. Photobiol. A: Chem.* 2019, 379: 39–46.

**13** Wang, W., Fang, J., Chen, H. Nano-confined g-$C_3N_4$ in mesoporous $SiO_2$ with improved quantum size effect and tunable structure for photocatalytic tetracycline antibiotic degradation. *J. Alloys Compd.* 2020, 819: 153064.

**14** Yang, G., Park, S.-J. Conventional and microwave hydrothermal synthesis and application of functional materials: a review. *Materials* 2019, 12 (7): 1177.

**15** Ashok, B., Hariram, N., Siengchin, S., Rajulu, A.V. Modification of tamarind fruit shell powder with in situ generated copper nanoparticles by single step hydrothermal method. *J. Bioresour. Bioprod.* 2020, 5 (3): 180–185.

**16** Li, T.-X., Miao, J.-W., Guo, E.-Y., Huang, H., Wang, J., Lu, Y.-P., Wang, T.-M., Cao, Z.-Q., Li, T.-J. Tungsten-containing high-entropy alloys: a focused review of manufacturing routes, phase selection, mechanical properties, and irradiation resistance properties. *Tungsten* 2021, 3 (2): 181–196.

**17** Choi, I.J., Kim, B.J., Lee, S.H., Jeong, B.J., Nasir, T., Cho, Y.S., Kim, N., Lee, J.-H., Yu, H.K., Choi, J.-Y. Fabrication of a room-temperature $NO_2$ gas sensor using morphology controlled CVD-grown tellurium nanostructures. *Sensors Actuat. B: Chem.* 2021, 333: 128891.

**18** Wang, Q., Lei, Y., Wang, Y., Liu, Y., Song, C., Zeng, J., Song, Y., Duan, X., Wang, D., Li, Y. Atomic-scale engineering of chemical-vapor-deposition-grown 2D transition metal dichalcogenides for electrocatalysis. *Energy Environ. Sci.* 2020, 13 (6): 1593–1616.

**19** Tang, L., Li, T., Luo, Y., Feng, S., Cai, Z., Zhang, H., Liu, B., Cheng, H.-M. Vertical chemical vapor deposition growth of highly uniform 2D transition metal dichalcogenides. *ACS Nano* 2020, 14 (4): 4646–4653.

**20** Meng, L., Zhou, Z., Xu, M., Yang, S., Si, K., Liu, L., Wang, X., Jiang, H., Li, B., Qin, P., Zhang, P., Wang, J., Liu, Z., Tang, P., Ye, Y., Zhou, W., Bao, L., Gao, H.-J., Gong, Y. Anomalous thickness dependence of Curie temperature in air-stable two-dimensional ferromagnetic 1T-$CrTe_2$ grown by chemical vapor deposition. *Nat. Commun.* 2021, 12 (1): 809.

**21** Mostafa, A.M., Mwafy, E.A. Synthesis of ZnO and Au@ZnO core/shell nano-catalysts by pulsed laser ablation in different liquid media. *J. Mater. Res. Technol.* 2020, 9 (3): 3241–3248.

**22** Yao, J.D., Zheng, Z.Q., Yang, G.W. Production of large-area 2D materials for high-performance photodetectors by pulsed-laser deposition. *Progress Mater. Sci.* 2019, 106: 100573.

**23** Zuñiga-Ibarra, V.A., Shaji, S., Krishnan, B., Johny, J., Sharma Kanakkillam, S., Avellaneda, D.A., Martinez, J.A.A., Roy, T.K.D., Ramos-Delgado, N.A. Synthesis and characterization of black $TiO_2$ nanoparticles by pulsed laser irradiation in liquid. *Appl. Surf. Sci.* 2019, 483: 156–164.

**24** Menazea, A.A., Mostafa, A.M., Al-Ashkar, E.A. Effect of nanostructured metal oxides (CdO, $Al_2O_3$, $Cu_2O$) embedded in PVA via Nd:YAG pulsed laser ablation on their optical and structural properties. *J. Mol. Struct.* 2020, 1203: 127374.

**25** Babinchak, W.M., Surewicz, W.K. Liquid–liquid phase separation and its mechanistic role in pathological protein aggregation. *J. Mol. Biol.* 2020, 432 (7): 1910–1925.

**26** Yuan, C., Levin, A., Chen, W., Xing, R., Zou, Q., Herling, T.W., Challa, P.K., Knowles, T.P.J., Yan, X. Nucleation and growth of amino acid and peptide supramolecular polymers through liquid–liquid phase separation. *Angew. Chem. Int. Ed.* 2019, 58 (50): 18116–18123.

**27** Ray, S., Singh, N., Kumar, R., Patel, K., Pandey, S., Datta, D., Mahato, J., Panigrahi, R., Navalkar, A., Mehra, S., Gadhe, L., Chatterjee, D., Sawner, A.S., Maiti, S., Bhatia, S., Gerez, J.A., Chowdhury, A., Kumar, A., Padinhateeri, R., Riek, R., Krishnamoorthy, G., Maji, S.K. α-Synuclein aggregation nucleates through liquid–liquid phase separation. *Nat. Chem.* 2020, 12 (8): 705–716.

**28** Ianiro, A., Wu, H., van Rijt, M.M.J., Vena, M.P., Keizer, A.D.A., Esteves, A.C.C., Tuinier, R., Friedrich, H., Sommerdijk, N.A.J.M., Patterson, J.P. Liquid–liquid phase separation during amphiphilic self-assembly. *Nat. Chem.* 2019, 11 (4): 320–328.

**29** Michael, J., Qifeng, Z., Danling, W. Titanium carbide MXene: synthesis, electrical and optical properties and their applications in sensors and energy storage devices. *Nanomater. Nanotechnol.* 2019, 9: 1–9.

**30** Wang, L., Hasanzadeh Kafshgari, M., Meunier, M. Optical properties and applications of plasmonic-metal nanoparticles. *Adv. Funct. Mater.* 2020, 30 (51): 2005400.

**31** Li, Q., Nie, L., Zhang, K., Li, Y., Chen, S., Zhu, G. Experimental investigation on aero-heating of rudder shaft within laminar/turbulent hypersonic boundary layers. *Chin. J. Aeronaut.* 2019, 32 (5): 1215–1221.

**32** Cui, F., Yue, Y., Zhang, Y., Zhang, Z., Zhou, H.S. Advancing biosensors with machine learning. *ACS Sens.* 2020, 5 (11): 3346–3364.

**33** Haleem, A., Javaid, M., Singh, R.P., Suman, R., Rab, S. Biosensors applications in medical field: a brief review. *Sens. Int.* 2021, 2: 100100.

**34** Morales, M.A., Halpern, J.M. Guide to selecting a biorecognition element for biosensors. *Bioconjug. Chem.* 2018, 29 (10): 3231–3239.

**35** Wu, Q., Miao, W.-s., Zhang, Y.-d., Gao, H.-j., Hui, D. Mechanical properties of nanomaterials: a review. *Nanotechnol. Rev.* 2020, 9 (1): 259–273.

**36** Crucho, J., Picado-Santos, L., Neves, J., Capitão, S. A review of nanomaterials' effect on mechanical performance and aging of asphalt mixtures. *Appl. Sci.* 2019, 9 (18): 3657.

**37** Gottardo, S., Mech, A., Drbohlavová, J., Małyska, A., Bøwadt, S., Riego Sintes, J., Rauscher, H. Towards safe and sustainable innovation in nanotechnology: state-of-play for smart nanomaterials. *NanoImpact* 2021, 21: 100297.

**38** Aljaberi, M.H.K., Al-Baghdadi, M.A.R.S., Al-Waily, M., Mohammadi-Aghdam, M., Goudarzi, T. Numerical investigation of mechanical behavior for lattice structure with effect of different nanomaterial types. *IOP Conf. Ser. Mater. Sci. Eng.* 2021, 1094 (1): 012172.

**39** Zeng, S., Ke, Y., Liu, Y., Shen, Y., Zhang, L., Li, C., Liu, A., Shen, L., Hu, X., Wu, H., Wu, W., Liu, Y. Synthesis and antidiabetic properties of chitosan-stabilized selenium nanoparticles. *Colloids Surf. B: Biointerfaces* 2018, 170: 115–121.

**40** Huo, X., Zhang, Y., Jin, X., Li, Y., Zhang, L. A novel synthesis of selenium nanoparticles encapsulated PLGA nanospheres with curcumin molecules for the inhibition of amyloid β aggregation in Alzheimer's disease. *J. Photochem. Photobiol. B: Biol.* 2019, 190: 98–102.

**41** Qian, F., Li, X., Tang, L., Lai, S.K., Lu, C., Lau, S.P. Selenium quantum dots: preparation, structure, and properties. *Appl. Phys. Lett.* 2017, 110 (5): 053104.

**42** Boroumand, S., Safari, M., Shaabani, E., Shirzad, M., Faridi-Majidi, R. Selenium nanoparticles: synthesis, characterization and study of their cytotoxicity, antioxidant and antibacterial activity. *Mater. Res. Express* 2019, 6 (8): 0850d8.

**43** Shah, C.P., Kumar, M., Pushpa, K.K., Bajaj, P.N. Acrylonitrile-induced synthesis of polyvinyl alcohol-stabilized selenium nanoparticles. *Cryst. Growth Des.* 2008, 8 (11): 4159–4164.

**44** Fardsadegh, B., Jafarizadeh-Malmiri, H. Aloe vera leaf extract mediated green synthesis of selenium nanoparticles and assessment of their In vitro antimicrobial activity against spoilage fungi and pathogenic bacteria strains. *Green Process Synth.* 2019, 8: 399–407.

**45** Mellinas, C., Jiménez, A., Garrigós, M.D.C. Microwave-assisted green synthesis and antioxidant activity of selenium nanoparticles using Theobroma Cacao L. bean shell extract. *Molecules* 2019, 24 (22): 4048.

**46** Kokila, K., Elavarasan, N., Sujatha, V. Diospyros montana leaf extract-mediated synthesis of selenium nanoparticles and their biological applications. *New J. Chem.* 2017, 41 (15): 7481–7490.

**47** Sharma, G., Sharma, A., Bhavesh, R., Park, J., Ganbold, B., Nam, J.-S., Lee, S.-S. Biomolecule-mediated synthesis of selenium nanoparticles using dried Vitis vinifera (raisin) extract. *Molecules* 2014, 19 (3): 2761–2770.

**48** Anu, K., Singaravelu, G., Murugan, K., Benelli, G. Green-synthesis of selenium nanoparticles using garlic cloves (Allium sativum): biophysical characterization and cytotoxicity on vero cells. *J. Clust. Sci.* 2016, 28 (1): 551–563.

**49** Krishnan, M., Ranganathan, K., Maadhu, P., Thangavelu, P., Kundan, S., Arjunan, N. Leaf extract of Dillenia indica as a source of selenium nanoparticles with larvicidal and antimicrobial potential toward vector mosquitoes and pathogenic microbes. *Coatings* 2020, 10 (7): 626.

**50** Geoffrion, L.D., Hesabizadeh, T., Medina-Cruz, D., Kusper, M., Taylor, P., Vernet-Crua, A., Chen, J., Ajo, A., Webster, T.J., Guisbiers, G. Naked selenium nanoparticles for antibacterial and anticancer treatments. *ACS Omega* 2020, 5 (6): 2660–2669.

**51** Guisbiers, G., Wang, Q., Khachatryan, E., Arellano-Jimenez, M.J., Webster, T.J., Larese-Casanova, P., Nash, K.L. Anti-bacterial selenium nanoparticles produced by UV/VIS/NIR pulsed nanosecond laser ablation in liquids. *Laser Phys. Lett.* 2015, 12 (1): 016003.

**52** Andrei, V., Roh, I., Yang, P. Nanowire photochemical diodes for artificial photosynthesis. *Sci. Adv.* 2023, 9: eade9044.

**53** Sheng, S.-Z., Wang, J.-L., Zhao, B., He, Z., Feng, X.-F., Shang, Q.-G., Chen, C., Pei, G., Zhou, J., Liu, J.-W., Yu, S.-H. Nanowire-based smart windows combining electro- and thermochromics for dynamic regulation of solar radiation. *Nat. Commun.* 2023, 14 (1): 3231.

**54** Su, J., Musgrave, C.B., Song, Y., Huang, L., Liu, Y., Li, G., Xin, Y., Xiong, P., Li, M. M.-J., Wu, H., Zhu, M., Chen, H.M., Zhang, J., Shen, H., Tang, B.Z., Robert, M., Goddard, W.A., Ye, R. Strain enhances the activity of molecular electrocatalysts via carbon nanotube supports. *Nat. Catal.* 2023, 6 (9): 818–828.

**55** Lin, Y., Cao, Y., Ding, S., Zhang, P., Xu, L., Liu, C., Hu, Q., Jin, C., Peng, L.-M., Zhang, Z. Scaling aligned carbon nanotube transistors to a sub-10 nm node. *Nat. Electron.* 2023, 6 (7): 506–515.

**56** Gates, B., Mayers, B., Grossman, A., Xia, Y. A sonochemical approach to the synthesis of crystalline selenium nanowires in solutions and on solid supports. *Adv. Mater.* 2002, 14 (23): 1749–1753.

**57** Zhang, H., Zhou, M., Tan, S., Li, G., Zhang, S., Hou, J. Carbothermal chemical vapor deposition route to Se one-dimensional nanostructures and their optical properties. *J. Phys. Chem. B* 2005, 109 (21): 10653–10657.

**58** Xia, Z.-M., Liu, Y.-N., Huang, Z., Qin, L.-Z., Lin, H., Li, Q. A facile green approach for synthesis of selenium nanowires with visible light photocatalytic properties. *J. Nanosci. Nanotechnol.* 2019, 19 (1): 156–162.

**59** Ren, L., Zhang, H., Tan, P., Chen, Y., Zhang, Z., Chang, Y., Xu, J., Yang, F., Yu, D. Hexagonal selenium nanowires synthesized via vapor-phase growth. *J. Phys. Chem. B* 2004, 108: 4627–4630.

**60** Yan, W., Burgos-Caminal, A., Das Gupta, T., Moser, J.-E., Sorin, F. Direct synthesis of selenium nanowire mesh on a solid substrate and insights into ultrafast photocarrier dynamics. *J. Phys. Chem. C* 2018, 122 (43): 25134–25141.

**61** Lei, W.B., Xu, W.Q., Qi, R.J., Huang, R. Facile synthesis and growth mechanism of trigonal selenium nanowires. *Chalcogenide Lett.* 2020, 17 (12): 615–621.

**62** Li, X., Li, Y., Li, S., Zhou, W., Chu, H., Chen, W., Li, I.L., Tang, Z. Single crystalline trigonal selenium nanotubes and nanowires synthesized by sonochemical process. *Cryst. Growth Des.* 2005, 5 (3): 911–916.

**63** Zhang, J., Zhang, S.-Y., Chen, H.-Y. CTAB-controlled synthesis of one-dimensional selenium nanostructures. *Chem. Lett.* 2004, 33 (8): 1054–1055.

**64** Zhang, S.Y., Liu, Y., Ma, X., Chen, H.Y. Rapid, large-scale synthesis and electrochemical behavior of faceted single-crystalline selenium nanotubes. *J. Phys. Chem. B* 2006, 110: 9041–9047.

**65** Chen, H., Shin, D.-W., Nam, J.-G., Kwon, K.-W., Yoo, J.-B. Selenium nanowires and nanotubes synthesized via a facile template-free solution method. *Mater. Res. Bull.* 2010, 45 (6): 699–704.

**66** Ren, Y., Niu, M., Gu, W., Fang, Y. Facile synthesis of trigonal selenium nanotubes in ethanol at low temperature. *Mater. Lett.* 2012, 82: 148–151.

**67** Filippo, E., Manno, D., Serra, A. Characterization and growth mechanism of selenium microtubes synthesized by a vapor phase deposition route. *Cryst. Growth Des.* 2010, 10 (11): 4890–4897.

**68** Chen, Y.-t., Zhang, W., Fan, Y.-q., Xu, X.-q., Zhang, Z.-x. Hydrothermal preparation of selenium nanorods. *Mater. Chem. Phys.* 2006, 98 (2–3): 191–194.

**69** Chen, H., Yoo, J.-B., Liu, Y., Zhao, G. Green synthesis and characterization of se nanoparticles and nanorods. *Electron. Mater. Lett.* 2012, 7 (4): 333–336.

**70** Kannan, S., Mohanraj, K., Prabhu, K., Barathan, S., Sivakumar, G. Synthesis of selenium nanorods with assistance of biomolecule. *Bull. Mater. Sci.* 2014, 37 (7): 1631–1635.

**71** Velayati, M., Hassani, H., Sabouri, Z., Mostafapour, A., Darroudi, M. Biosynthesis of Se-Nanorods using Gum Arabic (GA) and investigation of their photocatalytic and cytotoxicity effects. *Inorg. Chem. Commun.* 2021, 128: 108589.

**72** Velayati, M., Hassani, H., Sabouri, Z., Mostafapour, A., Darroudi, M. Green-based biosynthesis of Se nanorods in chitosan and assessment of their photocatalytic and cytotoxicity effects. *Environ. Technol. Innov.* 2022, 27: 102610.

**73** Jiang, Z.Y., Xie, Z.X., Xie, S.Y., Zhang, X.H., Huang, R.B., Zheng, L.S. High purity trigonal selenium nanorods growth via laser ablation under controlled temperature. *Chem. Phys. Lett.* 2003, 368: 425–429.

**74** Chen, Y.-t., Zhang, W., Zhang, F.-b., Zhang, Z.-x., Zhou, B.-z., Li, H.-L. A novel route to controlled synthesis of selenium nanowires. *Mater. Lett.* 2004, 58 (22–23): 2761–2763.

**75** Ma, Y., Qi, L., Shen, W., Ma, J. Selective synthesis of single-crystalline selenium nanobelts and nanowires in micellar solutions of nonionic surfactants. *Langmuir* 2005, 21: 6161–6164.

**76** Xie, Q., Dai, Z., Huang, W., Zhang, W., Ma, D., Hu, X., Qian, Y. Large-scale synthesis and growth mechanism of single-crystal Se nanobelts. *Cryst. Growth Des.* 2006, 6 (6): 1514–1517.

**77** Wang, D., Yu, H., Guan, Y., Zhu, M., Liu, S., Yao, J. A novel chemical reduction method to fabricate tunable selenium nanosphere and nanobelt based on cellulose nanocrystals. *Mater. Lett.* 2018, 230: 44–47.

**78** Wang, Q., Li, G.D., Liu, Y.L., Xu, S., Wang, K.J., Chen, J.S. Fabrication and growth mechanism of selenium and tellurium nanobelts through a vacuum vapor deposition route. *J. Phys. Chem. C* 2007, 111: 12926–12932.

**79** Xing, C., Xie, Z., Liang, Z., Liang, W., Fan, T., Ponraj, J.S., Dhanabalan, S.C., Fan, D., Zhang, H. 2D nonlayered selenium nanosheets: facile synthesis, photoluminescence, and ultrafast photonics. *Adv. Opt. Mater.* 2017, 5 (24): 1700884.

**80** Fan, T., Xie, Z., Huang, W., Li, Z., Zhang, H. Two-dimensional non-layered selenium nanoflakes: facile fabrications and applications for self-powered photo-detector. *Nanotechnology* 2019, 30 (11): 114002.

**81** Qin, J., Qiu, G., Jian, J., Zhou, H., Yang, L., Charnas, A., Zemlyanov, D.Y., Xu, C.-Y., Xu, X., Wu, W., Wang, H., Ye, P.D. Controlled growth of a large-size 2D selenium nanosheet and its electronic and optoelectronic applications. *ACS Nano* 2017, 11 (10): 10222–10229.

**82** Joannopoulos, J.D., Schlüter, M., Cohen, M.L. Electronic structure of trigonal and amorphous Se and Te. *Phys. Rev. B* 1975, 11 (6): 2186–2199.

**83** Phillips, W., Buchenau, U., Nücker, N., Dianoux, A. J., Petry, W. Dynamics of glassy and liquid selenium. *Phys. Rev. Lett.* 1989, 63 (21): 2381–2384.

**84** Yannopoulos, S.N., Andrikopoulos, K.S. Raman scattering study on structural and dynamical features of noncrystalline selenium. *J. Chem. Phys.* 2004, 121 (10): 4747–4758.

**85** Tugarova, A.V., Mamchenkova, P.V., Dyatlova, Y.A., Kamnev, A.A. FTIR and Raman spectroscopic studies of selenium nanoparticles synthesised by the bacterium Azospirillum thiophilum. *Spectrochim. Acta Part A: Mol. Biomol. Spectrosc.* 2018, 192: 458–463.

**86** Prasad, K.S., Selvaraj, K. Biogenic synthesis of selenium nanoparticles and their effect on As (III)-induced toxicity on human lymphocytes. *Biol. Trace Element Res.* 2014, 157 (3): 275–283.

**87** Hassanien, R., Abed-Elmageed, A.A.I., Husein, D.Z. Eco-friendly approach to synthesize selenium nanoparticles: photocatalytic degradation of sunset yellow azo dye and anticancer activity. *ChemistrySelect* 2019, 4 (31): 9018–9026.

**88** Khalid, A., Tran, P.A., Norello, R., Simpson, D.A., O'Connor, A.J., Tomljenovic-Hanic, S. Intrinsic fluorescence of selenium nanoparticles for cellular imaging applications. *Nanoscale* 2016, 8 (6): 3376–3385.

**89** Abbaszadeh, S., Allec, N., Wang, K., Karim, K.S. Low dark-current lateral amorphous-selenium metal–semiconductor–metal photodetector. *IEEE Electron Device Lett.* 2011, 32 (9): 1263–1265.

**90** Abbaszadeh, S., Karim, K.S., Karanassios, V. Measurement of UV from a microplasma by a microfabricated amorphous selenium detector. *IEEE Trans. Electron Devices* 2013, 60 (2): 880–883.

**91** Bubon, O., DeCrescenzo, G., Zhao, W., Ohkawa, Y., Miyakawa, K., Matsubara, T., Kikuchi, K., Tanioka, K., Kubota, M., Rowlands, J.A., Reznik, A. Electroded avalanche amorphous selenium (a-Se) photosensor. *Curr. Appl. Phys.* 2012, 12 (3): 983–988.

**92** Qin, D., Tao, H., Zhao, Y., Lan, L., Chan, K., Cao, Y. Field effect transistor from individual trigonal Se nanowire. *Nanotechnology* 2008, 19 (35): 355201.

**93** Wang, Z., Li, H., Tang, Z., Liu, Z., Ruan, Z., Ma, L., Yang, Q., Wang, D., Zhi, C. Hydrogel electrolytes for flexible aqueous energy storage devices. *Adv. Funct. Mater.* 2018, 28 (48): 1804560.

**94** Ito, H., Oka, M., Ogino, T., Takeda, A., Mizushima, Y. Selenium thin film solar cell. *Jpn. J. Appl. Phys.* 1982, 21: 77.

**95** Nakada, T., Kunioka, A. Polycrystalline thin-film TiO2/Se solar cells. *Jpn. J. Appl. Phys.* 1985, 24 (7): 536–538.

**96** Todorov, T.K., Singh, S., Bishop, D.M., Gunawan, O., Lee, Y.S., Gershon, T.S., Brew, K.W., Antunez, P.D., Haight, R. Ultrathin high band gap solar cells with improved efficiencies from the world's oldest photovoltaic material. *Nat. Commun.* 2017, 8 (1): 682.

**97** Di, Y., Zeng, Q., Huang, C., Tang, D., Sun, K., Yan, C., Wang, Y., Ke, S., Jiang, L., Hao, X., Lai, Y., Liu, F. Thermal-evaporated selenium as a hole-transporting material for planar perovskite solar cells. *Solar Energy Mater. Solar Cells* 2018, 185: 130–135.

**98** Sholkamy, E., Ahmad, M., Manal Yaser, M., Ali, A., Mehanni, M. Anticancer activity of biostabilized selenium nanorods synthesized by Streptomyces bikiniensis strain Ess_amA-1. *Int. J. Nanomed.* 2015, 10: 3389–3401.

**99** Vahdati, M., Tohidi Moghadam, T. Synthesis and characterization of selenium nanoparticles-lysozyme nanohybrid system with synergistic antibacterial properties. *Sci. Rep.* 2020, 10 (1): 510.

**100** Hashem, A.H., Khalil, A.M.A., Reyad, A.M., Salem, S.S. Biomedical applications of mycosynthesized selenium nanoparticles using penicillium expansum ATTC 36200. *Biol. Trace Element Res.* 2021, 199 (10): 3998–4008.

# VII

# Borophene, Stanene, Arsenene, and Antimonene

*Shancheng Yan*

## 7.1 Introduction

Due to the advent of graphene, the search for other graphene-like elements has become the research craze of the moment, driving the study of atomic layers composed of group III–VI elements such as P, B, Sn, As, and Sb. The most classical graphene-like material is phosphorene, which has a high hole mobility comparable to graphene. Happily, unlike graphene, both zigzag and armchair phosphorene nanosheets have a direct bandgap, and the size of the bandgap decreases monotonically with increasing bandwidth due to quantum confinement effects. This also means that phosphorene has excellent potential for use in field effect transistors (FETs) and optical sensor devices. Similar two-dimensional (2D) semiconductor materials, such as silicene and germanene, have some application limitations, such as zero bandgap, lack of environmental stability, or inability to be synthesized on a large scale, which are technical bottlenecks that must be resolved urgently.

Recently, the discovery of borophene [1–5], stanene [6–11], arsenene [12], and antimonene [13–15] has made single-element semiconductors of the IV–VI group a hot research topic, and borophene is an emerging single-element 2D semiconductor material with many interesting properties. It has many interesting properties. It has high carrier mobility, thermal conductivity, optical anisotropy, and superconductivity. Meanwhile, borophene has many structural morphologies with tunable bandgaps and maintains good stability in acidic and alkaline environments [16]. These features give it great potential for application in energy storage, sensing [17], and electronic information, and it is even expected to surpass graphene in some aspects.

On the other hand, stanene comprises Sn atoms arranged in a honeycomb lattice, which is isoelectronic with graphene. The low-energy electronic excitation can be regarded as linear in the honeycomb atomic arrangement. It is worth mentioning that stanene's low-flexure structure is more stable than graphene and reflects the large gap of quantum spin Hall states. Stanene has many essential properties, such as thermoelectric properties, topological superconductivity, and quantum anomalous Hall effect, and still possesses time-reversal symmetry when the self-selected orbital coupling (SOC) is considered. Similar to graphene, stanene is a Dirac material with ultra-high carrier mobility but has the

*Single Element Semiconductors: Properties and Devices*, First Edition. Yi Shi and Shancheng Yan.

same zero bandgap limitation as graphene. However, after doping it, stanene is expected to be a promising candidate as a material for optoelectronic applications. Arsenene and antimonene are single-atom layers of As and Sb, respectively, and are also recently proposed as emerging 2D materials. The optical bandgaps of the two are 1.6 eV and 1.5 eV, respectively. Although they are nonzero bandgap materials, the carrier mobility of arsenene is not ideal, and the carrier mobility of antimonene is also relatively average. However, their curved rhombic electronic structures make them environmentally stable, and they maintain good pristine properties even after being left in the air for 30 days.

In this chapter, we discuss the synthesis of borophene, stanene, arsenene, and antimonene and delve into these materials' properties and related applications. The outline of this chapter is as follows: Section 7.2 discusses the synthesis of borophene and its properties and applications, Section 7.3 discusses the preparation of stanene with its properties and applications, and Section 7.4 discusses the preparation of arsenene and antimonene with their properties and applications. Concluding remarks are given in Section 7.5.

## 7.2 Borophene

Borophene, an emerging 2D semiconductor of group III and the lightest Dirac material, has been widely studied for its excellent anisotropy, optical transparency, thermal properties, and superconductivity. This section describes its preparation methods, properties, and possible applications.

### 7.2.1 Preparation of Borophene

The preparation of borophene is challenge due to the unique structure of B and the complex B–B bonding. Zhang et al. [18] prepared a few layers of borophene using solvent thermally-assisted liquid-phase stripping. They utilized solvents with low surface tension, such as acetone, *N*,*N*-dimethylformamide, acetonitrile, ethanol, and *N*-methylpyridone, which made the stripping process more efficient. The preparation process can be roughly divided into three stages: firstly, block B was put into acetone solution with vigorous stirring to make it into a dispersion, after which it was heated and kept at 200°C for 24 h. Finally, the obtained solution was ultrasonicated and then centrifuged to remove the unpeeled particles to obtain the final product, borophene. The scanning electron microscope (SEM) image of the borophene prepared using the method is shown in Figure 7.1a, and the parallel lamellar structure can see the parallel layered structure, which confirms the successful stripping of few-layered borophene from block B. These borophenes have an average transverse scale of 5.05 mm, the few-layered flakes usually have larger sizes, and the surface is partially oxidized before being stripped off. Zhang et al. [19] succeeded in stripping borophene using the mechanical resonance method, and they put the amorphous B into ultrapure water in which silver nanowires were used as stabilizers, followed by 10 min of sonication. The mixture was then vibrated in a ball mill for 6 h. The final product was obtained after centrifugation. The transmission electron microscopy (TEM) image of borophene is shown in Figure 7.1b. The sample has a thin single-crystal structure with a thickness of less than 4 nm. No warpage occurs in this kind of borophene

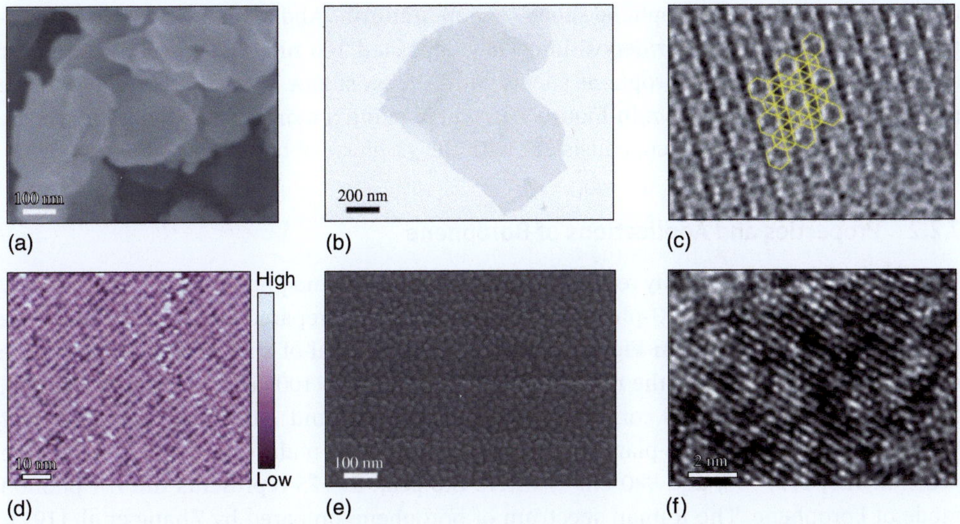


**Figure 7.1** (a) SEM image of the borophene. *Source:* Reprinted with permission from Ref. [18]. Copyright 2020, Royal Society of Chemistry. (b) TEM image of borophene prepared using mechanical resonance method. (c) HRTEM images of few-layer borophene. *Source:* Reprinted with permission from Ref. [19]. Copyright 2020, American Chemical Society. (d) Scanning tunneling microscopy (STM) image of the borophene. *Source:* Reprinted with permission from Ref. [20]. Copyright 2022, SpringerLink. (e) SEM image of borophene on the surface of carbon cloth. (f) HRTEM image of the prepared borophene. *Source:* Reprinted with permission from Ref. [21]. Copyright 2021, American Chemical Society.

structure, and high-resolution transmission electron microscopy (HRTEM) observed the lattice structure with alternating triangular lattice and hexagonal cavities. The density of hexagonal holes is $\eta = 1/9$, as shown in Figure 7.1c. It can be seen that the mechanical resonance stripping method is less costly and easy to operate and successfully prepared α-flake borophene. The various preparation methods mentioned previously are top-down, and there are many bottom-up methods to prepare borophene, including molecular beam epitaxy (MBE), chemical vapor deposition, and electron beam epitaxy. Using MBE, Chen et al. [20] synthesized large-size bilayer borophene on the Cu surface, and the final uniform and large piece of borophene formed is shown in Figure 7.1d. At the initial stage of growth, where a monolayer (ML) of borophene covers the Cu surface, excess B atoms evaporate to the surface to form a new structure. When the coverage of B reaches twice that of the ML, the entire surface is almost bilayered with borophene, and uniform and large-area bilayered single-crystal borophene is formed after high-temperature annealing. In this method, the interaction between the metal substrate and B plays a decisive role in the growth of borophene, and the adsorption capacity of the first layer of B on the Cu substrate $-0.41$ eV atom$^{-1}$ ($-0.148$ eVÅ$^{-2}$), which is about twice that of Ag. Tai et al. [21], on the other hand, successfully prepared borophene using chemical vapor deposition with a boron source of sodium borohydride. They used a two-zone chemical vapor deposition technique to grow borophene on carbon cloth, adding hydrogen to stabilize its structure. The SEM image of borophene on the surface of carbon cloth is shown in Figure 7.1e, which shows that the distribution of borophene sheets is more uniform. Abdi et al. also prepared borophene using chemical vapor deposition. They deposited 100 nm of Al on a silicon wafer and later synthesized the borophene sheets with Al's assistance. The HRTEM image of the prepared borophene is shown in Figure 7.1f, where parallel atomic bands with an average distance of 4.15 Å can be seen, consistent with the $\chi$3-phase of borophene.

### 7.2.2 Properties and Applications of Borophene

Raman spectra can powerfully reveal borophene's structure and phonon properties, broadly categorized into two types, $\chi$3-phase and $\beta_{12}$-phase [22], as prepared by Abdi et al. [23]. The Raman spectra are shown in Figure 7.2a,b. There are a total of six peaks in the $\chi$3-phase of borophene, which are in the range of 300, 375, 430, 780, 1100, and 1240 $cm^{-1}$. The 780, 1100, and 1240 $cm^{-1}$ peaks correspond to the A3g, A2g, and A1g vibrational modes of borophene formed due to in-plane stretching. The $\beta_{12}$-phase mainly has four major Raman peaks at 125, 175, 900, and 980 $cm^{-1}$, where the peak at 125 represents the ZA phonon mode of borophene. The Raman spectrum of borophene prepared by Zhang et al. [19] is shown in Figure 7.2c with five major Raman peaks. It is worth mentioning that the peak of borophene located at 1097 $cm^{-1}$ is several times higher than the bulk B intensity, which is attributed to intense vibrations within the sheet-mounted borophene, leading to its in-plane stretching enhancement.

Meanwhile, the difference between the peaks at 572 and 652 also implies noticeable molecular structure differences between the sheeted borophene and bulk B. The photoluminescence properties of borophene have been studied by Huo et al. [24]. The photoluminescence spectra of borophene and graphene heterojunction are shown in Figure 7.2d. It can be seen that it has an apparent characteristic peak at 492 nm, and there is a slight blue shift compared with the theoretical characteristic peak at 498 nm, which may be due to the interaction in the heterojunction.

Borophene is a promising 2D material for high-performance energy storage device applications. Abdi [23] prepared a new 2D borophene-based supercapacitor with borophene nanosheets and graphene working electrodes. Figure 7.3a compares the electrochemical performance of the pristine graphene working electrode with that of the supercapacitor spin-coated with borophene, and it can be seen that the area of the J-V curve of the spin-coated borophene device has increased by 2.25 $cm^2$. Figure 7.3b compares the V-T curve at a current of 1 $Ag^{-1}$. The discharge time of the new supercapacitor is nearly doubled compared to that of the pristine graphene, which also indicates that the new supercapacitor's discharge time is nearly doubled compared to that of the pristine graphene, nearly doubled, which also indicates the high specific capacity of the device.

Meanwhile, the device's capacitance is as high as 0.75 $F{\cdot}g^{-1}$ (350 $mF{\cdot}cm^{-2}$). At a scan rate of 5 $mV{\cdot}s^{-1}$, the capacitance reaches 0.58 $F{\cdot}g^{-1}$ (270 $mF{\cdot}cm^{-2}$), or so, the sCap of this device is higher than that of other known typical 2D materials, which makes it extremely valuable for applications in energy storage. Not only in the field of energy storage, borophene's performance in electrocatalytic hydrogen precipitation is also awe-inspiring. Tafel slope is one of the key indicators of electrocatalytic performance, and Tai et al. [21] measured the Tafel slopes of borophene grown at five different temperatures, 600, 650,

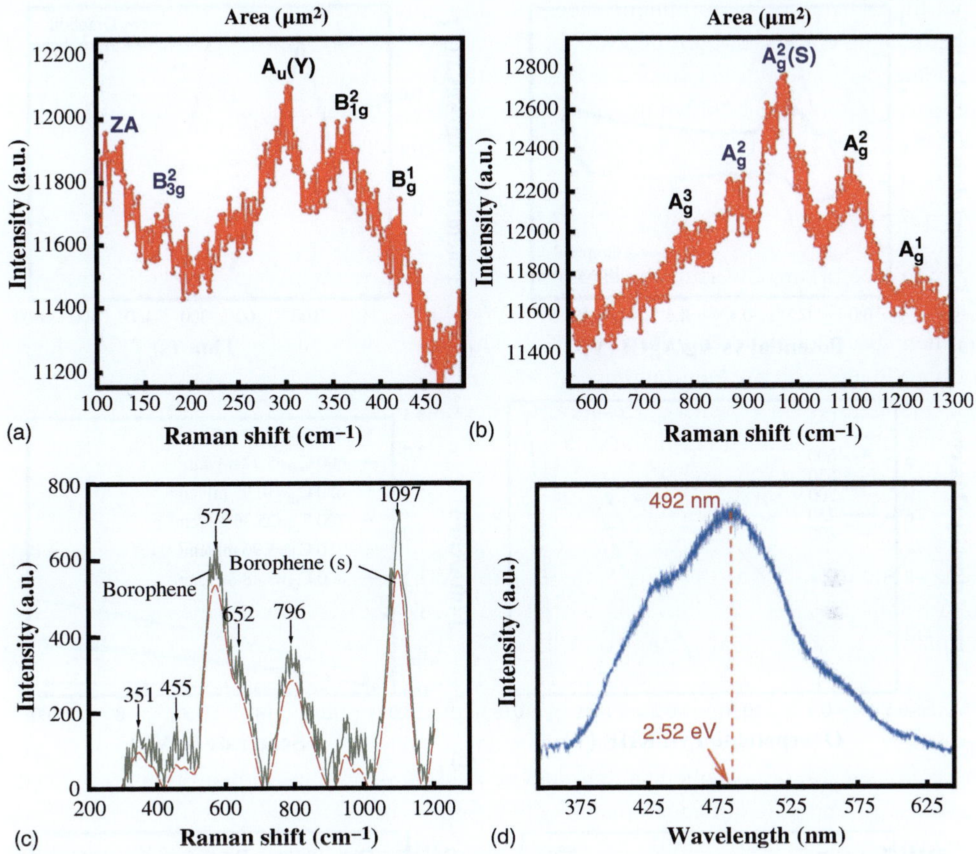


**Figure 7.2** (a, b) Raman spectra of the borophene. *Source:* Reprinted with permission from Ref. [23]. Copyright 2022, American Chemical Society. (c) Raman spectrum of borophene. *Source:* Reprinted with permission from Ref. [19]. Copyright 2020, American Chemical Society. (d) Photoluminescence spectra of borophene and graphene heterojunction. *Source:* Reprinted with permission from Ref. [24]. Copyright 2020, SpringerLink.

700, 750, and 800°C, as shown in Figure 7.3c. At 700°C, the Tafel slope of the sample is only 69 mV/dec, while at 800°C, it is as high as 196 mV/dec, which indicates the good electrocatalytic performance of borophene. Also, they calculated the electrochemical double layer capacitance (Cdl) of borophene, and the cyclic voltammetry scan data are shown in Figure 7.3d. The voltage range was 0.1–0.2 V, and the scan rate was 5–100 Mv/s. As a result, the Cdl of borophene was calculated to be 26.36 mF/cm$^2$ when the growth condition was 700°C for 30 min, which proved the high electrocatalytic performance of borophene. In addition to these applications, borophene can also be used as a humidity sensor. Huo et al. [24] investigated the sensitivity of the prepared borophene–graphene flexible sensor under different RH environments. The switching RH test results of the sensor are shown in Figure 7.3e, where the sensor's sensitivity continued to increase as the water vapor concentration increased from 0 to 85%, with the sensitivity exceeding 4200%. The response and recovery curves of the sensor were then tested at a water vapor

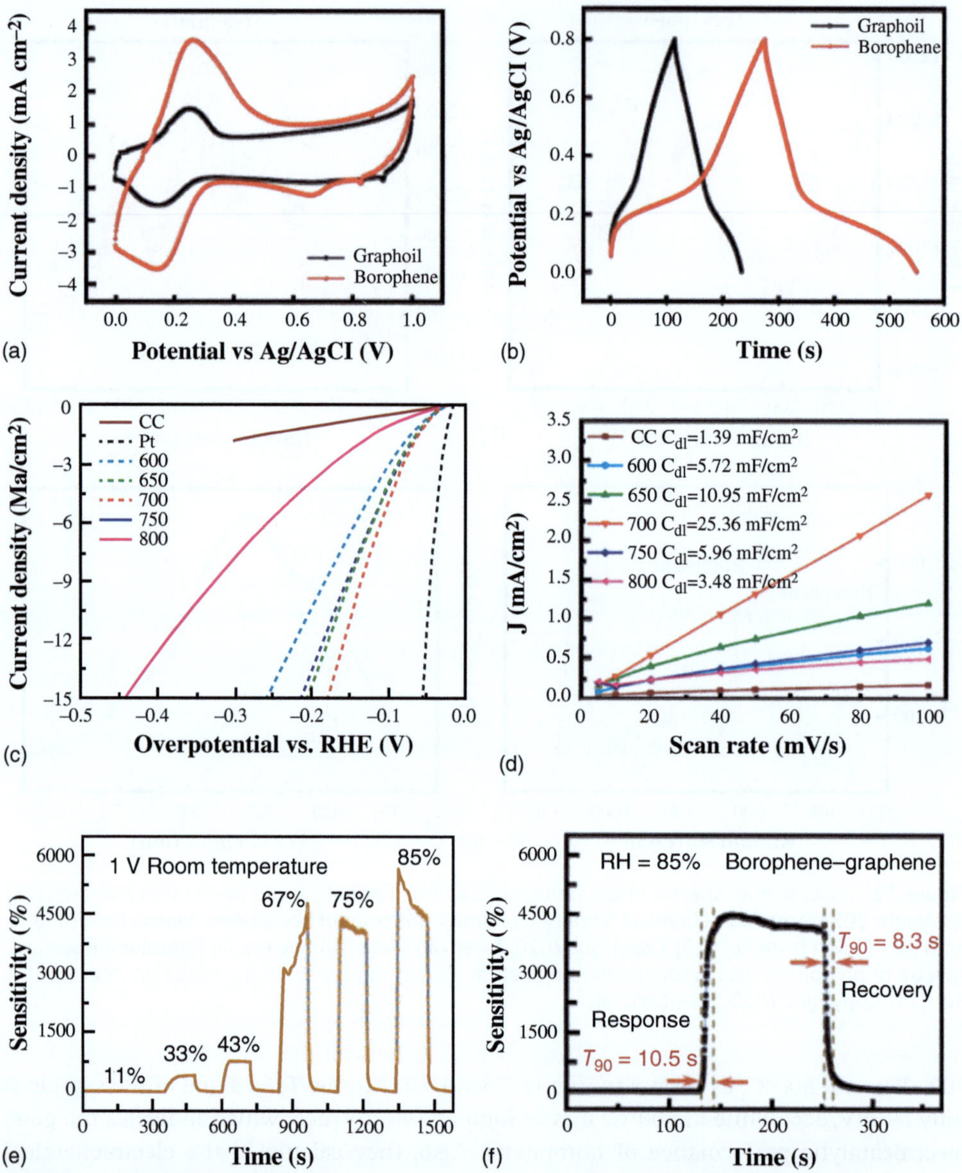


**Figure 7.3** (a, b) A comparison between the electrochemical performance of the supercapacitors based on the pristine graphite and the borophene-coated one. *Source:* Reprinted with permission from Ref. [23]. Copyright 2022, American Chemical Society. (c) Tafel slopes of borophene grown at different temperatures. (d) Cyclic voltammetry scan data of borophene. *Source:* Reprinted with permission from Ref. [21]. Copyright 2021, American Chemical Society. (e) Switching RH test results of the sensor. (f) Response and recovery curve of the heterostructure sensor at 85% RH. *Source:* Reprinted with permission from Ref. [24]. Copyright 2020, SpringerLink.

concentration of 85%, and the results are shown in Figure 7.3f, with a response time of 10.5 s and a recovery time of 8.3 s. These performances are much higher than those of other reported sensors for 2D materials.

## 7.3 Stanene

As a 2D material approximating graphene, stanene possesses excellent quantum effects, superconducting properties, and giant magnetoresistive properties, among others. This section will study its preparation method, properties, and applications in depth.

### 7.3.1 Preparation of Stanene

Among the methods of stanene preparation, MBE is the most common. Zhu et al. [25] used MBE to prepare 2D stanene. They used $Bi_2Te_3$ bulk single crystals and thin films as the substrate, and firstly, they grew $Bi_2Te_3$ thin films on silicon wafers with a thickness of about 40 nm. Afterward, high-purity tin was evaporated and deposited on the substrate at a rate of 0.4 layers/min, and the STM image of the stanene is shown in Figure 7.4a. Where the $Bi_2Te_3$ substrate is a step, the Sn atoms are deposited and grown in a bi-atomic layer structure, resulting in a complete and uniform stanene film. The film is not flat on the nanoscale, and there is a 0.06 nm difference in height, which the in-plane pressure may cause. The heights between the first, second, and third layers of stanene are about 0.35 nm, which is consistent with the spatial distances between the diatomic layers along the (111) direction in α-Sn and confirms that the Sn film is grown in a bi-atomic layer structure. Deng et al. [26] then used a modified MBE technique to grow stanene on Cu(111) at low temperatures. They evaporated tin at 800°C under ultravacuum conditions, with a vacuum pressure of less than $5 \times 10^{-10}$ mbar during the deposition process, while ensuring that the temperature of the Cu substrate was kept at a low temperature of 200 K. The stanene films were grown at low temperatures. The STM image of the prepared stanene is shown in Figure 7.4b, where the atomic density of the sample is 0.9 ML, and most of the substrate is uniformly covered by stanene. The thickness of the film is about 0.18 nm, which is determined as a single atomic layer, and the honeycomb lattice of stanene is completely flat. Whereas stanene synthesized once was almost always of low-flexural structure, the ones synthesized this time using low-temperature MBE were in a super-flat state, perhaps since the Sn atom is strongly coupled to the substrate, which reduces the energy of the interaction between the two. High-quality stanene was also prepared in an ultravacuum using a Co/Cu substrate by Chen et al. [27], who used an electron beam to evaporate the Sn and deposited it on a Co/Cu substrate at a low temperature of 80 K, thereby forming a single atomic layer of stanene. Figure 7.4c shows an STM image of stanene deposited on the substrate, which shows that the stanene is not continuously grown and that the film terminates at the step edges of the Co nano-islands. An atomically resolved image of stanene is shown in Figure 7.4d, which shows that it has a well-defined honeycomb lattice. The RT growth of the double-layered Co nano-islands on Cu is pseudocrystalline, which

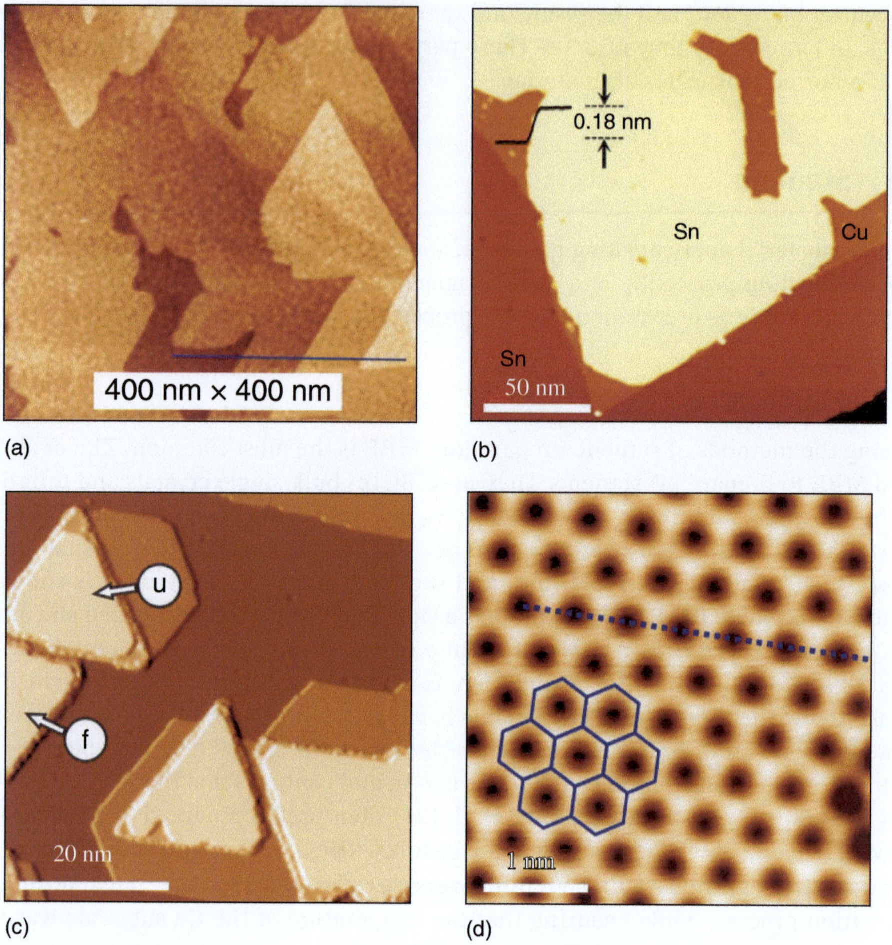


**Figure 7.4** (a) STM image of the stanene. *Source:* Reprinted with permission from Ref. [25]. Copyright 2015, SpringerLink. (b) STM image of the prepared stanene. *Source:* Reprinted with permission from Ref. [26]. Copyright 2021, SpringerLink. (c) STM image of stanene deposited on the substrate. (d) Atomically resolved image of stanene. *Source:* Reprinted with permission from Ref. [27]. Copyright 2023, American Chemical Society.

follows the growth of the atomic lattice of Cu(111), which also results in the stanene on Co/Cu(111) having the same (2 × 2) structural periodicity as that of the stanene on Cu(111). Growing high-quality stanene is challenging, but research on new methods never stops.

### 7.3.2 Properties and Applications of Stanene

Like graphene, massless Dirac fermions are supposed to exist in stanene, and the electronic structure of stanene [28] is illustrated in Figure 7.5a. The figure shows a large gap at the Γ point, and a linear dispersive Dirac cone appears at the K point without considering

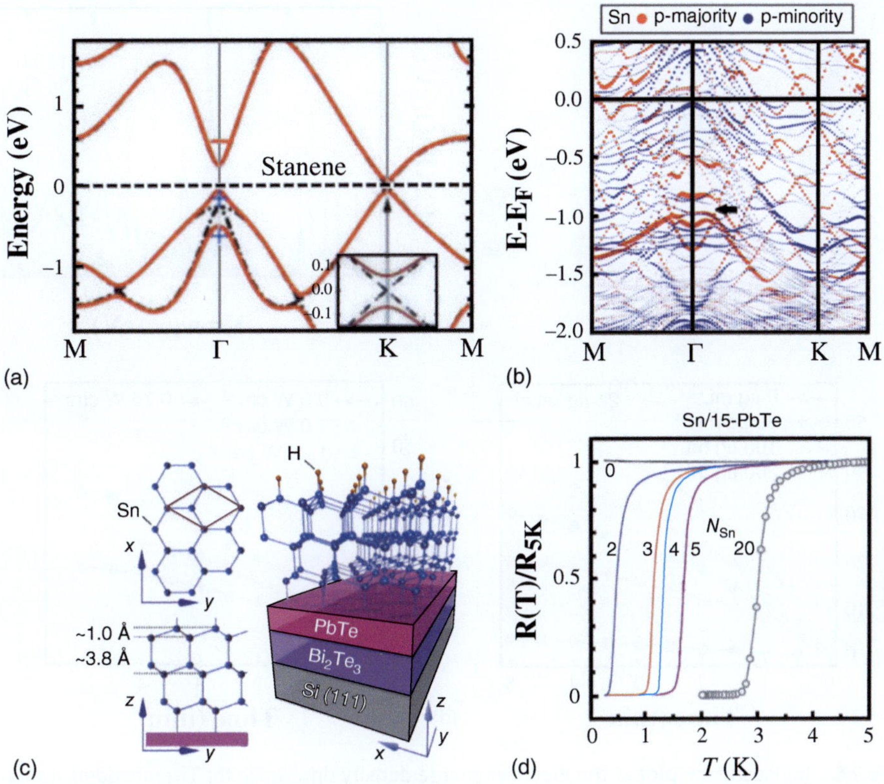


**Figure 7.5** (a) Electronic structure of stanene. *Source:* Reprinted with permission from Ref. [28]. Copyright 2019, WILEY-VCH. (b) p-orbital energy band of the Sn projected energy band structure. *Source:* Reprinted with permission from Ref. [27]. Copyright 2023, American Chemical Society. (c) Illustration of stanene lattice and $Sn/PbTe/Bi_2Te_3$ sandwich structure. (d) Normalized resistance of stanene with the increasing number of layers grown on substrates consisting of 15-PbTe/5-$Bi_2Te_3$/Si(111). *Source:* Reprinted with permission from Ref. [29]. Copyright 2018, SpringerLink.

the SOC. Figure 7.5b illustrates the p-orbital energy band of the Sn-projected energy band structure, where the SOC gap that opens below the Fermi energy level around −1.0 eV can be seen, and the SOC gap at Γ is about 0.25 eV. As an indirect bandgap semiconductor, stanene has a large bandgap at the Γ-point, which is drastically reduced and undergoes parity exchanges when the SOC opens. Superconductivity, a quantum phenomenon of broad interest, is also one of the many interesting properties that stanene is predicted to possess. Sn, as the material in which the Meiners effect was first discovered, stanene is also believed to be supposed to possess the spin Hall effect at room temperature. Stanene's stable superconductivity has been recently confirmed by Liao et al. [29], who prepared a three-layer heterostructure of $PbTe/Bi_2Te_3/Si$ triple heterostructure, as shown in Figure 7.5c. Figure 7.5d demonstrates the resistance change with the number of layers for the few-layer a-stanene on the topological insulator $Bi_2Te_3$. Its superconductivity starts to appear at two layers, and the transition temperature increases with the number of layers. Stanene's stable, super properties make it a promising candidate for constructing topological material devices.

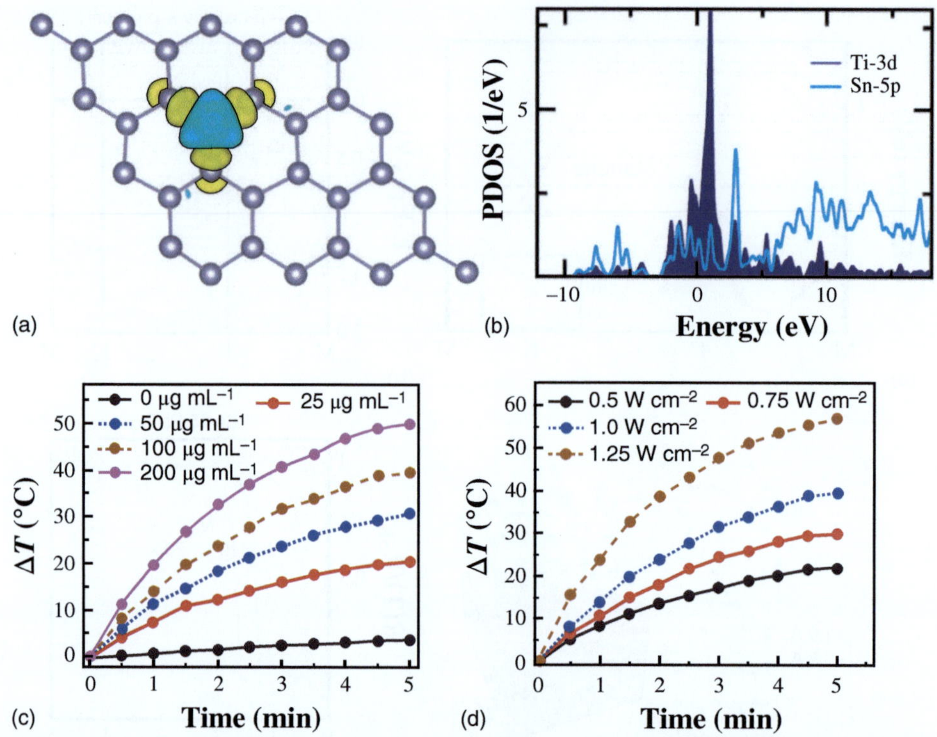


**Figure 7.6** (a) Isosurface plot of the electron charge density difference for Ti-embedded stanene ML with the isovalue of ±0.0002 a.u. (b) PDOS of the titanium and tin atoms of the Ti-embedded stanene. *Source:* Reprinted with permission from Ref. [30]. Copyright 2020, SpringerLink. (c) Temperature change profiles of SnNSs@PEG solutions at various concentrations (0–200 μg·mL$^{-1}$) under NIR laser irradiation (808 nm, 1 W·cm$^{-2}$). (d) Temperature change profiles of SnNSs@PEG solution (100 μg·mL$^{-1}$) exposed to different power densities (0.5–1.25 W·cm$^{-2}$) of NIR laser (808 nm). *Source:* Reprinted with permission from Ref. [31]. Copyright 2021, SpringerLink.

Zhou et al. [30] embedded a single layer of stanene in Ti atoms using DFT calculations, which allows the application of this structure to gas sensors. The results showed that, except $SO_2$, the other gases, including $CO_2$, $NO_2$, and $NH_3$, were weakly physisorbed on the original ML stanene. In contrast, adding Ti atoms into the ML stanene could change the adsorption type to chemisorption with a more robust adsorption capacity. Figure 7.6a illustrates the electron charge density difference isosurface plot of this structure, and Figure 7.6b shows its projected density of states (PDOS) plot, in which Sn and Ti atoms overlap with each other, illustrating the covalent nature of the interactions at the interface, which leads to the formation of more stable covalent bonds. It can be hypothesized that this structure can be used as a gas sensor with excellent performance. Ouyang et al. [31] prepared stanene using low-temperature and liquid-phase exfoliation techniques for biomedical applications. The calculated photothermal conversion efficiency of stanene modified using poly(ethylene glycol) was as high as 48.6%. Figure 7.6c demonstrates the temperature change curve of a 0~200 $\mu g \cdot mL^{-1}$ solution of poly(ethylene glycol)-modified stanene under 808 nm laser irradiation, and Figure 7.6d shows the concentration of 100 μg·mL$^{-1}$ of the solution under 808-nm near-infrared (NIR) laser irradiation at a power density of 0.5~1.25 W · cm$^{-2}$. The photothermal performance of this solution showed a clear concentration and power dependence, with the highest temperature increment ($\Delta T$) of the solution as high as 49.8°C when the concentration of the solution was 200 μg·mL$^{-1}$ and the power density of the laser was 1.0 W·cm$^{-2}$. In conclusion, stanene can be used in multimodal imaging, making photothermal therapy effective for tumor elimination and broadening its application in cancer phototherapy.

## 7.4 Arsenene and Antimonene

Arsenene and antimonene have also recently been discovered as graphene-like 2D semiconductors composed of As and Sb atoms, respectively, which have attracted much attention due to their excellent optical and electrical properties. This section will thoroughly study their preparation methods, properties, and applications.

### 7.4.1 Preparation of Arsenene and Antimonene

Current research on arsenene and antimonene preparation is still very rare, Tsai et al. [12] successfully synthesized multilayered arsenene using a plasma-assisted process, and after the plasma treatment, they annealed the InAs. During the annealing process, the As atoms were pushed to the surface of the substrate since InAs would react preferentially with the nitrogen ions in the plasma, forming fewer layers of arsenene. The TEM image of the final product obtained is shown in Figure 7.7a, which shows that InN and arsenene are present on the InAs substrate, with the arsenene being in the top layer. It was also found that the thickness of the arsenene could be somewhat controlled by varying the exposure time of the plasma and that a semiconductor-to-metal transition would occur in the thickness range of 14–21 nm. Tsai et al. [32] also synthesized a few-layer antimonene for the first time using a similar approach, using an InSb substrate as a template and annealing it at 450°C for half an hour to an hour after immersion in $N_2$ plasma. The TEM image of the prepared few-layer antimonene is shown in Figure 7.7b, with the presence of InN and a ~5-nm-thick antimonene film on the InSb substrate. The film has a layered, stepped structure with the InN layer underneath it acting as a buffer so that the lattice of antimonene matches the InSb lattice. The growth mechanism of antimonene synthesized using the plasma-assisted process is similar to that of previously synthesized arsenene, in which Sb atoms are squeezed to the surface during the annealing process to form antimonene films with fewer layers. In addition to plasma-assisted processes, epitaxial growth strategy has been a common means of preparing 2D materials. Ji et al. [33] used an epitaxial growth strategy to prepare high-quality single-crystal antimonene. They used mica as the substrate and evaporated the Sb powder at 660°C. The optical images of the deposited samples are shown in Figure 7.7c. The single-crystal antimonene has several polygonal shapes, namely triangular, hexagonal, rhombic, and trapezoidal, consistent with epitaxial growth. Wu et al. [34] used MBE to grow a ML of antimonene on a $PdTe_2$ substrate. Figure 7.7d illustrates an STM image of antimonene, and the inset shows

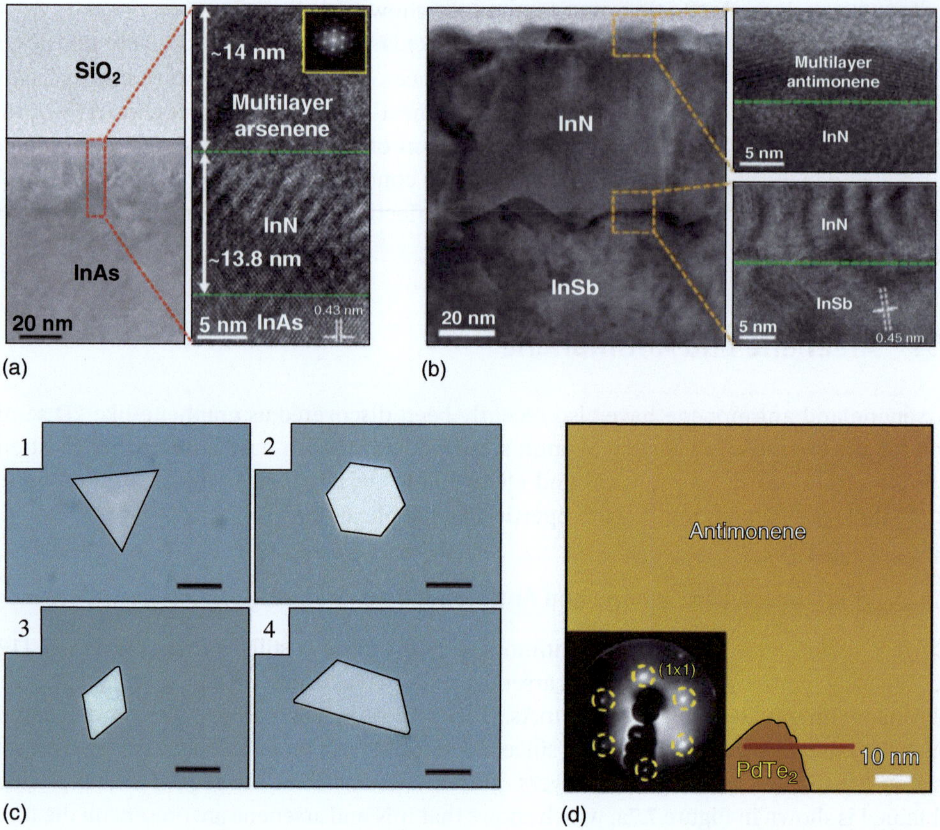


**Figure 7.7** (a) TEM images of arsenene. *Source:* Reprinted with permission from Ref. [12]. Copyright 2016, American Chemical Society. (b) TEM image of the prepared few-layer antimonene. *Source:* Reprinted with permission from Ref. [32]. Copyright 2016, Royal Society of Chemistry. (c) Optical images of the deposited samples. *Source:* Reprinted with permission from Ref. [33]. Copyright 2016, SpringerLink. (d) STM image of antimonene. *Source:* Reprinted with permission from Ref. [34]. Copyright 2017, WILEY-VCH.

six diffraction points from the ML of antimonene and $PdTe_2$ substrate, which not only have almost the same lattice constants but also have symmetric lattice distributions on the surface. So far, there are not many studies on the preparation of arsenene and antimonene, and more preparation methods are yet to be thoroughly investigated.

### 7.4.2 Properties and Applications of Arsenene and Antimonene

Arsenene and antimonene, as emerging 2D materials, must study their structure and phonon properties. The Raman spectra of arsenene prepared by Tsai et al. [12] are shown in Figure 7.8a, where the individual curves represent arsenene prepared by different plasma immersion times. The peaks at 205 $cm^{-1}$ represent the arsenene's $E_g$ vibrational mode, and the peak at 260 $cm^{-1}$ attachment represents the $A_{1g}$ vibrational mode of the arsenene. The $E_g$ mode in the arsenene is enhanced with the increase of immersion time, and the peak

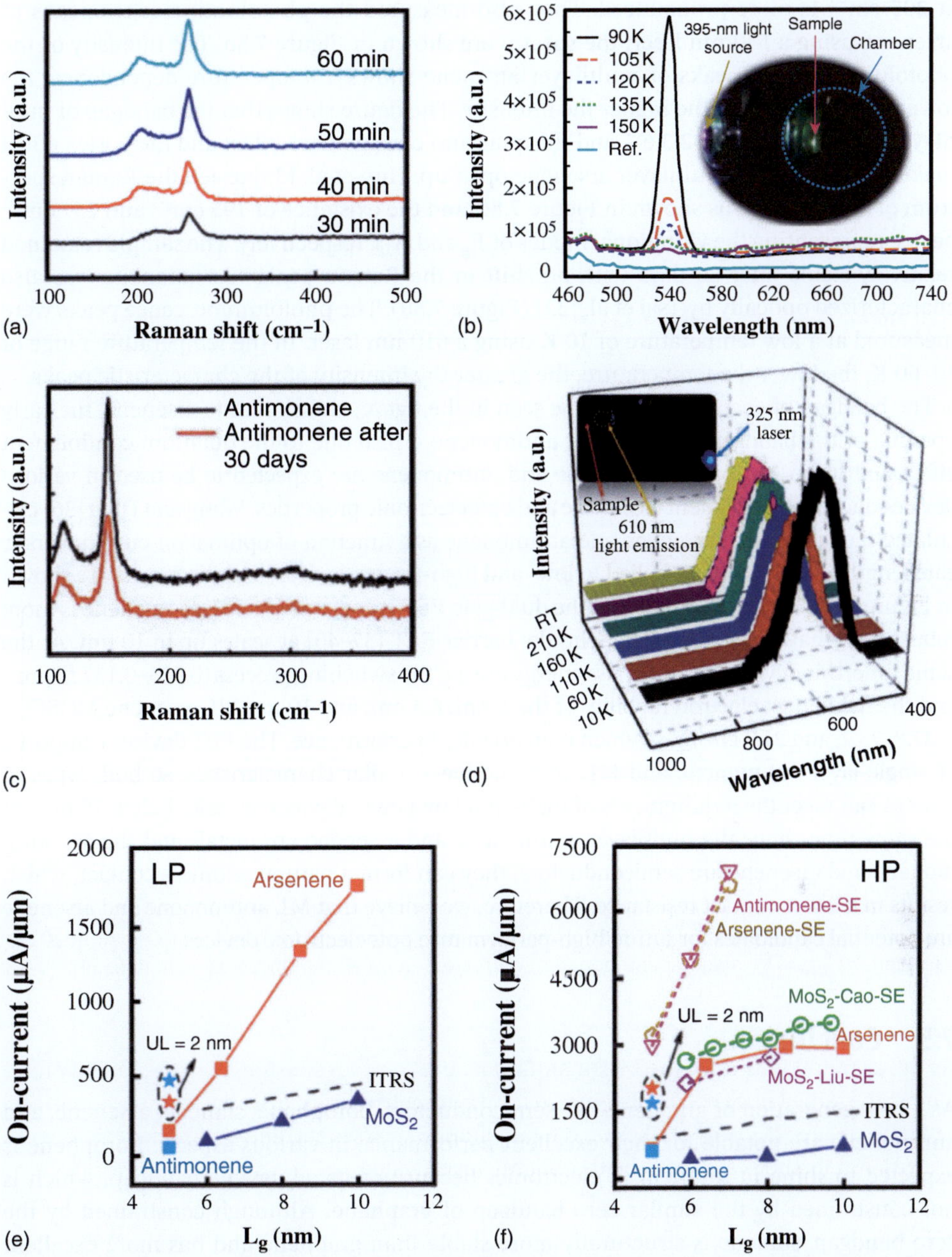


**Figure 7.8** (a) Raman spectra of arsenene and (b) photoluminescence spectra of arsenene. *Source:* Reprinted with permission from Ref. [12]. Copyright 2016, American Chemical Society. (c) Raman spectra of antimonene. *Source:* Reprinted with permission from Ref. [35]. Copyright 2017, American Chemical Society. (d) Photoluminescence spectra of antimonene. *Source:* Reprinted with permission from Ref. [32]. Copyright 2016, Royal Society of Chemistry. (e, f) Optimal on-current versus the gate length for the ML arsenene (red and orange), antimonene (cyan and magenta), $MoS_2$ (purple and green) double-gated n-metal oxide semiconductor field effect transistor (DG n-MOSFETs), and ML $MoS_2$ SBFETs with Ti electrodes (blue) in the (e) LP and (f) HP applications. *Source:* Reprinted with permission from Ref. [36]. Copyright 2015, SpringerLink.

at 205 $cm^{-1}$ is more pronounced. They also measured the photoluminescence peaks of arsenene using a 540-nm laser; the spectra are shown in Figure 7.8b. The intensity of the photoluminescence peaks of multilayer arsenene shows a temperature dependence; the lower the temperature, the higher the intensity. The figure shows that the bandgap of multilayer arsenene is about 2.3 eV, and the quantum confinement effect and the vortex effect make the bandgap of multilayer arsenene open up. Huo et al. [35] tested the Raman spectrum of antimonene, as shown in Figure 7.8c, and the existence of 195 $cm^{-1}$ and 257 $cm^{-1}$ peaks, representing the vibrational modes of $E_g$ and $A_{1g}$, respectively. The sample remained relatively stable after 30 days with no shift in the Raman peaks. Antimonene was also characterized optically by Tsai et al. [32] (Figure 7.8d). The photoluminescence peaks were measured at a low temperature of 10 K using a 610-nm laser. In the temperature range of 10–60 K, the lower the temperature, the greater the intensity of the characteristic peaks.

The bandgap of about 2.03 eV can be seen in the figure, and similar to arsenene, the early opening of the bandgap of multilayer antimonene is also due to the quantum confinement effect and turbo lamination. Arsenene and antimonene are expected to be used in various devices due to their excellent electronic and optoelectronic properties. Wang and Ding [36] calculated the bandgap of arsenene and antimonene as a function of optimal on-current versus gate length for FETs when applied to low- and high-power devices, and the results are shown in Figure 7.8e,f. The performance of the dual-gate FET composed of a ML of arsenene is more robust than that of the ML $MoS_2$ Schottky barrier FET [37–40] at scales up to 10 nm. At the same time, this device consumes less energy during the switching process (0.023–0.137 fJ/μm), and the field-effect electron mobility of the 5 nm, 8.5 nm, and 10 nm ML arsenene MOSFET is 27.8, 23.2, and 21.2 $cm^2/V{\cdot}s$, which is an excellent performance. The FET devices composed of single-layer antimonene and ML arsenene have similar characteristics, so both types of devices can meet the requirements of high- and low-power devices at scales below 10 nm. At the same time, since the multilayer antimonene and arsenene are metals and the ML antimonene and arsenene are semiconductors, they can form a natural seamless contact, which results in a lower contact resistance. Therefore, we believe that ML antimonene and arsenene are potential candidates for future high-performance optoelectronic devices.

## 7.5 Conclusion

As a new generation of graphene-like semiconductors, borophene, stanene, arsenene, and antimonene are notable for their excellent performance in various aspects. Borophene is expected to shine in the field of electronics because of its adjustable bandgap, which is not constrained by the similar zero bandgap of graphene. Although constrained by the zero bandgap, stanene is structurally more stable than graphene and has more excellent thermoelectric properties. It is structurally more stable than graphene and has better thermoelectric properties. Arsenene and antimonene also have suitable bandgap and strong stability, which are expected to be used in future high-performance electronic devices. However, in general, research on them has been scarce so far, and most of them are still at the level of theoretical calculations. More preparation methods and more creative applications still need to be discovered by researchers, and this class of single-element emerging 2D semiconductors has the potential to be widely used in various fields.

## References


1 Ranjan, P., Lee, J.M., Kumar, P., Vinu, A. Borophene: new sensation in Flatland. *Adv. Mater.* 2020, 32 (34): 2000531.

2 Ou, M., Wang, X., Yu, L., Liu, C., Tao, W., Ji, X., Mei, L. The emergence and evolution of borophene. *Adv. Sci.* 2021, 8 (12): 2001801.

3 Rahman, A., Rahman, M.T., Chowdhury, M.A., Bin Ekram, S., Uddin, M.M.K., Islam, M.R., Dong, L. Emerging 2D borophene: synthesis, characterization, and sensing applications. *Sens. Actuat. A: Phys.* 2023, 359: 114468.

4 Hou, C., Tai, G., Liu, Y., Wu, Z., Liang, X., Liu, X. Borophene-based materials for energy, sensors and information storage applications. *Nano Res. Energy* 2023, 2: e9120051.

5 Folorunso, O., Hamam, Y., Sadiku, R., Ray, S.S., Adekoya, G.J. Synthesis methods of borophene, graphene-loaded polypyrrole nanocomposites and their benefits for energy storage applications: a brief overview. *FlatChem* 2021, 26: 100211.

6 Dhungana, D.S., Grazianetti, C., Martella, C., Achilli, S., Fratesi, G., Molle, A. Two-dimensional silicone–stanene heterostructures by epitaxy. *Adv. Funct. Mater.* 2021, 31 (30): 2102797.

7 Ogikubo, T., Shimazu, H., Fujii, Y., Ito, K., Ohta, A., Araidai, M., Kurosawa, M., Le Lay, G., Yuhara, J. Continuous growth of germanene and stanene lateral heterostructures. *Adv. Mater. Interfaces* 2020, 7 (10): 1902132.

8 Li, J., Lei, T., Wang, J., Wu, R., Zhao, J., Zhao, L., Guo, Y., Qian, H., Ibrahim, K. Anisotropic electronic structure and interfacial chemical reaction of stanene/Bi2Te3. *J. Phys. Chem. C* 2020, 124 (8): 4917–4924.

9 Bechstedt, F., Gori, P., Pulci, O. Beyond graphene: clean, hydrogenated and halogenated silicene, germanene, stanene, and plumbene. *Progress Surf. Sci.* 2021, 96 (3): 100615.

10 Sharma, N., Chi, C.-H., Swaminathan, N., Dabur, D., Wu, H.-F. Introducing stanene oxyboride nanosheets as white light emitting probe for selectively identifying <5 μm microplastic pollutants. *Sens. Actuat. B: Chem.* 2021, 348: 130617.

11 Kumar, N., Lan, Y.-S., Chen, C.-J., Lin, Y.-H., Huang, S.-T., Jeng, H.-T., Hsu, P.-J. Self-assembly of magnetic Co atoms on stanene. *Phys. Rev. Mater.* 2022, 6 (6): 066001.

12 Tsai, H.-S., Wang, S.-W., Hsiao, C.-H., Chen, C.-W., Ouyang, H., Chueh, Y.-L., Kuo, H.-C., Liang, J.-H. Direct synthesis and practical bandgap estimation of multilayer arsenene nanoribbons. *Chem. Mater.* 2016, 28 (2): 425–429.

13 Gibaja, C., Rodriguez-San-Miguel, D., Ares, P., Gómez-Herrero, J., Varela, M., Gillen, R., Maultzsch, J., Hauke, F., Hirsch, A., Abellán, G., Zamora, F. Few-layer antimonene by liquid-phase exfoliation. *Angew. Chem. Int. Ed.* 2016, 55 (46): 14345–14349.

14 Fortin-Deschênes, M., Moutanabbir, O. Recovering the semiconductor properties of the epitaxial group V 2D materials antimonene and arsenene. *J. Phys. Chem. C* 2018, 122 (16): 9162–9168.

15 Wang, Y., Huang, P., Ye, M., Quhe, R., Pan, Y., Zhang, H., Zhong, H., Shi, J., Lu, J. Many-body effect, carrier mobility, and device performance of hexagonal arsenene and antimonene. *Chem. Mater.* 2017, 29 (5): 2191–2201.

16 Duo, Y., Xie, Z., Wang, L., Mahmood Abbasi, N., Yang, T., Li, Z., Hu, G., Zhang, H. Borophene-based biomedical applications: status and future challenges. *Coord. Chem. Rev.* 2021, 427: 213549.

**17** Taşaltın, N., Taşaltın, C., Güngör, S., Karakuş, S., Gürol, İ., Teker, M. Volatile organic compound detection performance of borophene and PANI:β borophene nanocomposite-based sensors. *J. Mater. Sci.: Mater. Electron.* 2022, 33 (31): 24173–24181.

**18** Zhang, F., She, L., Jia, C., He, X., Li, Q., Sun, J., Lei, Z., Liu, Z.-H. Few-layer and large flake size borophene: preparation with solvothermal-assisted liquid phase exfoliation. *RSC Adv.* 2020, 10 (46): 27532–27537.

**19** Zhang, Z., Zhou, M., Zhang, T., Yang, M., Yang, Q., Yu, J., Zhang, Y. Few-layer borophene prepared by mechanical resonance and its application in terahertz shielding. *ACS Appl. Mater. Interfaces* 2020, 12 (17): 19746–19754.

**20** Chen, C., Lv, H., Zhang, P., Zhuo, Z., Wang, Y., Ma, C., Li, W., Wang, X., Feng, B., Cheng, P., Wu, X., Wu, K., Chen, L. Synthesis of bilayer borophene. *Nat. Chem.* 2021, 14 (1): 25–31.

**21** Tai, G., Xu, M., Hou, C., Liu, R., Liang, X., Wu, Z. Borophene nanosheets as high-efficiency catalysts for the hydrogen evolution reaction. *ACS Appl. Mater. Interfaces* 2021, 13 (51): 60987–60994.

**22** Sheng, S., Wu, J.-B., Cong, X. Zhong, Q., Li, W., Hu, W., Gou, J., Cheng, P., Tan, P.-H., Chen, L., Wu, K. Raman spectroscopy of two-dimensional borophene sheets. *ACS Nano* 2019, 13 (4): 4133–4139.

**23** Abdi, Y., Mazaheri, A., Hajibaba, S., Darbari, S., Rezvani, S. J., Cicco, A. D., Paparoni, F., Rahighi, R., Gholipour, S., Rashidi, A., Byranvand, M. M., Saliba, M. A two-dimensional borophene supercapacitor. *ACS Mater. Lett.* 2022, 4 (10): 1929–1936.

**24** Hou, C., Tai, G.A., Liu, B., Wu, Z., Yin, Y. Borophene-graphene heterostructure: preparation and ultrasensitive humidity sensing. *Nano Res.* 2020, 14 (7): 2337–2344.

**25** Zhu, F.-F., Chen, W.-J., Xu, Y., Gao, C.-L., Guan, D.-D., Liu, C.-H., Qian, D., Zhang, S.-C., Jia, J.-F. Epitaxial growth of two-dimensional stanene. *Nat. Mater.* 2015, 14 (10): 1020–1025.

**26** Deng, J., Xia, B., Ma, X., Chen, H., Shan, H., Zhai, X., Li, B., Zhao, A., Xu, Y., Duan, W., Zhang, S.-C., Wang, B., Hou, J.G. Epitaxial growth of ultraflat stanene with topological band inversion. *Nat. Mater.* 2018, 17 (12): 1081–1086.

**27** Chen, C.-J., Chao, Y.-C., Lin, Y.-H., Zhuang, Y.-H., Lai, Y.-M., Huang, S.-T., MacDonald, A. H., Shih, C.-K., Wang, B.-Y., Su, J.-J., Hsu, P.-J. Single-atomic-layer stanene on ferromagnetic Co nanoislands with topological band structures. *ACS Nano* 2023, 17 (8): 7456–7465.

**28** Lyu, J.K., Zhang, S.F., Zhang, C.W., Wang, P.J. Stanene: a promising material for new electronic and spintronic applications. *Annalen der Physik* 2019, 531 (10): 1900017.

**29** Liao, M., Zang, Y., Guan, Z., Li, H., Gong, Y., Zhu, K., Hu, X.-P., Zhang, D., Xu, Y., Wang, Y.-Y., He, K., Ma, X.-C., Zhang, S.-C., Xue, Q.-K. Superconductivity in few-layer stanene. *Nat. Phys.* 2018, 14 (4): 344–348.

**30** Zhou, J., Liu, D., Wu, F., Yang, L., Xiong, Y., Abbasi, A. A DFT study on the possibility of embedding a single Ti atom into the perfect stanene monolayer as a highly efficient gas sensor. *Theor. Chem. Acc.* 2020, 139 (3): 1–11.

**31** Ouyang, J., Zhang, L., Li, L., Chen, W., Tang, Z., Ji, X., Feng, C., Tao, N., Kong, N., Chen, T., Liu, Y.-N., Tao, W. Cryogenic exfoliation of 2D stanene nanosheets for cancer theranostics. *Nano-Micro Lett.* 2021, 13 (1): 1–18.

**32** Tsai, H.-S., Chen, C.-W., Hsiao, C.-H., Ouyang, H., Liang, J.-H. The advent of multilayer antimonene nanoribbons with room temperature orange light emission. *Chem. Commun.* 2016, 52 (54): 8409–8412.

**33** Ji, J., Song, X., Liu, J., Yan, Z., Huo, C., Zhang, S., Su, M., Liao, L., Wang, W., Ni, Z., Hao, Y., Zeng, H. Two-dimensional antimonene single crystals grown by van der Waals epitaxy. *Nat. Commun.* 2016, 7 (1): 13352.

**34** Wu, X., Shao, Y., Liu, H., Feng, Z., Wang, Y.L., Sun, J.T., Liu, C., Wang, J.O., Liu, Z.L., Zhu, S.Y., Wang, Y.Q., Du, S.X., Shi, Y.G., Ibrahim, K., Gao, H.J. Epitaxial growth and air-stability of monolayer antimonene on PdTe2. *Adv. Mater.* 2016, 29 (11): 1605407.

**35** Huo, C., Sun, X., Yan, Z., Song, X., Zhang, S., Xie, Z., Liu, J., Ji, J., Jiang, L., Zhou, S., Zeng, H. Retraction of "few-layer antimonene: large yield synthesis, exact atomical structure, and outstanding optical limiting." *J. Am. Chem. Soc.* 2017, 139 (9): 3568–3568.

**36** Wang, Y., Ding, Y. Electronic structure and carrier mobilities of arsenene and antimonene nanoribbons: a first-principle study. *Nanoscale Res. Lett.* 2015, 10 (1): 1–10.

**37** Pizzi, G., Gibertini, M., Dib, E., Marzari, N., Iannaccone, G., Fiori, G. Performance of arsenene and antimonene double-gate MOSFETs from first principles. *Nat. Commun.* 2016, 7 (1): 12585.

**38** Cao, W., Kang, J., Sarkar, D., Liu, W., Banerjee, K. 2D semiconductor FETs—projections and design for sub-10 nm VLSI. *IEEE Trans. Electron Devices* 2015, 62 (11): 3459–3469.

**39** Liu, L., Lu, Y., Guo, J. On monolayer MoS2 field-effect transistors at the scaling limit. *IEEE Trans. Electron Devices* 2013, 60 (12): 4133–4139.

**40** Ni, Z., Ye, M., Ma, J., Wang, Y., Quhe, R., Zheng, J., Dai, L., Yu, D., Shi, J., Yang, J., Watanabe, S., Lu, J. Performance upper limit of sub-10 nm monolayer MoS2 transistors. *Adv. Electron. Mater.* 2016, 2 (9): 1600191.

# Index

Note: Page numbers in *italics* refers to figures.



*Single Element Semiconductors: Properties and Devices*, First Edition. Yi Shi and Shancheng Yan.